T0206297

Physics and Applications of Dilute Nitrides

Optoelectronic Properties of Semiconductors and Superlattices

A series edited by M.O. Manasreh, Department of Electrical and Computer Engineering, University of Arkansas, Fayetteville, AR

Physics and Applications of Dilute Nitrides

Edited by

Irina A. Buyanova and Weimin M. Chen

CRC Press
Taylor & Francis Group
Boca Raton London New York

CRC Press is an imprint of the
Taylor & Francis Group, an **informa** business

A TAYLOR & FRANCIS BOOK

First published 2004 by Taylor & Francis Group

Published 2020 by CRC Press
Taylor & Francis Group
6000 Broken Sound Parkway NW, Suite 300
Boca Raton, FL 33487-2742

First issued in paperback 2020

© 2004 by Taylor & Francis Group, LLC
CRC Press is an imprint of Taylor & Francis Group, an Informa business

No claim to original U.S. Government works

ISBN 13: 978-0-367-57830-5 (pbk)
ISBN 13: 978-1-59169-019-1 (hbk)

Visit the Taylor & Francis Web site at
http://www.taylorandfrancis.com

and the CRC Press Web site at
http://www.crcpress.com

Library of Congress Cataloging-in-Publication Data

Physics and applications of dilute nitrides / edited by Irina A. Buyanova and Weimin M. Chen.
 p. cm. -- (Optoelectronic properties of semiconductors and superlattices)
 Includes bibliographical references.
 ISBN 1-59169-019-6 (alk paper)
 1. Nitrides. 2. Semiconductors--Materials. I. Buyanova, Irina A. II. Chen, W. M. (Weimin M.)
III. Series.
 TK7871.15.N57P48 2004
 621.3815'2--dc22   2004000206

CONTENTS

ABOUT THE SERIES

The series *Optoelectronic Properties of Semiconductors and Superlattices* provides a forum for the latest research in optoelectronic properties of semiconductor quantum wells, superlattices, and related materials. It features a balance between original theoretical and experimental research in basic physics, device physics, novel materials and quantum structures, processing, and systems — bearing in mind the transformation of research into products and services related to dual-use applications. The following subfields, as well as others at the cutting edge of research in this field, will be addressed: long-wavelength infrared detectors, photodetectors (MWIR-visible-UV), infrared sources, vertical-cavity surface-emitting lasers, wide-band-gap materials (including blue-green lasers and LEDs), narrow-band-gap materials and structures, low-dimensional systems in semiconductors, strained quantum wells and superlattices, ultrahigh-speed optoelectronics, and novel materials and devices.

The main objective of this book series is to provide readers with a basic understanding of new developments in recent research on optoelectronic properties of semiconductor quantum wells and superlattices. The volumes in this series are written for advanced graduate students majoring in solid-state physics, electrical engineering, and materials science and engineering, as well as researchers involved in the fields of semiconductor materials, growth, processing, and devices.

PREFACE

Because of their high scientific interest and technological importance, dilute nitrides have been the topic of intense scientific research and technological developments ever since they were discovered. During the last decade, we have witnessed impressive progress in the science and technology of dilute nitrides, leading to a great number of scientific reports and device concepts. At the moment, there is an urgent need to provide a comprehensive and up-to-date review on the current status of our knowledge and the future perspectives of this rapidly developing field. The aim of the present book is to meet this demand by providing the readers with an overview of our present understanding of the physical properties of dilute nitrides, as well as of the great potential of these materials for various device applications, in particular in the areas of high-efficiency multijunction solar cells and near-infrared lasers operating within the fiber-optic communications window of 1.3–1.55 μm, where dilute nitrides are currently considered as among the most promising material systems. This book — the work of leading experts in the field — presents a broad and in-depth description of the basic electronic, optical, and vibrational properties of dilute nitrides; the properties of impurities and defects; recent developments in fabrication techniques; and potential device applications in optoelectronics and photonics. The book includes contributions from different points of view on some key issues that are currently under intense debate, with the hope of stimulating further research so that the controversies can be resolved quickly, leading to rapid advances and breakthroughs in our understanding.

This book is intended for postgraduate students, researchers, and engineers in both academia and industries active in the fields of solid-state physics, semiconductor materials science and engineering, and optoelectronic device applications. It will also be of interest to senior university students in physics and electrical engineering and for short-course instructors and consultants specializing in the field of semiconductor optoelectronics and photonics, and more generally in solid-state physics, materials science, and electrical engineering.

Dilute nitrides (or nitrogen-containing, anion-mixed III-V ternary and quaternary alloys) such as GaAsN, GaPN, or InGaAsN are derived from conventional III-V semiconductors such as GaAs, GaP, or InGaAs by the insertion of nitrogen into the group V sublattice. They were initiated in the beginning of the 1990s, inspired by the success in the fabrication of high-quality III-V alloys based on (Al, Ga, In)(As, P), as well as the breakthrough in growth of wide-band-gap III-nitrides. The main motivation by that time was to close the gap between the nitrides and the arsenides, and thus to fabricate light emitters covering the entire visible spectral range based on the direct band-gap materials. However, very soon it was realized that dilute nitrides in fact belonged to a novel class of highly mismatched semiconductors with strikingly unusual physical properties. These properties were markedly different from conventional semiconductor alloys and were found attractive for device applications in previously unforeseen areas.

Unlike conventional III-V semiconductor alloys, where the band-gap energy of the alloys can be reasonably approximated as a weighted linear average of the band gaps of their parental compounds, dilute nitrides exhibit a huge bowing of the band-gap energy due to a strong mismatch in electronegativity and atomic size between N and the replaced anion atom. For example, although GaN has a band-gap nearly 2 eV higher than that of GaAs, the band-gap of the ternary alloy GaN_xAs_{1-x} decreases rapidly with increasing N content. Just 1% N substitution decreases the room-temperature band-gap from ≈ 1.42 eV to ≈ 1.25 eV. Moreover, the giant band-gap bowing is accompanied by a reduction in the lattice constant of dilute nitrides, as opposed to conventional III-V alloys, where a reduction in band-gap energy is typically achieved by inserting an element that increases the lattice constant. Such a unique combination of these remarkable fundamental properties with the additional freedom in strain compensation and engineering provides unprecedented new possibilities and dimensions for widely extended band-structure engineering. The availability of GaIn-NAs(Sb) has opened the door for novel optoelectronic devices based on GaAs technologies operating within the important wavelength window of 1.3–1.55 µm for data- and tele-communications, as well as for innovative designs of efficient solar cells. Some of these devices are already making their way into the market. Dilute nitrides also hold great promise in many other potentially extremely important application areas, for example THz devices for sensing and medicine, optoelectronic integrated circuits (OEICs) for signal processing and defense, and the development of III-V materials grown on silicon with the promise of further electronic/photonic integration.

Besides the new possibilities in device applications provided by this new materials system, many intriguing physical properties distinguishably different from conventional semiconductor alloys have been discovered. In addition to the giant bowing in band-gap energy, dilute nitrides were shown to exhibit unusual splitting of the conduction-band states into two subbands, sublinear pressure dependence of the band-gap energy, a reduced temperature variation of the band-gap energy to a degree that is strongly dependent on host lattices, a strong enhancement in electron effective mass of Ga(In)AsN, a significant change in the character of the conduction-band edge, and high-lying conduction-band states, etc. These new findings have stimulated intense research to understand the underpinning physics and determine the exact role of N in modifying the electronic band-structure and optical properties of dilute nitrides. The extensive theoretical and experimental investigations have led to many new findings and physical insights as well as controversies, including the exact physical mechanism for the giant band-gap bowing. The research in dilute nitrides has also provided and will continue to provide valuable insight into both the physics of impurities as well as the physics of the largely unexplored class of highly mismatched semiconductors.

We would like to express our thanks to all contributors for being willing to share their experience and expertise with the interested readers. The excellent initiative and support by Prof. M. Omar Manasreh and the efficient collaboration with Taylor & Francis are gratefully acknowledged. We hope that this book of collected review articles from leading experts provides an up-to-date account of physics and applications of dilute nitrides and will further stimulate basic and applied research in the fascinating field of dilute nitrides and related highly mismatched semiconductors.

Irina A. Buyanova

Weimin M. Chen

Linköping, Sweden

November 2003

CONTRIBUTING AUTHORS

P.R.C. Kent and A. Zunger, National Renewable Energy Laboratory, Golden, CO

W. Walukiewicz, W. Shan, J. Wu, and K.M. Yu, Lawrence Berkeley National Laboratory, Berkeley, CA

E.P. O'Reilly, A. Lindsay, and S. Fahy, NMRC, University College, Lee Maltings, Prospect Row, Cork, Ireland

W.M. Chen and I.A. Buyanova, Department of Physics and Measurement Technology, Linköping University, Linköping, Sweden

P.J. Klar, Department of Physics and Materials Science Center, Philipps-University, Marburg, Germany

S. Tomic, University of Surrey, Surrey, U.K.

A. Polimeni and M. Capizzi, INFM and Dipartimento di Fisica, Università di Roma, "La Sapienza," Roma, Italy

A. Mascarenhas and S. Yoon, National Renewable Energy Laboratory, Golden, CO

S. Zhang and S.-H. Wei, National Renewable Energy Laboratory, Golden, CO

C. W. Tu, Department of Electrical and Computer Engineering, University of California, La Jolla, CA

M. Kondow, Hitachi Ltd., Central Research Laboratory, Tokyo, Japan

A. Adams and R. Fehse, University of Surrey, Surrey, U.K.

D.J. Friedman, J.F. Geisz, and A.J. Ptak, National Renewable Energy
Laboratory, Golden, CO

J.S. Harris, Solid State and Photonics Lab, Stanford University,
Stanford, CA

CHAPTER 1

An Atomistic View of the Electronic Structure of Mixed-Anion III-V Nitrides

P.R.C. KENT and A. ZUNGER

National Renewable Energy Laboratory, Golden, CO

1-591-69019-6/04/$0.00+$1.50

1.1 INTRODUCTION

Mixed-anion nitrides exhibit a duality in their electronic and transport properties, showing both homogeneous (bulklike) and heterogeneous (fluctuating) behavior. One observes homogeneous, bulklike characteristics, such as resonances within the continua, rigid shift of the conduction band with temperature and pressure, the appearance of new bulklike absorption edge such as E_+, and a split E_1. Significantly, the popular "band anticrossing" (BAC) model of Shan et al. [1] only addresses this homogeneous behavior. On the other hand, one also notes characteristics of heterogeneous localization centers and alloy fluctuations, such as a distribution of various nitrogen pairs and clusters whose levels are within the forbidden energy gap, absorption vs. emission Stokes shifts, emission blueshift upon increased excitation power, band tails with long decay times, and asymmetric line shapes. These heterogeneous localization characteristics are particularly apparent in photoluminescence (vs. absorption) and are pertinent to the design of photoemitting devices such as LEDs and lasers.

Theoretical approaches to the electronic structure of alloys can be divided into isomorphous and polymorphous models. In an isomorphous ("single shape") model, one considers a single (or very few) local atomic environment that spans the entire alloy structure. Clearly, explicit consideration of lattice relaxation, localization, and charge-transfer effect are excluded in such high-symmetry models. The most popular isomorphous alloy model applied to the dilute III-V nitrides is the BAC [1], in which the alloy is constructed from a single substitutional nitrogen impurity and embedded in the host, e.g., GaAs, with a composition-dependent coupling to the conduction-band minimum (CBM). This model treats only the perturbed host states (PHS), but in describing the alloy in terms of a single impurity motif, it ignores fluctuations due to inhomogeneities. Although this model permits remarkable fitting to measured bulklike absorption quantities such as composition, pressure, and temperature-dependent band gaps $E_g(x)$, $E_g(p)$, and $E_g(T)$, the entire phenomenology of the evident alloy fluctuation behavior remains experimentally unexplained. Furthermore, the dimension of alloy state evolution with composition is lost because the composition dependence underlying the model is, by construction, smooth.

Polymorphous ("many shapes") alloy models focus on the central property that distinguishes disordered random alloys from ordered compounds, namely, the existence of a plurality of local atomic environments [2–6]. Critically, these models are able to address both the bulklike features addressed by the isomorphous models, and, in addition, properly address the characteristics due to heterogeneous localization centers and alloy fluctuations. Our approach [7, 8] describes the *transition* between the two types of behavior as a function of concentration and pressure. In $GaP_{1-x}N_x$, for example, the Ga atom can be surrounded by five distinct near-neighbor structures, $P_m N_{4-m}$ (with $0 \leq m \leq 4$); nitrogen pairs can have arbitrary separation in the alloy; etc. Even a random distribution of impurities creates clusters by chance. Once different local environments are acknowledged, the phenomena of localization and fluctuations follows naturally, for some impurity clusters may induce a large enough perturbation with respect to the bulk to split a level into the gap. Different clusters then create different levels, leading to an inhomogeneous distribution. As in the isomorphous alloy models, perturbed host states (PHS) are allowed. However, unlike such models, cluster states (CS) are also included, and these two types of states can interact. The plurality of different CS are responsible for most of the characteristics of dilute nitrides observed in photoluminescence (PL).

We have developed a polymorphous model of dilute III-V nitrides [7, 8] based upon large-scale atomistic supercell calculations and the empirical pseudopotential method (EPM). This model provides a straightforward explanation of a wide variety of experimentally observed phenomena, whether observed in PL or absorption (photoluminescence excitation, PLE). Our calculations are also able to address the issue of short-range ordering in quaternary alloys, such as InGaAsN, with only simple extensions.

The isomorphous (band anticrossing) and polymorphous (atomistic EPM) models disagree on several issues, including:

1. Our polymorphous model includes alloy fluctuations via the existence
    of many different local environments around nitrogen atoms, giving
    rise to pairs, triplets, etc. These fluctuations are entirely absent from
    the two-level band anticrossing model. The fluctuations lead to exci-
    ton localization, Stokes-like shift between absorption and emission,
    blueshift with increasing temperature at low temperatures, and the
    creation of a localized-to-delocalized transition at the amalgamated
    point x_c.
2. Our model shows clearly that more than two levels are perturbed by
    nitrogen incorporation, e.g., the L_{1c}, X_{1c} host states. This leads to the
    formation of a wide band of states above E_- and below E_+ that is
    entirely absent from the two-level band anticrossing model. The
    mixing-in of L-character is evident experimentally, both in the reso-
    nant Raman scattering of Cheong et al. [9] and in the BEEM exper-
    iment of Kozhevnikov et al. [10], which directly detected the L-
    character.
3. For GaPN, the two-level BAC is forced to attribute alloy bowing to
    the downward shift of the impurity states, rather than to the proper
    alloy conduction band that lies above it.
4. With increasing temperature, the band-gap significantly redshifts, as
    states are thermally depopulated in both GaAs [11, 12] and-gap [13].
    This depopulation is not accounted for in the BAC.
5. Exchange of discrete levels with continua due to pressure: multiple PL
    lines emerge from the conduction continua upon application of pressure
    to dilute alloys [14], and with the addition of nitrogen, select high-
    energy PL lines disappear [14] due to partial CS delocalization. The
    BAC cannot account for either of these effects.

The BAC and supercell calculations agree on the following points

1. The E_- level moves down and the E_+ level moves up as x_N increases.
2. The pressure coefficient (reduced with respect to the bulk) saturates as
    x_N increases.
3. N incorporation raises the electron effective mass at the bottom of the
    CBM, and if one moves farther up in energy into the conduction band,
    the mass increases even further.
4. The E_1 transition at ~3 eV is unaffected by nitrogen.
5. Incorporation of N in GaP reverses the pressure coefficient of GaP to
    a strongly positive value and makes the interband transition direct.

We give an outline of the calculations in Sections 1.2 and 1.3. In Section
1.4 we present a detailed comparison with experiment. In Section 1.5
we present results for the InGaAsN quaternary. Conclusions are given in
Section 1.6.

1.2   THEORY OF DILUTE III-V NITRIDES

Numerous theoretical methods have been applied to study the anomalous effects seen in III-V nitrides [1, 7, 8, 15–24], yielding sometimes conflicting results. The primary reason for the lack of consensus between different methods is the unprecedented difficulty of theoretically modeling the properties of these alloys. Critically, some approximations that can yield predictions in good agreement with experiments for conventional III-V alloys are no longer valid in III-V-N solid solutions [25, 26]. Photoluminescence (PL) [27, 28], pressure [29], and theoretical evidence [7, 8] demonstrate that fluctuations in nitrogen content, including the random formation of pairs, triplets, and clusters of nitrogen atoms, must be explicitly included for accurate modeling and understanding of these materials. Many common theoretical methods are therefore not applicable, because they consider only a single structural motif throughout the alloy volume (isomorphous models). This includes the band anticrossing model (BAC), the virtual-crystal approximation (VCA), and coherent potential approximation (CPA), even though it may provide a good fit to *some* properties. Additionally, some further common approximations that work in ordinary III-V alloys are not applicable here. For example, effective-mass approaches [17] are not suitable for modeling nitrogen localized states, since they couple host bands from across the entire Brillouin zone. Models that ignore atomic relaxation resulting from the N-P or N-As size mismatch [22–24, 30] are also problematic, since there is large relaxation that couples to the electronic structure; density functional calculations cannot yet be applied to sufficiently large supercells. Tight-binding methods suffer from small basis sets and are unable to reproduce the details of the full electronic structure.

An accurate theoretical description of these features is afforded by the atomistic supercell, where one places As, Ga, and N atoms on lattice sites in any desired configuration (random, clustered, etc.), relaxes the atomic positions, and solves the Schrödinger equation for the final configuration. The latter can be solved using density functional calculations in the local density approximation (LDA-DFT) (Wei et al. [19]), tight binding (Lindsay and O'Reilly [17]), or EPM. The LDA-DFT approach has not been abandoned, and it yields good information about wavefunction localization, but it is computationally difficult to obtain good statistics on different nitrogen configurations or treat ultradilute alloys. Tight-binding methods can provide only limited wavefunction information, but may be applied to large supercells. The EPM is unique in being able to study very large — hence potentially very dilute or impuritylike — systems while yielding LDA-error-free real-space wavefunctions, and it enables routine study of 10^5 atom systems.

1.2.1 Large Supercell EPM Calculations

A brief overview of the key components of EPM calculations for nitrides is given below. Further details can be found in a work by Kent and Zunger [8]

The first step in any atomistic electronic-structure calculation consists of describing the microstructure and nanostructure of the system under investigation. We do so by generating supercells and then distributing cations and anions at the corresponding atomic sites of these supercells. This distribution must be consistent with the investigated atomic-ordering, if any, as well as with the intended alloy concentration. Furthermore, when studying disordered alloys, the atomic distribution must be random and the supercells must be as large as possible in order to reproduce the different possible chemical environments. "Isomorphous" alloy models, such as the VCA, and band anticrossing describe the alloy via a single motif, or atomic distribution. Here, we use a "polymorphous" approach where the alloy consists of many different motifs, or local environments.

Once the atomic configuration has been decided, atomic relaxation must be taken into account to be able to accurately reproduce the electronic properties of nitride alloys. Neglecting or poorly approximating atomic relaxation leads to incorrect predictions of electronic and optical properties [26]. The relaxed atomic positions are obtained from the valence-force-field method (VFF) [31, 32]. In this method, the forces between atoms are calculated using a simple "balls and springs" Hamiltonian, which includes both bond-stretching and bond-bending terms. The method is accurate because, in the bulk, all atoms are fully coordinated, unlike surface atoms or point defects. The parameters of the VFF method that are typically used are fit to experimental elastic constants [32] for non-nitrides, and fit to LDA results [33] for nitrides. The method yields a good agreement with first-principles results for the atomic coordinates of mixed-anion nitride alloys [8, 16].

The crystal potential $V(r)$ is written as a superposition of screened atomic pseudopotentials $v_\alpha(r)$, where α = Ga, In, N, P, As, etc. These modern pseudopotentials differ from the traditional empirical pseudopotentials [34, 35] in that (1) they are specified continuously at all reciprocal lattice vectors (and hence can be applied to large unit cells), (2) the all-important band offsets between different materials are fitted, (3) the measured bulk effective masses and LDA-calculated deformation potentials are carefully fitted, (4) the potential depends on the local environment, and (5) the potential explicitly depends on strain.

Finally, we can use the EPM technique to calculate the optical and electronic properties of very large supercells [36]. The eigenfunctions and eigenvalues of this Hamiltonian are determined by using the folded-spectrum method [37]. This numerical technique produces single-particle eigensolutions in a given energy window without having to obtain (and orthogonalize to) lower energy eigensolutions. As a result, the overall method scales linearly in computational time with the number N of atoms in the supercell, while conventional band-structure methods — that require both self-consistency of the crystal potential and knowledge of all occupied levels — exhibit a time scaling of N^3. The key EPM approximation is that there is no self-consistency beyond the GaAs and GaN binaries. The approximation is checked for accuracy by comparing the resultant alloy bowing and band offsets relative to LDA-DFT for simple superlattices [18, 26].

1.3 ELECTRONIC STRUCTURE EVOLUTION OF DILUTE N III-VS

In this section we concentrate, for reasons of clarity, on the electronic structure evolution of GaAsN. We have previously shown the mechanism to be common to GaPN [8] and InGaAsN [38], and expect it to apply equally well to InAsN, InPN, GaSbN, InSbN, and derivative quaternaries.

1.3.1 Electronic Structure Of Dilute Impurities

The fundamental physics of dilute nitride impurities in GaAs is characterized by the formation of nitrogen localized near band-gap CS. Historically, only the $a_1(N)$ level, resonant 150–180 meV above the conduction-band minimum (CBM), has been identified [39], but small clusters of nitrogen atoms create other levels. The CS result from the differences in atomic size and orbital energies between the nitrogen and arsenic atom it substitutes. Our empirical pseudopotential calculated $a_1(N)$ level is at $E_c + 150$ meV and $E_c + 180$ meV for 4,096 and 13,824 atom cells, respectively, in close agreement with experiment.

To consider the role of small nitrogen aggregates formed during growth, we have considered a number of prototypical clusters: pairs, triplets, clusters of multiple nitrogens around a single gallium, and directed chains of nitrogen atoms. Many other clusters are possible, particularly in higher nitrogen concentration alloys, even on the basis of random statistics. A full description of different clusters is given in a work by Kent and Zunger [8]. Here, for clarity, we concentrate on a single type of cluster.

In Figure 1.1, we show the calculated energy levels for a Ga-centered tetrahedron with its four vertices occupied by $As_{4-p}N_p$, with $0 \leq p \leq 4$. Note that $p = 1$ corresponds to an isolated impurity, and $p = 2$ to a first nearest-neighbor N-N pair. We see that the levels become deeper as p increases, consistent with the fact that, on an absolute scale, the CBM of GaN is ~0.5 eV below that of GaAs. The induced CS are highly nitrogen localized, evidenced by the wavefunction isosurface (inset). We also considered (Figure 1.1) extended [1,1,0]-oriented chains of increasing length, motivated

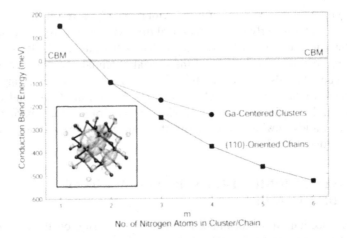

Figure 1.1 Energy levels of Ga-centered nitrogen clusters and [1,1,0]-directed nitrogen chains in GaAs, calculated in 4096 atom cells. (Inset) Wavefunction isosurface of cluster state at a $Ga(N_4)$ cluster, showing strong nitrogen localization.

by the comparatively deep nature of even a [1,1,0]-oriented pair ($p = 2$, above). These extended chains consist of 3, 4, 5, etc. nitrogen atoms. We observed that each additional atom in the chain produced successively deeper levels. In general, we find that an increased local concentration of nitrogen atoms, of any orientation, induces deep, dipole allowed levels. It is not necessary that the nitrogens be immediate neighbors. Small nitrogen aggregates therefore can contribute to below band-gap PL, even at low impurity concentrations.

1.3.2   Evolution of Electronic Structure With Composition

Nitrogen introduces two types of states in the dilute limit: (1) the perturbed host states (PHS) residing within the continuum such as $a_1(X_{1c})$, $a_1(L_{1c})$, and $a_1(\Gamma_{1c})$, and (2) the cluster states (CS) residing inside (or near) the band gap, e.g., the pair and higher cluster states (Figure 1.1). We next address the question of how the PHS and CS evolve as the nitrogen composition increases.

We perform calculations as a function of nitrogen concentration by randomly distributing up to 20 nitrogen atoms onto the anion sites of GaAs in a 1,000-atom supercell, and 13,824-atom supercells for convergence checks, repeating this for 15 randomly selected configurations at each composition. The ensuing energy levels are then collected and analyzed for their degree of localization by computing for each level ϕ_i the average distance $R_a^{(i)}$ from each nitrogen site α at which 20% of the wavefunction is enclosed. Through this measure we have classified each level as either "localized" or "quasi-localized."

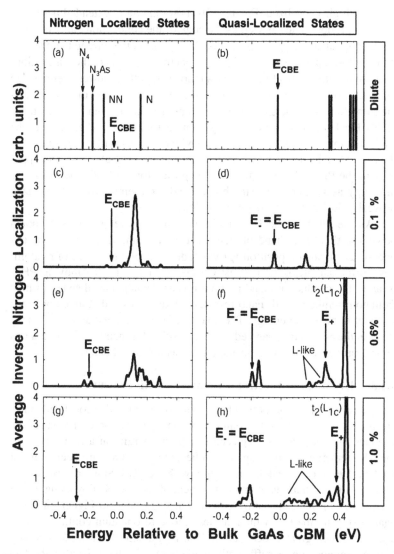

Figure 1.2 Spectral dependence of average nitrogen localization for (left) nitrogen localized "cluster states" and (right) quasi-localized "perturbed host states" of GaAsN for selected nitrogen compositions. The vertical arrows show the position of the alloy conduction band edge E_{CBE}.

Figure 1.2 depicts the spectral dependence of the average localization $\Sigma_\alpha(1/R_\alpha^{(i)})$ for localized and quasi-localized levels of GaAsN. Figure 1.2a shows the resonant localized single-impurity $a_1(N)$ state, located within the conduction band, and selected pair, triplet ($GaAs[N_3]$), and quadruplet ($GaAs[N_4]$) cluster states appearing inside the band gap. These wavefunctions are highly localized. Figure 1.2b shows the more extended perturbed X, L, and Γ host states as well as the edge of the conduction band, denoted by the bold arrow "E_{CBE}" (also called "E_-" [1]). As the nitrogen concentration increases, panels d, f, h, and j show that the edge E_{CBE} of the conduction-band minimum (vertical heavy arrow) moves rapidly to lower energies due to anticrossing and repulsion with higher energy members of the PHS. At the same time, the energy of the CS are pinned and remain fixed, as these highly localized states do not strongly interact with each other. Indeed, the wavefunctions of the CS do not change with composition. At the same time, the $t_2(L_{1c})$ band appears constant in energy, at $E_c + 0.4$ eV, while the upper edge of the PHS (also called "E_+" [1]) appears for $x \approx 0.6\%$ and moves up in energy as x_N increases. This broad band represents mostly delocalized or weakly localized a_1 perturbed host states.

As the edge of the PHS move rapidly to lower energies ("optical bowing"), this broad band of states sweeps past the discrete CS one-by-one. At a critical composition x_c (which depends on the degree of randomness in the samples), the deepest CS are overtaken by the moving PHS. Near x_c, the conduction-band minimum is an "amalgamated state" formed from both semilocalized (Fano-resonance-like) states and more delocalized states of comparable energy. The duality of semilocalized and delocalized states at the conduction band edge is responsible for many of the anomalous optical properties of dilute nitride alloys, as discussed below.

1.3.3   Evolution of Electronic Structure With Pressure

An interesting question is to what extent localized and delocalized states are mixed. Klar et al. [40] found that upon application of pressure, the conduction band edge is displaced to higher energies at a rapid pace, so the CS reemerge into the gap. This reflects the low pressure coefficient of the CS, due to their weak hybridization with the PHS. Similarly, Buyanova et al. [41] found that quantum confinement of the GaPN alloy using a GaP barrier can displace the alloy conduction band edge to higher energies, again exposing the CS, which are less prone to quantum confinement on account of their greater localization. One would expect that for sufficiently high N composition or sufficiently high-energy CS, the CS will hybridize with the delocalized host states and become hostlike. At this point, they will acquire a similar degree of delocalization as the host. Upon application of pressure or quantum confinement, these states will *not* emerge into the band gap, moving instead with the conduction band edge.

We have calculated the energy levels vs. pressure of a supercell containing ⁓64,000 atoms with selected nitrogen clusters placed in it [42]. We consider two limiting nitrogen compositions: a highly dilute alloy where the CS are still in the gap, below the conduction band edge (Figures 1.3a and 1.3b), and a well developed, post-amalgamated alloy, where the CS have already been overtaken by the PHS and reside above the conduction band edge (Figures 1.3c and 1.3d). We apply pressure to both cases. We denote with "D" and "L" whether the state is "delocalized" or "localized," respectively.

- *Isolated nitrogen in the dilute limit:* In the highly dilute alloy, an *isolated* nitrogen impurity (Figure 1.3a) appears at low pressure as a localized $a_1(N)$ level above the delocalized conduction band edge $a_1(\Gamma_{1c})$, as observed by Wolford et al. [39] and by Liu et al. [43]. As pressure is applied, the $a_1(\Gamma_{1c})$ and $a_1(N)$ levels anticross, leading to the emergence of the localized $a_1(N)$ level into the gap, with its characteristically small pressure coefficient ($a_P = 12$ meV/GPa at $P \approx 4$ GPa).

- *Nitrogen triplet in the dilute limit:* As an example of a highly localized CS, we consider an N-N-N triplet in the C_{2c} geometry (a triplet of atoms aligned along the [1,1,0] axis). At low pressure, this CS appears in the dilute alloy as an ultralocalized level, 250 meV below the conduction band edge (Figure 1.3b). As pressure is applied, the localized state remains in the gap: the delocalized conduction band edge moves rapidly to higher energies, whereas the triplet state moves with a very low pressure coefficient, owing to its highly localized character.

- *Impurities in well-developed alloy:* A 1.5% random alloy has various N clusters in it, which are formed by chance. In our example, the conduction band edge has now descended by 360 meV from the dilute limit (1.5 eV, Figure 1.3a, to 1.14 eV in the 1.5% alloy, Figure 1.3c). All CS are by now resonances above the conduction band edge, as evidenced by the fact that upon application of pressure (Figure 1.3c), no states emerge into the band gap. To test specifically this point, we have deliberately placed in the well-developed alloy the same N-N-N triplet that produced a deep localized gap state in the ultradilute limit (Figure 1.3b). We find (Figure 1.3d) that this state now lies in the conduction band. To probe if the N-N-N CS is localized or not, we apply pressure to this cell (Figure 1.3d). Inspection of the lowest 200 meV above the conduction band edge shows no ultralocalized state. We see that whereas, in the dilute limit (Figures 1.3a and 1.3b), high pressure inevitably leads to the existence of a localized NNN level in the gap, in the strongly post-amalgamated limit (Figure 1.3d), the CS has sufficiently hybridized with the host conduction states that its pressure coefficient is no longer different from the host states. Consequently, all conduction states are displaced in energy in tandem, and no level emerges into the gap (Figure 1.3d).

In Figure 1.4, we describe schematically the pressure behavior of CS for dilute alloys (parts a, b), and concentrated alloys (c, d). One can expect that there are intermediate cases between the ultradilute limit (where all CS exist in the gap at high pressure) and the well-developed alloy (no CS

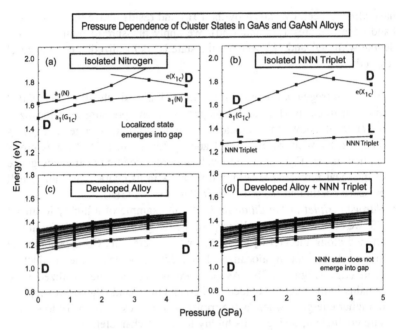

Figure 1.3 Calculated pressure dependence of cluster states in GaAsN. "D" and "L" denote delocalized and localized states, respectively. (a) Isolated nitrogen in GaAs, (b) N-N-N triplet in GaAs, (c) the well-developed 1.5% GaAsN alloy, (d) the 1.5% alloy containing the N-N-N triplet.

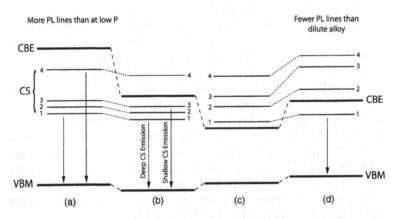

Figure 1.4 Schematic description of the displacement of cluster states (CS) and conduction band edge (CBE) in nitride alloys with pressure and composition. Note that upon application of pressure, the shallow CS emerge (or stay) in the gap in the dilute alloy (a, b), while in the concentrated alloy only the deepest CS emerge into the gap (c, d).

in the gap). In these intermediate cases, the weakly hybridized CS, above but still close to the conduction band edge (e.g., state 4 in Figures 1.4a and 1.4b or state 1 in Figures 1.4c and 1.4d), will eventually emerge into the gap upon application of pressure (since their pressure coefficients are sufficiently distinct from those of the host conduction states). Thus the number of PL lines will *increase* with pressure (Figure 1.4a vs. Figure 1.4b). In contrast, the strongly hybridized CS that are further above the conduction band edge will not emerge into the gap upon application of pressure (e.g., states 2, 3, 4 in Figure 1.4c and Figure 1.4d). We expect that the CS that are deepest in the dilute alloy (farthest from the CBM, lowest energy PL), such as state 1 in Figure 1.4c, will be overtaken by the conduction band edge last, and thus reemerge upon application of pressure, Figure 1.4d. The behavior illustrated in Figure 1.4 contradicts the impurity band model [44] that suggests the CS will broaden into a band at high nitrogen concentrations, and thus will not reemerge as *narrow* PL lines with application of pressure. Likewise, the BAC [1] does not predict the behavior illustrated, since the CS do not feature in this model.

1.4 EXPERIMENTAL CONSEQUENCES AND VERIFICATION

Based on our calculated results, we have divided the evolution of the electronic structure into three distinct regions, corresponding to increasing nitrogen content. The distinct experimental features observable in each region are interpreted in terms of the duality of CS and PHS.

1.4.1 Pre-amalgamation: Ultradilute Nitrogen Concentrations

In the ultradilute regime (nitrogen concentration $x < 0.01\%$) one observes:

Localized, single impurity levels near the band gap
Anomalously small pressure dependence of single impurity states

1.4.1.1 *Localized, single impurity levels near the band gap*

In conventional isovalent alloys such as GaAs:P or GaAs:In (where the underlining indicates the substituted host atom), the ensuing perturbation potential $V_{As} - V_P$ or $V_{Ga} - V_{In}$ is too weak to create a bound state in the gap. In contrast, absorption and photoluminescence excitation (PLE) of GaP:N and GaAs:N show the "N_x center" due to anion-substitutional isolated nitrogen. In GaP:N, this level appears as an impurity-bound exciton at $E_{CBM} - 33$ meV *below* the conduction-band minimum (CBM) [27, 45–47], whereas in GaAs:N it appears as a *sharp resonance* at $E_{CBM} + 180$ meV [39, 43, 48, 49] above the CBM.

1.4.1.2   Anomalously small pressure dependence of single impurity states

Shallow, effective-mass-like impurity levels (GaAs:Zn or GaAs:Si) are constructed from the wavefunction of the single nearest host crystal state. Consequently, when pressure is applied, such impurity levels change their energy at the same rate as the energetically nearest host crystal state [50]. In contrast, the impurity levels in dilute GaP:N and GaAs:N have anomalously small pressure coefficients. In GaP:N, the energy of the impurity-bound exciton is almost pressure independent [51, 52], whereas the X_{1c} CBM of the GaP host crystal descends at a rate of -14 meV/GPa. In GaAs:N, the nitrogen level moves with pressure to higher energies at a much slower rate (^{a}40 meV/GPa [43, 49]) than the G_{1c} CBM of GaAs [53] ($+110$ meV/GPa). Such small pressure coefficients usually indicative of localization, whereby the wavefunction is constructed from many bands of the host crystal, rather than from the nearest host crystal state [54]. Examination of calculated wavefunctions [8, 55] finds this to be the case.

1.4.2   Intermediate Nitrogen Concentrations

In the intermediate concentration regime (up to $\approx 1\%$ nitrogen), one observes:

> Sharp photoluminescence (PL) lines due to impurity clusters
> Redshift between absorption/PLE and emission
> Composition-pinning of the impurity-pair energy levels
> Selective delocalization of localized states upon application of pressure

1.4.2.1   Sharp photoluminescence (PL) lines due to impurity clusters

Even random substitution of impurities onto the atomic sites of a host crystal creates, by chance, impurity pairs and higher-order clusters. In conventional isovalent III-V alloys, such pairs give rise to broad resonances, within the valence and conduction continua [2–6], but no gap levels. In contrast, in GaPN and GaAsN, the N-N pairs form discrete levels inside the band-gap extending in GaP down to $E_{CBM} - 160$ meV [47, 56–58] and in GaAs down to $E_{CBM} - 10$ meV [43, 49, 59] or $E_{CBM} - 80$ meV [60–62]. Such clusters do not appear to create deep levels in ordinary, non-nitride, alloys. These cluster states are found in EPM calculations (Section 1.3).

1.4.2.2   Redshift between absorption/PLE and emission

In high-structural-quality, random, direct-gap III-V alloys, absorption and emission occur at the same energy. In contrast, already at a concentration of 0.05–0.1% nitrogen in GaAs, the emission lines are redshifted with respect to absorption [63]. At higher concentrations, the shift increases in energy [60, 64].

As the concentration of nitrogen increases further, one observes composition-pinning of the impurity-pair energy levels.

1.4.2.3   Composition-pinning of the impurity-pair energy levels

The sharp emission lines from the pair levels remain initially at a fixed energy as the nitrogen composition increases both in GaP:N [44] and in GaAs:N (0.05-0.1% [63]). This surprising pinning suggests that the impurities do not interact with each other. This behavior is characteristic of deep transition metal impurities in semiconductors [54, 65], but not of hydrogenic impurities (Si:P, As), which readily broaden into bands and shift in energy as their concentration increases [66]. The pinning behavior is captured in our calculations (Figure 1.2 and Section 1.3).

As the concentration increases further, the PL from pair states becomes asymmetric, with a sharp high-energy cutoff and a low-energy tail [28, 58, 67–69], where the carriers have anomalously long lifetimes [69–71]. At yet higher concentrations, all of the pair/cluster lines disappear into a single, broad emission line [28, 40, 44]. This behavior contrasts with conventional alloys, where the emission line is featureless at all alloy compositions.

1.4.2.4   Selective delocalization of localized states on application of pressure

Under hydrostatic pressure, multiple PL lines (CS) emerge from the conduction continua in dilute alloys into the band-gap [43]. However, a careful accounting of PL lines in 0.25–0.4% superlattices [14] demonstrates that certain PL lines are absent in the higher concentration alloys, even though they would be expected to be well inside the band-gap at high pressure. EPM calculations [42], Figure 1.3, reveal that the disappearance of certain lines under pressure can be attributed to partial localization of CS with increased nitrogen concentration, resulting in higher pressure coefficients.

1.4.3   Post-Amalgamation: High Nitrogen Concentration Alloys

Once all of the sharp lines of pairs/clusters disappear, additional unexpected effects remain:

1.4.3.1   Band-gap shows huge and composition-dependent optical bowing

In conventional $A_xB_{1-x}C$ isovalent III-V alloys, the band-gap $E_g(x)$ changes with respect to the composition-weighted average of the constituents with constant bowing coefficient (usually <1eV). In $GaP_{1-x}N_x$ and $GaAs_{1-x}N_x$, the bowing is huge and composition dependent, being largest at small x: ≈ 26 eV at $x < 1\%$ eV, and ≈ 16 eV at $x > 1\%$ [72].

1.4.3.2 Electron mass is anomalously heavy but decreases with concentration

In conventional alloys, the mass changes monotonically with composition [53]. The reduction of the band-gap upon N addition (bowing) will reduce the effective mass, whereas mixing of L- and X-character in the Γ-like CBM due to the impurity potential will increase the mass. The balance between these effects will depend on the nitrogen concentration. In conventional alloys, the second effect is absent. Small amounts ($\approx 1\%$) of nitrogen increase the 0.066 m_e mass of pure GaAs to ≈ 0.4 m_e [73] or to 0.12–0.19 m_e [74], but subsequent addition of nitrogen appears to reduce the electron mass [73]. As the Fermi energy moves further into the conduction band, the effective mass becomes higher [75]. In GaP, 2.5% nitrogen creates a large mass of ≈ 0.9 m_e [68] compared with the X-band effective masses ($m_\parallel^* \approx 0.25$ m_e, $m_\perp^* \approx 4.8$ m_e [53]).

1.4.3.3 Reduction in band-gap with increased temperature slows down with nitrogen addition

Band gaps are always reduced as temperature is increased [53]. However, in conventional alloys the temperature coefficient is close to the concentration-weighted average over the constituents. This reduction in PL energy with increased temperature slows down dramatically with small addition of nitrogen to GaAs [11, 12] and-gap [13]. Furthermore, the intensity of the PL lines of conventional alloys decreases with increasing temperature, but this decrease is accelerated by nitrogen addition, especially at low temperatures [76].

1.4.3.4 Energy of the PL lines blueshifts as excitation power increases

The energy of the PL lines is blueshifted as the excitation power increases [70], indicating occupation of previously empty states (so excitation must now occupy higher energy states). This is known to occur in alloys containing localized, quantum dotlike clusters [77].

1.4.3.5 Emission decay time increases with decreasing emission energy

In other words, the states that are deeper in the gap (lower emission energies) have weaker dipole transition elements (or equivalently, less Γ character and more off-Γ character) [78].

1.5   QUATERNARY PHYSICS

The $(Ga_{1-y}In_y)(As_{1-x}N_x)$ alloys hold great promise for overcoming the poor temperature characteristics of conventional long-wavelength lasers [79–82], and they are a key candidate material for high-efficiency multijunction solar cells [83]. One technologically important feature of these quaternaries is that they can have a perfect lattice match to a GaAs or InP substrate with the appropriate ratio of indium y to nitrogen x concentrations, thus avoiding the growth problems encountered in strained samples, such as the ternary alloys GaAsN and GaPN. The relationship between the indium and nitrogen concentrations leading to a perfect lattice match of $(Ga_{1-y}In_y)(As_{1-x}N_x)$ with GaAs or InP has been indicated in a work by Ballaiche [84], assuming Vegard's rule.

Quaternaries also distinguish themselves from ternary alloys by the nonuniqueness of the number of bonds of each cation-anion type. The ratio between the number of A–C and B–C bonds in $(A_{1-x}B_x)C$ ternaries is simply $1-x{:}x$, while the number of GaAs, InAs, GaN, and InN bonds in $(Ga_{1-y}In_y)(As_{1-x}N_x)$ alloys is not only related to the compositions x and y but also depends on the possible short-range atomic ordering of the quaternary. Another way of saying this is that $Ga_{0.5}In_{0.5}As_{0.5}N_{0.5}$ can be viewed as either GaAs+InN or GaN+InAs, but the formula alone does not reveal which of the two cases is right. Short-range atomic ordering (SRO) in $(Ga_{1-y}In_y)(As_{1-x}N_x)$ may seriously affect its optical and electronic properties, since atomic ordering is known to alter the band-gap and electronic wavefunctions in the mixed-anion $Ga(As_{1-x}N_x)$ ternary system [85]. This additional question raises two questions [38]: What is the effect of the different kinds of atomic bonds with respect to the disorder $(Ga_{1-y}In_y)(As_{1-x}N_x)$ alloy, and (2) what are the consequences of any resulting short-range order on the optical properties?

The first question was answered by using Monte Carlo (MC) simulations for which the internal energy incorporates strain effects, as predicted by the valence-force-field approach and chemical bond energies. These MC calculations revealed that, in $(Ga_{1-y}In_y)(As_{1-x}N_x)$ alloys lattice-matched to GaAs, nitrogen atoms prefer to be surrounded by In atoms, whereas As prefers to bond with gallium atoms. In other words, the number of In–N (large cation–small anion) and Ga–As (small cation–large anion) bonds increases relative to the random system. The question was answered by performing simulations on different kind of large $(Ga_{0.94}In_{0.06})(As_{0.98}N_{0.02})$ supercells: one with the SRO just mentioned above, and another without. The most important results of these simulations is that SRO $(Ga_{1-y}In_y)(As_{1-x}N_x)$/GaAs is expected to increase the band-gap with respect to the random alloy case, and results in the emergence of a band tail of localized states at the conduction-band minimum due to different clusters of nitrogens surrounded by varying numbers of indium and gallium atoms. This prediction has now been confirmed in several X-ray absorption experiments [86, 87].

1.6 CONCLUSIONS

We have presented a polymorphous model of dilute III-V nitrides based upon large-scale atomistic supercell calculations and the empirical pseudopotential method (EPM). Our model recognizes the existence of a plurality of local atomic environments and is, therefore, able to address the bulklike features addressed by isomorphous models such as the "band anticrossing" (BAC) model [1], and, in addition, properly address the characteristics due to heterogeneous localization centers and alloy fluctuations.

The atomistic supercell empirical pseudopotential model provides a straightforward explanation of a wide variety of experimentally observed phenomena, whether observed in PL or absorption (PLE). In particular, our approach describes the transition between the two types of behavior as a function of concentration and pressure. The model is also able to address the issue of short-range ordering in quaternary alloys, such as InGaAsN, and is equally applicable to other ternary and quaternary nitride alloys.

Within an isomorphous description, the entire phenomenology of alloy fluctuation behavior observed experimentally remains unexplained.

References

1. Shan, W., Walukiewicz, W., Ager III, J.W., Haller, E.E., Geisz, J.F., Friedman, D.J., Olson, J.M., and Kurtz, S.R., "Band Anticrossing in GaInNAs Alloys," *Phys. Rev. Lett.* 82: 1221–1224 (1999).
2. Magri, R., Froven, S., and Zunger, A., "Electronic Structure and Density of States of the Random $Al_{0.5}Ga_{0.5}As$, $GaAs_{0.5}P_{0.5}$ and $Ga_{0.5}In_{0.5}$ as Semiconductor Alloys," *Phys. Rev. B* 44: 7947–7964 (1991).
3. Zunger, A. and Jaffe, J., "Structural Origin of Optical Bowing in Semiconductor Alloys," *Phys. Rev. Lett.* 51: 662–665 (1983).
4. Bernard, J. and Zunger, A., "Electronic Structure of ZnS, ZnSe, ZnTe, and Their Pseudobinary Alloys," *Phys. Rev. B* 36: 3199–3228 (1987).
5. Wei, S.H. and Zunger, A., "Disorder Effects on the Density of States of the II-VI Semiconductor Alloys $Hg_{0.5}Cd_{0.5}Te$, $Cd_{0.5}Zn_{0.5}Te$, and $Hg_{0.5}Zn_{0.5}Te$," *Phys. Rev. B* 43: 1662–1677 (1991).
6. Mader, K. and Zunger, A., "Short- and Long-Range-Order Effects on the Electronic Properties of III-V Semiconductor Alloys," *Phys. Rev. B* 51: 10462–10476 (1995).
7. Kent, P.R.C. and Zunger, A., "Evolution of III-V Nitride Alloy Electronic Structure: The Localized to Delocalized Transition," *Phys. Rev. Lett.* 86: 2613–2616 (2001).
8. Kent, P.R.C. and Zunger, A., "Theory of Electronic Structure Evolution in GaAsN and GaPN Alloys," *Phys. Rev. B* 64: 115208 (2001).
9. Cheong, H.M., Zhang, Y., Mascarenhas, A., and Geisz, J.F., "Observation of Nitrogen-Induced Levels in $GaAs_{1-x}N_x$ Using Resonant Raman Scattering," *Phys. Rev. B* 61: 13687–13690 (2000).

10. Kozhevnikov, M., Narayanamurti, V., Reddy, C.V., Xin, H.P., Tu, C.W., Mascarenhas, A., and Zhang, Y., "Evolution of GaAs$_{1-x}$N$_x$ Conduction States and Giant Au/GaAs$_{1-x}$N$_x$ Schottky Barrier Reduction Studied by Ballistic Electron Emission Spectroscopy," *Phys. Rev. B* 61: R7861–R7864 (2000).

11. Uesugi, K., Suemune, I., Hasegawa, T., Akutagawa, T., and Nakamura, T., "Temperature Dependence of Band-gap Energies of GaAsN Alloys," *Appl. Phys. Lett.* 76: 1285–1287 (2000).

12. Polimeni, A., Capizzi, M., Geddo, M., Fischer, M., Reinhardt, M., and Forchel, A., "Effect of Temperature on the Optical Properties of (InGa)(AsN)/GaAs Single Quantum Wells," *Appl. Phys. Lett.* 77: 2870–2872 (2000).

13. Yaguchi, H., Biwa, G., Miyoshi, S., Aroki, D., Arimoto, K., Onabe, K., Ito, R., and Shiraki, Y., "Temperature Dependence of Photoluminescence of GaP$_{1-x}$N$_x$ Alloys," *J. Crystal Growth* 189/190: 496–499 (1998).

14. Weinstein, B.A., Stambach, S.R., Ritter, T.M., Maclean, J., and Wallis, D.J., "Evidence for Selective Delocalization of N-Pair States in Dilute GaAs$_{1-x}$N$_x$," *Phys. Rev. B* 68: 035336 (2003).

15. Jones, E.D., Modine, N.A., Allerman, A.A., Kurtz, S.R., Wright, A.F., Tozer, S.T., and Wei, A.F., "Band Structure of In$_x$Ga$_{1-x}$As$_{1-y}$N$_y$ Alloys and the Effects of Pressure," *Phys. Rev. B* 60: 4430–4433 (1999).

16. Bellaiche, L., Wei, S.H., and Zunger, A., "Localization of Percolation in Semiconductor Alloys: GaAsN and GaPN," *Phys. Rev. B* 54: 17568–17576 (1996).

17. Lindsay, A. and O'Reilly, E.P., "Theory of Enhanced Bandgap Non-Parabolicity in GaN$_x$As$_{1-x}$ and Related Alloys," *Solid State Commun.* 112: 443–447 (1999).

18. Mattila, T., Wei, S.H., and Zunger, A., "Localization and Anticrossing of Electron Levels in GaAs$_{1-x}$N$_x$ Alloys," *Phys. Rev. B* 60: R11245–R11248 (1999).

19. Wei, S.H. and Zunger, A., "Giant and Composition-Dependent Optical Bowing Coefficient in GaAs:N Alloys," *Phys. Rev. Lett.* 76: 664–667 (1996).

20. Zhang, Y., Mascarenhas, A., Xin, H.P., and Tu, C.W., "Scaling of Band-Gap Reduction in Heavily Nitrogen Doped GaAs," *Phys. Rev. B* 63: 161303(R) (2001).

21. Wang, L.W., "Large Scale LDA Band-gap Corrected GaAsN Calculations," *Appl. Phys. Lett.* 78: 1565–1567 (2001).

22. Hjalmarson, H.P., Vogl, P., Wolford, D.J., and Dow, J.D., "Theory of Substitutional Deep Traps in Covalent Semiconductors," *Phys. Rev. Lett.* 44: 810–813 (1980).

23. Gil, B., Albert, J.P., Camassel, J., Mathieu, H., and Benoit á la Guillaume, C., "Model Calculation of Nitrogen Properties in III-V Compounds," *Phys. Rev. B* 33: 2701–2712 (1986).

24. Jaros, M. and Brand, S., "Electronic States Associated with the Substitutional Nitrogen Impurity in GaP$_x$As$_{1-x}$," *J. Phys. C* 12: 525–539 (1979).

25. Bellaiche, L., Wei, S.H., and Zunger, A., "Composition Dependence of Interband Transition Intensities in GaPN, GaAsN, and GaPAs Alloys," *Phys. Rev. B* 56: 10233–10240 (1997).

26. Bellaiche, L., Wei, S.H., and Zunger, A., "Band Gaps of GaPN and GaAsN Alloys," *Appl. Phys. Lett.* 70: 3558–3560 (1997).

27. Thomas, D.G., Hopfield, J.J., and Frosch, C.J., "Isoelectronic Traps due to Nitrogen in Gallium Phosphide," *Phys. Rev. Lett.* 15: 857–860 (1965).

28. Zhang, Y., Mascarenhas, A., Geisz, J.F., Xin, H.P., and Tu, C.W., "Discrete and Continuous Spectrum of Nitrogen-Induced Bound States in Heavily Doped GaAsN," *Phys. Rev. B* 63: 085205 (2001).

29. Tsang, M.S., Wang, J.N., Ge, W.K., Li, G.H., Fang, Z.L., Chen, Y., Han, H.X., Li, L.H., and Pan, Z., "Hydrostatic Pressure Effect on Photoluminescence from a GaN$_{0.015}$As$_{0.985}$/GaAs Quantum Well," *Appl. Phys. Lett.* 78: 3595–3597 (2001).

30. Zhang, Y. and Ge, W., "Behavior of Nitrogen Impurities in III-V Semiconductors," *J. Lumin.* 85: 247–260 (2000).

31. Keating, P., "Effect of Invariance Requirements on the Elastic Strain Energy of Crystals with Application to the Diamond Structure," *Phys. Rev. B* 145: 637–645 (1966).
32. Martin, R.M., "Elastic Properties of ZnS Structure Semiconductors," *Phys. Rev. B* 1: 4005–4011 (1970).
33. Kim, K., Lambrecht, W.R.L., Segall, B., and van Schilfgaarde, M., "Effective Masses and Valence-Band Splittings in GaN and AlN," *Phys. Rev. B* 56: 7363–7375 (1997).
34. Cohen, M.L. and Bergstresser, T.K., "Band Structures and Pseudopotential Form Factors for Fourteen Semiconductors of the Diamond and Zinc-blende Structures," *Phys. Rev.* 141: 789–796 (1966).
35. Chelikowsky, J.R. and Cohen, M.L., "Electronic Structure of Silicon," *Phys. Rev. B* 10: 5095–5107 (1974).
36. Wang, L.W., Bellaiche, L., Wei, S.H., and Zunger, A., "'Majority Representation' of Alloy Electronic States," *Phys. Rev. Lett.* 80: 4725–4728 (1998).
37. Wang, L. and Zunger, A., "Solving Schrödinger's Equation around a Desired Energy: Aapplication to Silicon Quantum Dots," *J. Chem. Phys.* 100: 2394–2397 (1994).
38. Kim, K. and Zunger, A., "Spatial Correlations in GaInAsN Alloys and Their Effects on Band-Gap Enhancement and Electron Localization," *Phys. Rev. Lett.* 86: 2609–2612 (2001).
39. Wolford, D.J., Bradley, J.A., Fry, K., and Thompson, J., "The Nitrogen Isoelectronic Trap in GaAs," in *Proceedings of the 17th International Conference of the Physics of Semiconductors* (New York: Springer, 1984), 627.
40. Klar, P.J., Grüning, H., Heimbrodt, W., Koch, J., Höhnsdorf, F., Stolz, W., Vicente, P.M.A., and Camassel, J., "From N Isoelectronic Impurities to N-Induced Bands in the GaN_xAs_{1-x} Alloy," *Appl. Phys. Lett.* 76: 3439–3441 (2000).
41. Buyanova, I.A., Yu Rudko, G., Chen, W.M., Xin, H.P., and Tu, C.W., "Radiative Recombination Mechanism in GaN_xP_{1-x} Alloys," *Appl. Phys. Lett.* 80: 1740–1742 (2002).
42. Kent, P.R.C. and Zunger, A., "Failure of Nitrogen Cluster States to Emerge into the Bandgap of GaAsN with Application of Pressure," *Appl. Phys. Lett.* 82: 559–561 (2003).
43. Liu, X., Pistol, M.E., Samuleson, L., Schwetlick, S., and Seifert, W., "Nitrogen Pair Luminescence in GaAs," *Appl. Phys. Lett.* 56: 1451–1453 (1990).
44. Zhang, Y., Fluegel, B., Mascarenhas, A., Xin, H.P., and Tu, C.W., "Optical Transitions in the Isoelectronically Doped Semiconductor GaP:N: An Evolution from Isolated Centers, Pairs, and Clusters to an Impurity Band," *Phys. Rev. B* 62: 4493–4500 (2000).
45. Thomas, D.G. and Hopfield, J.J., "Isoelectronic Traps due to Nitrogen in Gallium Phosphide," *Phys. Rev.* 150: 680–689 (1966).
46. Cohen, E., Sturge, M.D., Lipari, N.O., Altarelli, M., and Baldereschi, A., "Acceptorlike Excite S States of Excitons Bound to Nitrogen Pairs in GaP," *Phys. Rev. Lett.* 35: 1591–1594 (1975).
47. Cohen, E., and Sturge, M.D., "Excited States of Excitons Bound to Nitrogen Pairs in GaP," *Phys. Rev. B* 15: 1039–1051 (1977).
48. Perkins, J.D., Mascarenhas, A., Zhang, Y., Geisz, J.F., Friedman, D.J., Olson, J.M., and Kurtz, S.R., "Nitrogen-Activated Ttransitions, Level Repulsion, and Band-gap Reduction in $GaAs_{1-x}N_x$ with $x < 0.03$," *Phys. Rev. Lett.* 82: 3312–3315 (1999).
49. Liu, X., Pistol, M.E., and Samuelson, L., "Excitons Bound to Nitrogen Pairs in GaAs," *Phys. Rev. B* 42: 7504–7512 (1990).
50. Vogl, P., "Chemical Trends of Deep Impurity Levels in Covalent Semiconductors," *Festkörperprobleme* 21: 191–219 (1981).
51. Eremets, M.I., Krasnovskij, O.A., Struzhkin, VV., and Shirokov, A.M., "Bound Excitons in GaP under Pressures up to 10 GPa," *Semicond. Sci. Technol.* 4: 267–268 (1989).
52. Gil, B., Baj, M., Camassel, J., Mathieu, H., Benoit á la Guillaume, C., Mestres, N., and Pascual, J., "Hydrostatic-Pressure Dependence of Bound Excitons in GaP," *Phys. Rev. B* 29: 3398–3407 (1984).

53. Landolt-Börnstein, O. Madelung, Ed., *Numerical Data and Functional Relationships in Science and Technology*, Vol. 22a (Berlin: Springer-Verlag, 1987).
54. Zunger, A., *Solid State Physics*, Vol. 39 (Boston: Academic, 1986), 275.
55. Kent, P.R.C. and Zunger, A., "Nitrogen Pairs, Triplets, and Clusters in GaAs and GaP," *Appl. Phys. Lett.* 79: 2339 (2001).
56. Yaguchi, H., Miyoshi, S., Biwa, G., Kibune, M., Onabe, K., Shiraki, Y., and Ito, R., "Photoluminescence Excitation Spectroscopy of $GaP_{1-x}N_x$ Alloys: Conduction-Band-Edge Formation by Nitrogen Incorporation," *J. Crystal Growth* 170: 353–356 (1997).
57. Liu, X., Bishop, S.G., Baillargeon, J.N., and Cheng, K.Y., "Band-gap Bowing in $GaP_{1-x}N_x$ Alloys," *Appl. Phys. Lett.* 63: 208–210 (1993).
58. Xin, H.P. and Tu, C.W., "Effects of Nitrogen on the Band Structure of GaN_xP_{1-x} Alloys," *Appl. Phys. Lett.* 76: 1267–1269 (2000).
59. Schwabe, R., Seifert, W., Bugge, F., Bindemann, R., Agekyan, V.F., and Pogarev, S.V., "Photoluminescence of Nitrogen-Doped VPE GaAs," *Solid State Commun.* 55: 167–173 (1985).
60. Makimoto, T., Saitô, H., Nishida, T., and Kobayashi, N., "Exitonic Luminescence and Absorption in Dilute $GaAs_{1-x}N_x$ Alloy ($x < 0.3$)," *Appl. Phys. Lett.* 70: 2984–2986 (1997).
61. Makimoto, T., Saito, H., and Kobayashi, N., "Origin of Nitrogen-Pair Luminescence in GaAs Studied by Nitrogen Atomic-Layer-Doping in MOVPE," *Jpn. J. Appl. Phys.* 36: 1694–1697 (1997).
62. Saito, H., Makimoto, T., and Kobayashi, N., "Photoluminescence Characteristics of Nitrogen Atomic-Layer-Doped GaAs Grown by MOVPE," *J. Crystal Growth* 170: 372–376 (1997).
63. Grüning, H., Chen, L., Hartmann, T., Klar, P.J., Heimbrodt, W., Höhnsdorf, F., Koch, J., and Stolz, W., "Optical Spectroscopic Studies of N-Related Bands in Ga(N,As)," *Phys. Stat. Solidi B* 215: 39–45 (1999).
64. Buyanova, I.A., Pozina, G., Hai, P.N., Thinh, N.Q., Bergman, J.P., Chen, W.M., Xin, H.P., and Tu, C.W., "Mechanism for Rapid Thermal Annealing Improvements in Undoped GaN_xAs_{1-x}/GaAs Structures Grown by Molecular Beam Epitaxy," *Appl. Phys. Lett.* 77: 2325–2327 (2000).
65. Caldas, M.J., Fazzio, A., and Zunger, A., "A Universal Trend in the Binding Energies of Deep Impurities in Semiconductors," *Appl. Phys. Lett.* 45: 671–673 (1984).
66. Mott, N., *Metal-Insulator Transition* (London: Taylor & Francis, 1974).
67. Buyanova, I.A., Chen, W.M., Monemar, B., Xin, H.P., and Tu, C.W., "Effect of Growth Temperature on Photoluminescence of GaN/As/GaAs Quantum Well Structures," *Appl. Phys. Lett.* 75: 3781–3783 (1999).
68. Xin, H.P. and Tu, C.W., "Photoluminescence Properties of GaNP/GaP Multiple Quantum Wells Grown by Gas Source Molecular Beam Epitaxy," *Appl. Phys. Lett.* 77: 2180–2182 (2000).
69. Yaguchi, H., Miyoshi, S., Arimoto, H., Saito, S., Akiyama, H., Onabe, K., Shiraki, T., and Ito, R., "Nitrogen Concentration Dependence of Photoluminescence Decay Time in $GaP_{1-x}N_x$ Alloys," *Solid-State Electronics* 41: 231–233 (1997).
70. Buyanova, I.A., Chen, W.M., Pozina, G., Monemar, B., Xin, H.P., and Tu, C.W., "Mechanism for Light Emission in GaNAs/GaAs Structures Grown by Molecular Beam Epitaxy," *Phys. Stat. Solidi B* 216: 125–129 (1999).
71. Mariette, H., "Picosecond Spectroscopy in III-V Compounds and Alloy Semiconductors," *Physica B* 146B: 286–303 (1987).
72. Toivonen, J., Hakkarainen, T., Sopanen, M., and Lipsanen, H., "High Nitrogen Composition GaAsN by Nitrogen Pressure Metalorganic Vapor-Phase Epitaxy," *J. Cryst. Growth* 221: 456–460 (2000).
73. Zhang, Y., Mascarenhas, A., Xin, H.P., and Tu, C.W., "Formation of an Impurity Band and Its Quantum Confinement in Heavily Doped GaAs:N," *Phys. Rev. B* 61: 7479–7482 (2000).

74. Hai, P.N., Chen, W.M., Buyanova, I.A., Xin, H.P., and Tu, C.W., "Direct Determination of Electron Effective Mass in GaNAs/GaAs Quantum Wells," *Appl. Phys. Lett.* 77: 1843–1845 (2000).

75. Yu, K.M., Walukiewicz, W., Shan, W., Ager III, J., Wu, J., and Haller, E.E., "Nitrogen-Induced Increase of the Maximum Electron Concentration in Group III-N-V Alloys," *Phys. Rev. B* 61: R13337–R13340 (2000).

76. Onabe, K., Aoki, D., Wu, J., Yaguchi, H., and Shiraki, Y., "MOVPE Growth and Luminescence Properties of GaAsN Alloys with Higher Nitrogen Concentrations," *Phys. Stat. Solidi A* 176: 231–235 (1999).

77. Mattila, T., Wei, S.H., and Zunger, A., "Electronic Structure of 'Sequence Mutations' in Ordered GaInP$_2$ Alloys," *Phys. Rev. Lett.* 83: 2010–2013 (1999).

78. Takahashi, M., Moto, A., Tanaka, S., Tanabe, T., Takagishi, S., and Karatani, K., "Observation of Compositional Fluctuations in GaNAs Alloys Grown by Metalorganic Vapor-Phase Epitaxy," *J. Cryst. Growth* 221: 461–466 (2000).

79. Xin, H.P. and Tu, C.W., "GaInNAs/GaAs Multiple Quantum Wells Grown by Gas-Source Molecular Beam Epitaxy," *Appl. Phys. Lett.* 72: 2442–2444 (1998).

80. Kondow, M., Uomi, K., Niwa, A., Kitatani, T., Watahiki, S., and Yazawa, Y., "GaInNAs: A Novel Material for Long-Wavelength-Range Laser Diodes with Excellent High-Temperature Performance," *Jpn. J. Appl. Phys.* 35: 1273–1275 (1996).

81. Nakahara, K., Kondow, M., Kitatani, T., Yazawa, Y., and Uomi, K., "Continuous-Wave Operation of Long-Wavelength GaInNAs/GaAs Quantum Well Laser," *Electron. Lett.* 32: 1585–1586 (1996).

82. Sato, S., Osawa, Y., Saito, T., and Gujimara, I., "Room-Temperature Pulsed Operation of 1.3 mm GaInNAs/GaAs Laser Diode," *Electron. Lett.* 33: 1386–1387 (1997).

83. Kurtz, S.R., Allerman, A.A., Jones, E.D., Gee, J.M., Banas, J.J., and Hammons, B., "InGaAsN Solar Cells with 1.0 eV Band Gap, Lattice Matched to GaAs," *Appl. Phys. Lett.* 74: 729–732 (1999).

84. Bellaiche, L., "Band Gaps of Lattice-Matched (Ga,In)(As,N) Alloys," *Appl. Phys. Lett.* 75: 2578–2580 (1999).

85. Bellaiche, L., and Zunger, A., "Effects of Atomic Short Range Order on the Electronic and Optical Properties of GaAsN, GaInN, and GaInAs Alloys," *Phys. Rev. B* 57: 4425–4431 (1998).

86. Tournie, E., Pinault, M.A., and Guzman, A., "Mechanisms Affecting the Photoluminescence Spectra of GaInNAs after Post-Growth Annealing," *Appl. Phys. Lett.* 80: 4148–4150 (2002).

87. Ciatto, G., Boscherini, F., D'Acapito, F., Mobilio, S., Baldassarri, G., Polimeni, A., Capizzi, M., Gollub, D., and Forchel, A., "Atomic Ordering in (InGa)(AsN) Quantum Wells: An In K-Edge X-ray Absorption Investigation," *Nucl. Instruments Methods Phys. B* 200: 34–39 (2003).

CHAPTER 2

Band Anticrossing in III-N-V Alloys: Theory and Experiments

W. WALUKIEWICZ, W. SHAN, J. WU, and K.M. YU

Materials Sciences Division, Lawrence Berkeley National Laboratory, Berkeley, California

2.1   INTRODUCTION

Alloying is one of the most commonly used methods to tailor properties of semiconductor materials for specific applications. In many instances, the parameters of the semiconductor alloys are quite well described by a simple interpolation between properties of the endpoint materials using first- or

second-order polynomials. The procedure has been justified within a virtual-crystal approximation (VCA), in which the random alloy potential is
approximated by a periodic lattice of average atomic potential [1–3]. The
approximation works well for the cases where the properties of the endpoint
materials do not differ drastically. For example, alloying of a large variety
of elemental and compound semiconductors produces materials with energy
gaps well approximated by Equation 2.1 [1]:

$$E_G(x) = xE_A + (1-x)E_B - bx(1-x) \qquad (2.1)$$

where x is the composition, E_A and E_B are the band gaps of the endpoint
materials, and b is the bowing parameter describing deviation of the band-
gap dependence from the linear interpolation.

The approximation represented by Equation 2.1 works well for the
cases of alloys with the bowing parameters smaller than 1 eV. It applies to
a wide range of the elemental and compound semiconductor alloys. For
example $Al_xGa_{1-x}As$ and $GaAs_xP_{1-x}$ fall into this category of alloys with the
bowing parameters of 0.2 eV [4] and 0.21 eV [5], respectively. Replacing
Ga with Al atom in $Al_xGa_{1-x}As$ or P with As atom in $GaAs_xP_{1-x}$ introduces
only a small perturbation to the crystal lattice. Consequently, the average
VCA potential does not differ much from the actual potential at any lattice
site, and the alloy disorder effects can be accounted for using a simple
perturbation theory. This leads to only a small deviation of the energy gap
from the linear interpolation between the endpoint gaps.

However, it has been recognized quite early that, in many instances, the
band gaps of semiconductor alloys depend very strongly on the alloy composition. Most notably, it has been found that the band gap in GaN_xAs_{1-x} is
drastically reduced at low N contents [6]. Attempts to explain the composition
dependence of the band gap in these alloys using Equation 2.1 have led to
unrealistically large, composition-dependent bowing parameters [6, 7]. Similarly, large bowings of the fundamental band gap have been found in other
group $III-N_x-V_{1-x}$ alloys [8–11]. Understanding of the unusual properties of
these alloys has become a challenge for the existing theoretical methods used
to calculate the electronic band structure. It was obvious that the VCA could
not be used to consider the large perturbation of the crystal lattice associated
with the replacement of a column-V atom with small, highly electronegative
N atom. The band structure calculations of ordered alloys have predicted
large, N-induced band-gap reductions in such alloys, although they were not
accurate enough to be directly compared with experiments [12]. Theoretical
and experimental aspects of the III-N-V alloys have been extensively discussed and reviewed in a series of articles [13, 14].

The effects of N incorporation on the properties of III-V compound
semiconductors have been studied for almost four decades [15]. The early

studies were limited to very low impuritylike concentrations of nitrogen. However, even at these low concentrations, nitrogen has a pronounced effect on the material properties. For example, an efficient electroluminescence was observed in GaP, an indirect gap semiconductor doped with nitrogen. The emission was attributed to formation of acceptorlike highly localized states [16]. The origin of these states has been elucidated from the theory based on the tight-binding approximation [17]. It showed that incorporation of any isovalent impurity into a semiconductor material produces localized, acceptorlike states. The energy location of the state is determined by the nature and the strength of the local potential introduced by the isovalent impurity. Theoretical calculations clearly show that the energy levels of elements with high electronegativity are located at energies lower than those of more-metallic impurities. Thus replacement of As with P produces an energy level high in the conduction band of GaAs, whereas substitution of As with highly electronegative N results in a level located close to the conduction-band minimum.

The difference in the location of the energy levels of the isovalent impurities is the key to understanding of the different types of the semiconductor alloys. In the case of "well-matched alloys" with a small electronegativity difference between constituent elements, the energy levels of the isovalent impurities are not observable, as they are located deeply in the conduction band [17]. Increasing of the impurity content toward alloylike concentrations leads to a relatively weak interaction between the impurity and the extended band states, resulting in an incremental modification of the semiconductor band structure, mostly in the energy range away from the conduction-band minimum. On the other hand, in the cases where the electronegativity difference between constituent elements is fairly large, incorporation of highly electronegative impurities gives rise to localized energy levels close to the conduction-band edge. At impuritylike concentrations, such levels preserve their separate identity and can be observed as discrete levels. However, at higher, alloylike, concentrations, the strong hybridization between the localized and extended states drastically modifies the electronic structure close to the conduction-band minimum, resulting in large changes of the optical and electrical properties of such highly mismatched alloys (HMAs).

The difference between well-matched and highly mismatched alloys resembles the previously considered problem of the amalgamation- and persistence-type alloys [18]. In the amalgamation type, the electronic states of different constituents of the alloy are indistinguishable, and a single electronic structure gradually evolves with changing alloy composition. In the persistence-type alloys, the component materials preserve their separate identity over a composition range. The later type of behavior is most clearly observed in alloys with highly electronegative constituents, such as alkali halides. For example, separate states associated with Br and Cl ions were clearly observed in KBr-KCl, a classical example of persistence-type alloys [19].

In recent years, several approaches have been developed to consider the electronic properties of highly mismatched alloys. Some of them are discussed in Chapters 1, 3, and 8 of this book. In this chapter, we present the band-anticrossing (BAC) model of the electronic structure of highly mismatched alloys that considers an interaction between localized states of highly electronegative isovalent impurities and the extended states of the semiconductor matrix. A general theory of the electronic structure of HMAs based on coherent potential approximation is presented, and the effects of the BAC interaction on the basic band-structure parameters are discussed in Section 2.2. Experimental results and the theoretical interpretation of the optical properties of group III-N-V alloys are presented in Section 2.3. Experimental verifications of novel phenomena predicted by the BAC model are discussed in Section 2.4, with the emphasis on the enhancement of effective mass and decrease in mobility of carriers, as well as mutual passivation between substitutional dopants on the group-III sites and nitrogen atoms.

2.2   THEORY OF BAND ANTICROSSING

The electronic structure of highly mismatched alloys (e.g., GaN_xAs_{1-x}) can be described by considering the interaction between the localized states and extended states within the many-impurity Anderson model [20]. The total Hamiltonian of the system is the sum of three terms [21, 22]:

$$H = \sum_{\mathbf{k}} E_{\mathbf{k}}^c c_{\mathbf{k}}^+ c_{\mathbf{k}} + \sum_{j} E_j^d d_j^+ d_j + \frac{1}{\sqrt{N}} \sum_{j,\mathbf{k}} (e^{i\mathbf{k}\cdot j} \, V_{\mathbf{k}j} \, c_{\mathbf{k}}^+ d_j + h.c.) \quad (2.2)$$

The first term is the Hamiltonian of the electrons in the band states with energy dispersion $E_{\mathbf{k}}^c$. The second term corresponds to the electron localized on the jth impurity site with energy E_j^d. The third term describes the change in the single electron energy due to the dynamic mixing between the band states and the localized states. Following Anderson's scheme, the hybridization strength is characterized by the parameter $V_{\mathbf{k}j}$ defined by Anderson [20]:

$$V_{\mathbf{k}j} = \sum_{l} e^{i\mathbf{k}\cdot(l-j)} \int a^*(\mathbf{r} - l) H_{HF}(\mathbf{r}) \varphi_d(\mathbf{r} - j) d\mathbf{r} \quad (2.3)$$

where $a(\mathbf{r} - j)$ and $\varphi_d(\mathbf{r} - j)$ are the Wannier function belonging to the band and the localized wavefunction of the impurity on the jth site, respectively. $H_{HF}(\mathbf{r})$ is the single-electron energy described in the Hartree-Fock approximation [20].

The Fourier transform of the retarded Green's function, $G_{kk'}(E) = <<c_k \mid c_{k'}^+ >>$, satisfies the following equation of motion [23]:

$$E << c_k \mid c_{k'}^+ >>=<\left[c_k, c_{k'}^+\right]_+ > + << \left[c_k, H\right] \mid c_{k'}^+ >> \qquad (2.4)$$

In Equation 2.4, the bracket $<\ldots>$ represents the thermodynamic ensemble average. As follows from the commutation relations between the operators, an integral equation for $G_{kk'}$ has the form

$$G_{kk'} = \delta_{kk'} G_{kk}^{(0)} + \frac{1}{N} G_{kk}^{(0)} \sum_{k''j} \tilde{V} \cdot e^{i(k-k'')j} G_{k''k'} \qquad (2.5)$$

where $G_{kk'}^{(0)} = \delta_{kk'} (E - E_k^c + i0^+)^{-1}$ is the unperturbed Green's function, and the renormalized interaction parameter is given by

$$\tilde{V} = V_{kj} \cdot V_{k'j} / \left(E - E_j^d\right) \approx V^2 / \left(E - E_j^d\right) \qquad (2.6)$$

where V is the average value of V_{kj}, assuming weak dependencies on k and j.

For the single-impurity case, the Green's function in Equation 2.5 can be solved analytically, and the exact solution has been obtained by Anderson [20]. The hybridization term produces a profound effect on the electronic structure of the system. In general, one shall consider finite but dilute concentrations of impurities, $0 < x << 1$. In this case, the single-site coherent potential approximation (CPA) is adequate for the many-impurity system [24, 25]. In the CPA treatment, a configurational averaging is performed, neglecting correlations between positions of the impurities. Consequently, the space translational invariance of the average Green's function is partially restored, and k resumes its well-defined properties as a good quantum number. In momentum space, the diagonal Green's function in CPA can be written as [21, 24, 25]:

$$G_{kk}(E) = \left[E - E_k^c - \frac{V^2 x}{E - E^d - i\pi\beta V^2 \rho_0\left(E^d\right)}\right]^{-1} \qquad (2.7)$$

The new dispersion relations are determined by the poles of $G_{kk}(E)$. The solutions are given by an equivalent two-state-like eigenvalue problem:

$$\begin{vmatrix} E_k^c - E(k) & V\sqrt{x} \\ V\sqrt{x} & E^d + i\Gamma_d - E(k) \end{vmatrix} = 0, \qquad (2.8)$$

where $\Gamma_d = \pi\beta V^2 \rho_0(E^d)$ is the broadening of E^d in the single-impurity Anderson model [19]. V is the value of V_{kj} averaged over k and j, and ρ_0 is the unperturbed density of states (DOS) of E_k^c. In this approximation, the effective contribution of ρ_0 is represented by its value evaluated at E^d and multiplied by a prefactor β that is to be determined by experiments. If $\Gamma_d = 0$, Equation 2.8 is reduced to the BAC model with two restructured dispersions for the upper and lower conduction subbands [26]:

$$E_\pm(\mathbf{k}) = \frac{1}{2}\left\{ \left(E_k^c + E^d \right) \pm \sqrt{\left(E_k^c - E^d \right)^2 + 4V^2 x} \right\}. \qquad (2.9)$$

If the broadening Γ_d is nonzero but small, so that $2V\sqrt{x} >> \pi B V^2 \rho_0(E^d)$ and $|E_k^c - E^d| >> \pi B V^2 \rho_0(E^d)$, an approximate analytical solution for Equation 2.8 can be obtained:

$$\tilde{E}_\pm(\mathbf{k}) \approx E_\pm(\mathbf{k}) + i\Gamma_d \frac{\left[E_\pm(\mathbf{k}) - E_k^c \right]}{\left[E_\pm(\mathbf{k}) - E_k^c \right] + \left[E_\pm(\mathbf{k}) - E^d \right]} \equiv E_\pm(\mathbf{k}) + i\Gamma_\pm(\mathbf{k})$$

$$(2.10)$$

The real part, $E_\pm(\mathbf{k})$, is defined in Equation 2.9 by the BAC model. The imaginary part of the dispersion relations defines the hybridization-induced uncertainty of the energy. Note that the imaginary part in Equation 2.10 is proportional to the admixture of the localized states ($|d\rangle$) to the restructured wavefunctions ($|E_\pm(\mathbf{k})\rangle$) in the two-state-like perturbation picture described by Equation 2.8 [27],

$$\Gamma_\pm(\mathbf{k}) = \left| \langle d | E_\pm(\mathbf{k}) \rangle \right|^2 \cdot \Gamma_d \qquad (2.11)$$

Figure 2.1 shows the dispersion relations given by Equation 2.9 for $GaN_{0.005}As_{0.995}$ near the Brillouin zone center. The broadening of the dispersion relations is given by the imaginary part of Equation 2.10. The hybridization parameter $V = 2.7$ eV used in the calculations has been determined from the fitting of Equation 2.9 to the composition dependence of the band gap of GaN_xAs_{1-x} alloys [26]. Figure 2.2 shows that this single fitting parameter provides a very good agreement with the experimental data reported by various groups over the years [28–31].

The hybridization-induced modifications of the electronic structure have large effects on the properties of HMAs. For instance, the broadening parameter that defines a finite lifetime for the lowest conduction band $|E_-$

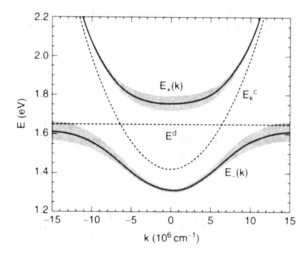

Figure 2.1 Conduction band restructuring for $GaN_{0.005}As_{0.995}$. The broadening of the dispersion curves of the newly formed subbands illustrates the energy uncertainties as defined in Equation 2.9. All the energies are referenced to the top of the valence band of GaAs.

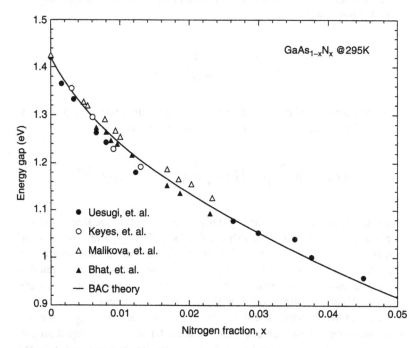

Figure 2.2 Comparison between the experimentally observed and calculated band-gap reduction of GaN_xAs_{1-x} as a function of N concentration. The calculations are based on the BAC model with $V = 2.7$ eV, $E^c = E_M = 1.42$ eV, and $E^d = E_N = 1.65$ eV.

(**k**)⟩ through the uncertainty principle imposes a limit to the mobility of free electrons that conduct current in the lowest conduction band:

$$\mu = \frac{e\tau(k_F)}{m_-^*(\mathbf{k}_F)} \approx \frac{e\,\hbar}{m_-^*(\mathbf{k}_F) \cdot \Gamma_-(\mathbf{k}_F)} \qquad (2.12)$$

The mobility is affected by the level broadening as well as by the BAC-induced enhancement of the density-of-states electron effective mass that can be calculated from the dispersion $E_-(\mathbf{k})$ in Equation 2.9 [32, 33],

$$m_-^*(\mathbf{k}_F) = \hbar^2 \left|\frac{k}{dE_-(\mathbf{k})/dk}\right|_{k=\mathbf{k}_F} = m_0^* \cdot \left[1 + \frac{V^2 x}{\left(E^d - E_-(\mathbf{k}_F)\right)^2}\right] \qquad (2.13)$$

where m_0^* is the electron effective mass of the unperturbed dispersion E_k^c. The Fermi wave vector $\mathbf{k}_F$ and Fermi energy $E_F = E_-(\mathbf{k}_F)$ are determined by the free-electron concentration (n) calculated from the restructured density of states [34],

$$n(E_F) = \int \frac{\rho(E)\,dE}{1 + \exp[(E - E_F)/k_B T]} \qquad (2.14)$$

The restructured density of states is given by the imaginary part of the Green's function as shown in the following expression:

$$\rho(E) = \frac{1}{\pi} \text{Im} \sum_{\mathbf{k}} G_{\mathbf{kk}}(E) = \frac{1}{\pi} \int \rho_0(E_k^c) \text{Im}\left[G_{\mathbf{kk}}(E)\right] dE_\mathbf{k}^c. \qquad (2.15)$$

The integration converges rapidly with $E_\mathbf{k}^c$ in a small range that is proportional to x. The calculated perturbed DOS for GaN$_x$As$_{1-x}$ with several small values of x is shown in Figure 2.3. Note that the anticrossing interaction leads to a dramatic redistribution of the electronic states in the conduction band. The most striking feature of the DOS curves is the clearly seen gap between E_+ and E_- that evolves with increasing N content.

In the Green's function calculation, the k dependence of $V_{\mathbf{kj}}$ is assumed to be weak on the momentum scale we are interested in. In Equation 2.6, the parameter $V_{\mathbf{kj}}$ is averaged over the impurity sites and in k space. In the simplest case, all the impurity atoms are of the same type, so that the j dependence of $V_{\mathbf{kj}}$ is removed. The k dependence of $V_{\mathbf{kj}}$ can be estimated

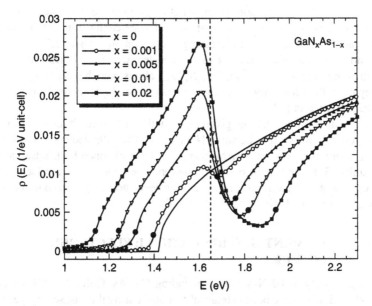

Figure 2.3 Density of states of GaN$_x$As$_{1-x}$ for a range of values of x as compared with the unperturbed DOS. The two black dots on each curve indicate the energy positions of the E$_-$ and E$_+$ subband edges. The vertical dashed line indicates the original position of the nitrogen localized level.

from Equation 2.3. Assuming that the Hartree-Fock energy varies slowly in space and can be replaced by a constant ε_{HF}, one obtains

$$V_k = \varepsilon_{HF} \cdot \sum_l e^{i\mathbf{k}\cdot\mathbf{l}} \int a^*(\mathbf{r}-\mathbf{l})\varphi_d(\mathbf{r})d\mathbf{r}. \qquad (2.16)$$

Due to the localized character of both $a(r)$ and $\varphi_d(r)$, the overlap integral in Equation 2.16 is essentially zero when they are located on two sites far apart from each other. In an attempt to model the k-dependence of V_k, the integral in Equation 2.16 is replaced by an exponentially decaying function $\approx \exp(-l/l_d)$, and one obtains

$$V_k = \varepsilon_{HF} \sum_l e^{i\mathbf{k}\cdot\mathbf{l}-l/l_d} = \frac{V_0}{\left(1+l_d^2 k^2\right)^2}. \qquad (2.17)$$

There is experimental evidence indicating that the values of V_k at the L point in GaN$_x$As$_{1-x}$ [35] and at the X point in GaN$_x$P$_{1-x}$ [36] are about three to four times smaller than the V_k at the Γ point. This ratio corresponds to a localized wavefunction decay length (l_d) on the order of the lattice constant. This result indicates that the off-zone-center conduction-band minima are affected by the anticrossing interaction only when their energies are close to the localized level. This is consistent with recent measurements of the optical properties of $In_yGa_{1-y}N_xAs_{1-x}$ alloys, which have shown that alloying with N has only very small effects on the high energy transitions at large k vectors [33].

The dispersion relations given by Equation 2.9 were obtained considering only an interaction between N states and the extended states close to the Γ minimum. A more general result has been obtained by Lindsay and O'Reilly [37] using the $k \cdot p$ approximation. The approach is applicable to a larger range of electron energies and has been successfully used to model optical gain in GaInNAs-based lasers [38].

2.3 EXPERIMENTAL VERIFICATION I: ELECTRONIC STRUCTURE

A large variety of III-N-V alloys including GaNAs, GaInNAs, AlGaNAs, GaNP, and InNP have been extensively studied during the past several years. Most of the samples for the spectroscopic studies discussed below were grown by either metalorganic vapor-phase epitaxy (MOVPE) with dimethylhydrazine as nitrogen source or gas-source molecular-beam epitaxy (MBE) using an RF plasma nitrogen-radical beam source. The nitrogen contents in those samples were determined using secondary-ion mass spectroscopy (SIMS) and indirectly from the change of the lattice constant measured by reflection [0,0,4] double-crystal X-ray diffraction. The AlGaNAs and some InNP samples were also synthesized using N^+ implantation [39–41]. AlGaAs epitaxial films were grown by MOVPE on GaAs substrates. N^+ implantation was carried out at room temperature using multiple energies in an attempt to create a uniform depth distribution of nitrogen in the implantation region. A detailed discussion on synthesis of III-N-V alloys via N-ion implantation will be given in the next section.

2.3.1 Interband Optical Transitions

The conduction-band splitting into two nonparabolic subbands predicted by the BAC model has been unambiguously observed in GaN_xAs_{1-x} and $Ga_{1-y}In_yN_xAs_{1-x}$ using photomodulation spectroscopy [26, 42, 43]. Figure 2.4 shows photoreflectance (PR) spectra recorded from GaN_xAs_{1-x} samples. The PR spectrum of GaAs ($x = 0$) exhibits two sharp derivativelike spectral features corresponding to the transition from the top of the valence band to the bottom of the conduction band (E_0 transition), and the transition between the spin-orbit split-off band and the conduction-band minimum ($E_0 + \Delta_0$ transition). For N-containing samples, in addition to the PR spectral

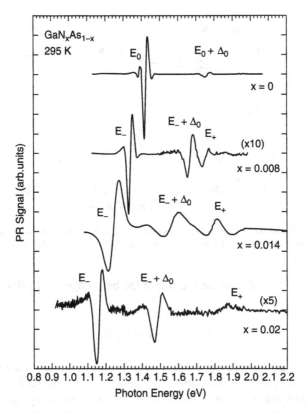

Figure 2.4 PR spectra of GaN$_x$As$_{1-x}$ samples with different N concentrations.

features related to the transition across the fundamental band gap (E_- transition) and the transition from the top of the spin-orbit split-off valence band to the bottom of the conduction band ($E_- + \Delta_0$ transition), an extra feature (E_+) appears at higher energies in the PR spectra. With increasing N concentration, the E_- and $E_- + \Delta_0$ transitions shift to lower energy, and the E_+ transition moves in the opposite direction. Shown in Figure 2.5 are the E_- and E_+ transition energies in Ga$_{1-y}$In$_y$N$_x$As$_{1-x}$ as a function of N concentration reported by several different groups [41, 43–45]. The non-linear dependence of the transition energies on N concentration can be well described by the BAC model using a coupling constant $V = 2.7$ eV.

It is also worth noting that, as shown in Figure 2.4, the spin-orbit splitting energy Δ_0 is equal to ≈0.34 eV for all the measured samples and does not depend on N content. The results demonstrate that incorporation of N into GaAs and GaInAs affects mostly the conduction band and has a negligible effect on the electronic structure of the valence band. Using Equation 2.9, one can obtain a simple relationship between the subband-

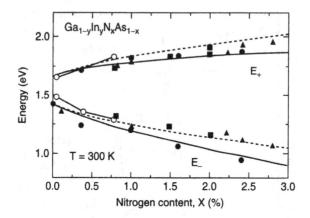

Figure 2.5 Dependence of E_- and E_+ transitions on N composition. The solid and dotted lines represent the BAC model predictions for $Ga_{1-y}In_yN_xAs_{1-x}$ ($y = 3x$) and GaN_xAs_{1-x}, respectively.

splitting $\alpha = E_+ - E_-$ and the redshift of the band edge $\beta = E_M(k = 0) - E_-$ ($k = 0$) [42]:

$$\alpha = 2\beta + E_N - E_M(0) \tag{2.18}$$

E_N is the energetic position of the N state relative to the top of the valence band. Since for small N concentrations E_N is constant, Equation 2.18 defines a universal relationship for a given material system specified by E_M. Implicit in the derivation of Equation 2.18 is an assumption that alloying with N affects the conduction band only and that the location of the valence band remains unchanged. Any N-induced change in location of the valence-band edge would manifest itself in a change of the slope in Equation 2.18 from two to larger (smaller) value for the downward (upward) shift of the valence-band edge.

Figure 2.6 shows the band splitting and the band-gap shift measured at room temperature in a number of $Ga_{1-y}In_yN_xAs_{1-x}$ samples ($x = 0$ to ~2.4%, $y = 0$ to ~8%). The values of the band-gap reductions in $Ga_{1-x}In_xN_yAs_{1-y}$ samples were corrected for the In-induced shift of E_M from $E_M = 1.42$ eV in GaAs, for the effects of In and N on the band gap are fully separable. The solid line in the figure represents the relationship given by Equation 2.18 predicted by the anticrossing model, with the valence band independent of the N concentration. It is in good overall agreement with the experimental data. Two lines with slopes of 2.2 and 1.8 in Figure 2.6 represent the cases where a downward and upward shift of the valence-band edge accounts for approximately 10% of the total band-gap shift. A linear fit to the data in Figure 2.6 yields a slope of 2.1, indicating a rather small 5% downward shift of the valence-band edge as a function of the N content.

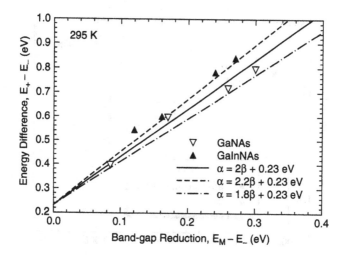

Figure 2.6 The relationship between the measured band splitting and band-gap shift for GaInNAs.

The band-anticrossing effects have also been observed in other group III-V materials alloyed with nitrogen, including GaN_xP_{1-x}, [46] InN_xP_{1-x} [9, 40, 41], $GaN_xSb_yAs_{1-x-y}$ [10], and InN_xSb_{1-x} [11]. Shown in Figure 2.7 is, for instance, the change of the band-gap energy of gas-source MBE-grown InN_xP_{1-x} as a function of N content. The band-gap energies of the samples were determined by absorption and PR measurements. The incorporation of nitrogen reduces the band gap in the same manner as in the case of $Ga_{1-x}In_xN_yAs_{1-y}$, shown in Figure 2.2. The best fit to the experimental data using the BAC model yields the energy position of $E_N \approx E_v$ + 2.0 eV for the localized N-level and the coupling constant of V = 3.5 eV for the InN_xP_{1-x} alloy system.

2.3.2 Effects of Hydrostatic Pressure

Another important prediction of the BAC model is the characteristic anticrossing behavior of the E_- and E_+ transitions under applied pressure. The pressure dependence of the optical transitions associated with the E_- and E_+ subband edges in an In-free $GaN_{0.015}As_{0.985}$ sample and a $Ga_{0.95}In_{0.05}N_{0.012}As_{0.988}$ sample are shown in Figure 2.8. The anticrossing behavior of two strongly interacting energy levels with distinctly different pressure dependencies is unmistakably observed. The E_- transition has a strong dependence at low pressures and gradually saturates at high pressures, whereas the E_+ transition has weak pressure dependence at low pressures and displays a much stronger dependence at high pressures. The solid lines through the experimental data in the figure are results of calculations using Equation 2.9. The best fits to the data yield the energy of the

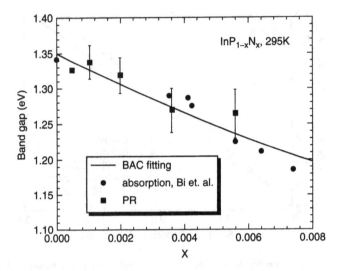

Figure 2.7 Band-gap energy of $InP_{1-x}N_x$ as a function of nitrogen concentration. The solid line is a fit based on the BAC model.

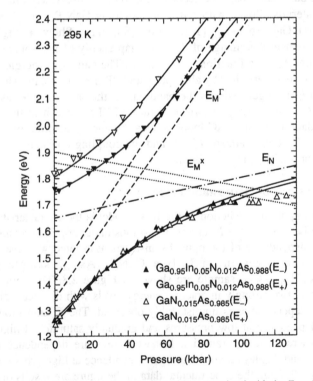

Figure 2.8 Effects of pressure on the optical transitions associated with the E_- and E_+ transitions in $GaN_{0.015}As_{0.985}$ and $Ga_{0.95}In_{0.05}N_{0.012}As_{0.988}$.

nitrogen state, $E_N = E_V + 1.65$ eV for both samples at atmospheric pressure, and it is independent of In concentration [26, 41]. These results prove that the effects of alloying with In on the band gap can be separated from the shifts produced by the interaction with N states, allowing for an independent determination of E_M from a given In concentration in $Ga_{1-y}In_yN_xAs_{1-x}$ alloys.

The observed change in the dependence of the E_- and E_+ transitions on pressure indicates that application of pressure gradually changes the character of the E_--subband edge from extended E_M-like to localized E_N-like, and the character of the E_+-subband edge from the localized-like to extended-like. To illustrate such a transformation, Figure 2.9 shows the theoretically calculated dispersion relation of the E_- (k) and $E_+(k)$ conduction subbands at three different pressures [43]. The interaction between the localized N states and the extended conduction-band states has a dramatic effect on the dispersion relations $E_\pm(k)$ of the two subbands. As the effect of the interaction is most pronounced for the extended states located close to E_N, the dispersion relations for the conduction subbands are strongly affected by applying pressure that shifts E_N with respect to the conduction band of the matrix. Application of pressure has the strongest effect on the lower subband, which narrows drastically at high pressures. Narrowing of the band indicates a gradual, pressure-induced transformation in the nature of the lowest subband from an extended to highly localized state. This transformation is associated with a pressure-induced enhancement of the effective mass and the density of states in the lower subband.

The large enhancement of the effective mass in the lower $E_-(k)$ subband as a function of electron energy and externally applied pressure can be demonstrated by the pressure dependence of the transitions between quantum confinement states in GaN_xAs_{1-x}/GaAs multiple quantum wells. Figure 2.10 shows the effect of pressure on the lowest confinement transition energy ($E_{n=1}$) in a $GaN_{0.016}As_{0.984}$/GaAs multiple quantum well (MQW) sample

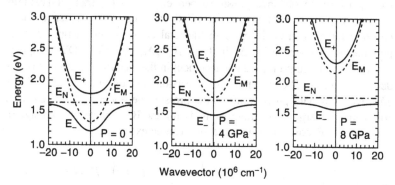

Figure 2.9 Calculated dispersion of the $E_-(k)$ and $E_+(k)$ subbands at three different pressures.

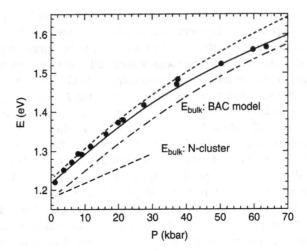

Figure 2.10 The first transition energy E_1 as a function of hydrostatic pressure for $x = 0.016$ and well width = 7 nm. Solid curve: calculated values with GaN_xAs_{1-x} electron effective mass given by Equation 2.19; Short dashed curve: calculated values assuming GaN_xAs_{1-x} electron effective mass equal to m^*_{GaAs}. Dot-dashed line: calculated pressure dependence of fundamental band gap in bulk $GaN_{0.016}A_{0.984}$ based on BAC model.

with 7-nm wells and 20-nm barriers. The pressure dependence of the fundamental band gap of bulk $GaN_{0.016}As_{0.984}$ (E_-) calculated based on the BAC model is also plotted in the figure. The results suggest that an increase in the total confinement energy $\Delta E = E_{n=1} - E_-$ in the MQW sample with applied pressure is primarily due to the very different pressure dependence of the band gap for the $GaN_{0.016}As_{0.984}$ wells and GaAs barriers. The increase in the well depth enhances the quantum confinement effects at higher pressures. However, the experimentally obtained pressure dependence of the $E_{n=1}$ transition exhibits a different picture: relatively strong nonlinear pressure dependence and a decrease rather than an increase in the total confinement energy ΔE with pressure. Such a decrease in ΔE in the $GaN_{0.016}As_{0.984}$/GaAs MQW sample is the result of pressure-induced enhancement in the electron effective mass, as described above. The flattening of dispersion relation and increase of the effective mass with pressure affect the pressure dependence of the electron confinement energy in quantum wells. Using the BAC model, the change in the electron effective mass with pressure can be expressed as [47]:

$$m^* \approx \hbar^2 \left. \frac{k}{dE_-(k)/dk} \right|_{k=0} = 2\,m^*_{GaAs} \left/ \left[1 - \frac{E_M(P) - E_N(P)}{\sqrt{\left(E_M(P) - E_N(P)\right)^2 + 4V^2x}} \right]_{k=0} \right.$$

$$(2.19)$$

Here $E_M(P)$ and $E_N(P)$ represent, respectively, the pressure dependence for the band gap of the GaAs matrix and the N level. It is evident from Equation 2.19 that incorporation of N into GaAs leads to a significant increase of the electron effective mass. The mass is further increased with the application of hydrostatic pressure. The effective mass given by Equation 2.19 has been used to calculate pressure dependence of the $E_{n=1}$ transition energy. The calculations represented by solid line in Figure 2.10 are in good agreement with the experimental data. The results confirm the large pressure-induced increase of the effective mass from 0.11 m_0 at ambient pressure to 0.28 m_0 at 70 kbar. This can be compared with a less than 30% change in the effective mass if a linear change, $m_e(P) \approx m_e(0)E_0(P)/E_0(0)$ for the conventional alloys such as AlGaAs [48], is used.

2.3.3   Effects of the Higher Conduction-Band Minima

Alternative interpretations of some of the observed effects discussed above, such as the appearance of E_+ transition and the pressure dependence of the E_- and E_+ transitions, were also proposed [49–56]. For instance, it has been argued that the observed changes in the conduction band structure are a result of interactions between states originating from the extended states of the Γ, L, and/or X conduction-band minima. It has been argued that incorporation of N breaks the crystal symmetry and splits the degenerate L and X minima into the a_1 and t_2 states. The a_1 states strongly interact with the states at the Γ minimum, leading to a downward shift of the conduction-band edge. The close proximity of the L minimum energy at $E_V + 1.705$ eV to the energy of the localized N-state $E_N \approx E_V + 1.65$ eV was invoked in the argument that interaction with either of these states could be responsible for the E_+ and E_- transitions [44]. It was also proposed that the impuritylike band of interacting nitrogen pairs and cluster states is responsible for the downward shift of the conduction-band edge in GaNAs alloys [55, 56].

Several groups have studied the N-induced effects on the L conduction-band edges by measuring the E_1 ($\Lambda_{4v,5v} - \Lambda_{6c}$) transition near the L points of the Brillouin zone in GaN_xAs_{1-x} using different experimental methods [33, 35, 57, 58]. Figure 2.11 shows PR spectra associated with the E_1 transitions from the $L_{4,5}$ valence-band edge to the L_6 conduction-band minima in several GaN_xAs_{1-x} samples. A small increase of the E_1 transition energy with increasing x has been observed relative to the E_1 transition at 2.925 eV in GaAs.

Figure 2.12 plots the change in the energy of the E_1 transition, as well as the shift of E_- and E_+ transitions as functions of the N concentration in GaNAs. It is clear from the figure that the E_1 transition shifts to higher energy at a much slower rate than the E_+ transition. On the other hand, the relative energy shift of the E_+ transition as a function of N concentration

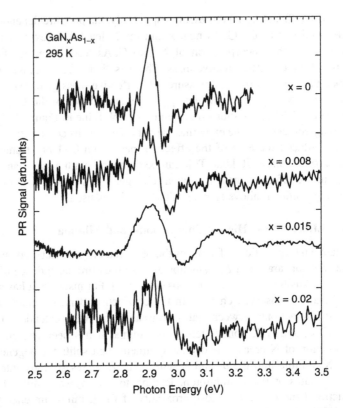

Figure 2.11 PR spectral features of the E_1 transition in GaN$_x$As$_{1-x}$ with different N concentration.

is, within the experimental uncertainties, the same as downward shift of the E_- transition, which represents the fundamental band-gap reduction in the GaNAs samples. This salient feature is a typical characteristic of two-level anticrossing interaction, in which the upward shift of the upper state and the downward shift of the lower state are exactly the same in magnitude. The much slower upward shift of the E_1 compared with the E_+ transition energy rules out the possibility that the E_+ can be associated with N-induced Γ_v–L_c transition. The slow, monotonic increase of the E_1 transition energy with N concentration and the lack of a splitting of the L-band edge are also in disagreement with the theoretical calculations [51] attributing the E_+ spectral feature to transitions from Γ_v to configuration-weighted average of nitrogenlike a_1(N) and L-like a_1(L$_c$) states.

To further elucidate the role of the higher energy minima, the effects of N on the band structure of Ga$_{1-y}$Al$_y$As alloys were investigated [58]. In these alloys, the Γ band-edge minimum shifts from about 0.5 eV below in GaAs to slightly over 0.5 eV above the X conduction-band minima in AlAs. The large relative energy shift is expected to strongly affect the strength of

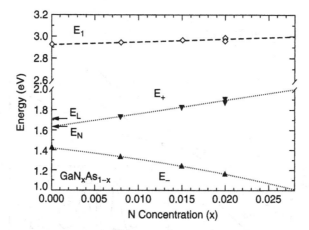

Figure 2.12 Measured energies of the E_1 transition, as well as the E_- and E_+ transitions in GaN$_x$As$_{1-x}$. The shift of E_1 transition is less than 20% of that of E_+ transition. The arrows mark the energy position of E_M and E_N with respect to the top of the valence band in GaAs matrix. The dotted and dashed lines are for guiding the eye.

the interaction between those two minima. Ga$_{1-y}$Al$_y$N$_x$As$_{1-x}$ alloys used in the study were synthesized by implanting nitrogen ions into MOVPE-grown Al$_y$Ga$_{1-y}$As epitaxial films on GaAs substrates followed by postimplantation thermal annealing. Energy positions of the experimentally observed E_+ and E_- transitions for a GaN$_{0.0085}$As$_{0.9915}$ and four Ga$_{1-y}$Al$_y$N$_x$As$_{1-x}$ samples are shown in Figure 2.13. The dependencies of the energies of the Γ, L, and X conduction-band minima on the Al content in Al$_y$Ga$_{1-y}$As are shown in the figure. The band-gap E_0 measured in the as-grown wafers was used to determine the Al concentration of the Ga$_{1-y}$Al$_y$As epitaxial films. The inset in Figure 2.13 shows a comparison of the PR spectra between an as-grown Al$_{0.35}$Ga$_{0.65}$As sample and an N$^+$-implanted Al$_{0.35}$Ga$_{0.65}$N$_x$As$_{1-x}$ sample.

As illustrated in Figure 2.13, the positions of the E_+ and E_- transitions can be again well explained assuming an anticrossing interaction between states at the Γ minimum and the localized nitrogen state E_N, whose energy depends on the Al content as $E_N = 1.65 + 0.61y$ eV. It is interesting to see that the E_+ transition lies above the Γ conduction-band minimum even at the indirect band-gap region of the Al$_y$Ga$_{1-y}$As matrix. The results demonstrate that there is no strong interaction between the X and Γ minima to give rise to the E_+ transition and that the Γ conduction-band minimum plays a dominant role in the band-anticrossing interaction. Also, there is no clear correlation between the location of the X conduction-band edge and the E_- and E_+ transition energies. For example, the X-edge lies close to the E_+ transition in the sample with $y = 0.35$, but as much as 0.17 eV above the E_+ transition energy in GaNAs. This would not be possible if an interaction between Γ an X were responsible for the E_+ band-edge shift.

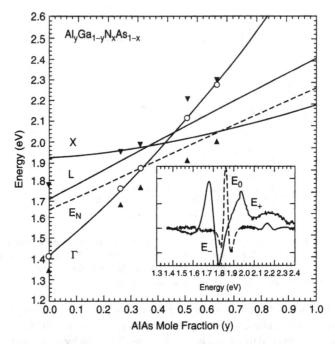

Figure 2.13 The E_- and E_+ transition energies measured for $Al_{1-y}Ga_yN_xAs_{1-x}$ samples. Open circles represent E_0 transitions in samples with $x = 0$. The dependence of the Γ, X, and L conduction-band edges on the AlAs mole fraction in AlGaAs alloys are shown by the solid lines. The dashed line represents the estimated change of the E_N position with Al content. The inset shows a comparison of PR spectra measured on the $Al_{0.35}Ga_{0.65}As$ samples with N (solid line) and without N (dashed line).

A very interesting and important case is represented by GaN_xP_{1-x} alloys, in which the lowest conduction-band-edge (X_c) minima are located slightly above the N level. Figure 2.14 shows a PR spectrum of a $GaN_{0.023}P_{0.977}$ sample taken over a wide photon energy range (1.45–3.35 eV). A PR spectrum of GaP is also shown in the figure for comparison. Note that, for the indirect optical transitions such as the $Γ_V$–X_c transition in GaP as marked by the arrow in the figure, no PR signal is detectable. However, in contrast, two PR spectral features are clearly observed in the $GaN_{0.023}P_{0.977}$ sample: the strong PR feature at the energy of 1.96 eV located below the indirect band gap (E_{gX}) of GaP, and the weak feature at the energy of 2.96 eV above the direct band gap ($E_Γ$) of GaP. Similar PR spectra can be observed in all the GaNP samples. The strong PR signals below the indirect band gap of GaP indicates a change in the nature of the optical transitions in GaNP, and it suggests a change in the nature of the band gap, from indirect in GaP to direct in GaNP. This N-induced transformation from indirect to direct gap can be understood in terms of the band-anticrossing model discussed above. It predicts formation of two subbands $E_-(k)$ and $E_+(k)$ as a result of the

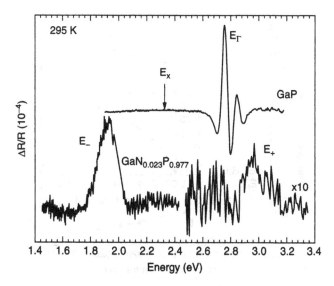

Figure 2.14 Comparison of PR spectra between $GaN_{0.023}P_{0.977}$ and GaP. The indirect transition associated with the fundamental band gap (E_{gx}) of GaP cannot be detected by PR measurement.

interaction of the localized N states and the extended conduction-band states of the GaP matrix. The band-gap energy is given by the energy of the lower subband edge, $E_-(0)$, relative to the top of the valence band. The PR spectrum of the $GaN_{0.023}P_{0.977}$ sample in Figure 2.14 is in good agreement with the prediction of the BAC model. The PR spectral feature on the lower energy side can be attributed to the E_- transition. The feature at higher energy can be assigned to the E_+ transition. It provides direct evidence that the N-induced transformation from indirect to direct gap semiconductor is a result of the BAC interaction between the N states and the higher lying Γ minimum.

Figure 2.15 shows the energy positions of the E_- and E_+ transitions determined by the photomodulation measurements for the GaN_xP_{1-x} samples used in this work. The dashed lines represents the BAC fitting using Equation 2.9 with well-known band-structure parameters of the GaP matrix; $E_M(\Gamma) = 2.78$ eV, and $E_N = 2.18$ eV. The best fit to the experimental data yields $V = 3.05$ eV. Based on the band-anticrossing model, the wavefunctions of the E_- states are combinations of the wavefunctions of the localized N states and the extended states at Γ_c minimum. The relative contributions of the different states can be readily estimated using the model. The inset in Figure 2.15 shows the calculated admixture of the Γ_c states to the states at the E_- edge. It is seen that for an alloy composition as low as $x = 0.031$, the wavefunction of the E_- conduction-band edge has already a significant 26% increase in the admixture of the Γ_c states that are optically coupled to

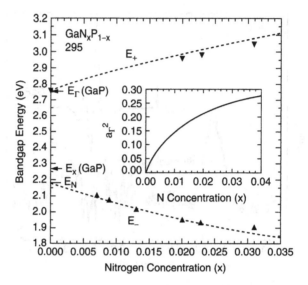

Figure 2.15 E_- and E_+ transition energies in GaNP samples as a function of N concentration. The dashed lines are the calculated variation of the E_- and E_+ edges with N content. The inset shows the relative contribution of the Γ_c states to the wavefunction of the E_- edge states.

the Γ_V states. This explains the strong direct optical transitions that are responsible for the absorption edge shown in Figure 2.15.

The role of N in the conduction-band structure of GaN$_x$P$_{1-x}$ alloys can be further examined by studying the effects of hydrostatic pressure on the photoluminescence (PL) transition energies. The results are shown in Figure 2.16 along with two previously reported extreme cases of the pressure dependencies of the indirect band gap ($\Gamma_V - X_c$) of GaP [59, 60]. A large reduction of the PL signal was observed at the energies corresponding to the onset of the indirect $\Gamma_V - X_c$ absorption [61]. The PL emission energies show a small upward shift with increasing pressure. The measured pressure coefficients are larger than the pressure coefficient of the highly localized N level, $dE_N/dP = 1.5$ meV/kbar, but much smaller than the pressure coefficient of $dE_{M\Gamma}/dP = 10$ meV/kbar for the $E_{M\Gamma}$ conduction-band edge in GaP [59, 62]. This is consistent with the fact that E_- is located much closer to the E_N energy level, as shown in Figure 2.15, so that its wavefunction has a predominantly N-like character.

The hydrostatic-pressure-induced shift of the fundamental band gap can be calculated by substituting the pressure-dependent energy levels $E_N(P)$ and $E_{M\Gamma}(P)$ into Equation 2.9. The solid lines in Figure 2.16 show the changes of the calculated fundamental band-edge energies. The results are in good agreement with the experimentally observed pressure dependence of the PL emission lines.

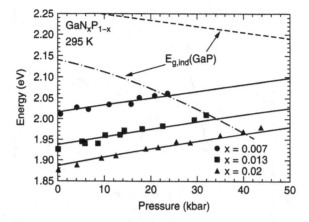

Figure 2.16 The pressure dependence of the PL emissions from the GaN$_x$P$_{1-x}$ samples with $x = 0.007$, 0.013, and 0.02. The solid lines represent the calculated changes of the PL energies using the BAC model. The dashed and dash-dotted lines are two extreme cases of the pressure dependence of the indirect band gap of GaP.

2.4 EXPERIMENTAL VERIFICATION II: BAND-ANTICROSSING-INDUCED NOVEL PHENOMENA

As has been mentioned in the previous section, thin films of GaN$_x$As$_{1-x}$, InN$_x$P$_{1-x}$, and Al$_y$Ga$_{1-y}$N$_x$As$_{1-x}$ have been successfully synthesized by N$^+$ implantation into GaAs, InP, and Al$_y$Ga$_{1-y}$As. Nitrogen implantation followed by rapid thermal processing was proven to be a practical and convenient method for the formation of diluted III-N$_x$-V$_{1-x}$ alloys [39–41]. In all three cases, the fundamental band-gap energy for the ion-beam-synthesized III-N$_x$-V$_{1-x}$ alloys was found to decrease with increasing N implantation dose in a manner similar to that observed in epitaxially grown GaN$_x$As$_{1-x}$ thin films. In GaN$_x$As$_{1-x}$, the highest value of x (fraction of "active" substitutional N on As sublattice) achieved using N$^+$ implantation and conventional rapid thermal annealing (RTA) technique was 0.006 with the highest N activation efficiency of $\approx 15\%$. It was observed that substitutional N$_{As}$ is thermally unstable at temperatures higher than 850°C. Transmission electron microscopy (TEM) suggested that formation of N-related voids resulted at RTA temperatures higher than 850°C due to the segregation of N [63].

Alternative means to improve N incorporation in N-ion-synthesized GaN$_x$As$_{1-x}$ alloys using the pulsed-laser melting (PLM) method has also been developed recently [64, 65]. The PLM technique involves the melting of the implant-damaged or amorphized layer induced by the near-surface absorption of intense (pulsed) laser radiation and the subsequent rapid epitaxial regrowth from the liquid. Epitaxy is seeded at the solid–liquid interface by the crystalline bulk in a manner very similar to liquid-phase

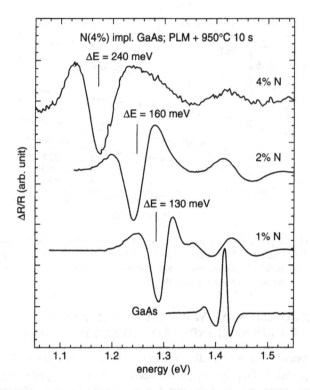

Figure 2.17 PR spectra measured from a series of samples implanted with increasing levels of N (x_{imp}) and processed by PLM at an energy fluence of 0.34 J/cm² and subsequent RTA at 950°C for 10 s.

epitaxy (LPE), but with the whole process occurring on a much shorter time scale, typically between 10^{-8} and 10^{-6} s [66, 67].

Figure 2.17 shows a series of PR spectra from samples implanted with increasing amounts of N processed by PLM with energy fluence of 0.34 J/cm² and subsequently RTA at 950°C for 10 s [64, 65]. Such PLM-RTA postimplantation treatments represent the "optimum" conditions giving rise to clear, sharp optical transitions. The PR spectra from the PLM-RTA samples shown in Figure 2.17 all exhibit distinct optical transitions across the fundamental band gap of the materials. The optical-transition energies of the various samples are indicated in the figure. A monotonic decrease in the band gap with increasing implanted N content x_{imp} can be clearly observed, indicating that an increasing fraction of N is incorporated in the As sublattice with increasing x_{imp}.

The amount of N incorporated in the As sublattice ("active" N) for the GaN_xAs_{1-x} layers formed by a combination of the N⁺ implantation and PLM-RTA method calculated using the BAC model is ~40–60% of the implanted value. This is over five times higher than that observed in samples synthesized by RTA alone [41]. Such drastic improvement can be attributed to the extremely short melt duration (~200 ns) and regrowth process that promotes N substitution in the As sublattice and inhibits the formation of nitrogen-related voids, which have been observed in samples formed by N+ implantation followed by RTA only [63]. In addition to the enhanced N incorporation, the GaN_xAs_{1-x} layers synthesized by N+ implantation followed by PLM-RTA were also found to be thermally stable up to annealing temperature >950°C. This new sample synthesis technique provided a convenient and reliable method for preparing various samples in the following studies.

2.4.1 Enhancement in Maximum Electron Concentration

The maximum achievable electron and/or hole concentration is an important criterion in the design of semiconductor devices. A universal rule that predicts the maximum free-carrier concentration achievable by doping has been developed and shown to be valid for a wide variety of semiconductor materials [68–70]. The rule is based on the amphoteric native-defect model that relates the type and concentrations of compensating native defects responsible for dopant compensation to the location of the Fermi level with respect to a common energy reference. According to this model, GaAs is predicted to exhibit limitations on the maximum free-electron concentration. Experimentally, the maximum electron concentration n_{max} in GaAs achievable under equilibrium conditions is limited to about 10^{18} to 10^{19} cm^{-3} [71].

As has been discussed earlier, the BAC model not only explains the band gap reduction in III-N_x-V_{1-x} alloys, but it also predicts that the N-induced modifications of the conduction band may have profound effects on the transport properties of this material system [43]. In particular, the downward shift of the conduction-band edge and the enhancement of the DOS effective mass in GaInNAs may lead to much enhanced maximum electron concentration n_{max}. Figure 2.18 shows the electron concentration in Se-doped MOVPE-grown $Ga_{1-3x}In_{3x}N_xAs_{1-x}$ films with $x = 0$ to 0.033 measured by Hall effect and the electrochemical capacitance-voltage (ECV) technique [32]. Since the Se atomic concentrations in these films are at least an order of magnitude higher than the free-electron concentration (in the range of 2–7 × 10^{20} cm^{-3}), the measured free-electron concentration shown in Figure 2.18 can be considered to be the maximum achievable free-electron concentration, n_{max}.

Figure 2.18 shows that the n_{max} increases strongly with the N-content x, with a maximum observed value of 7 × 10^{19} cm^{-3} for $x = 0.033$. This value is ~20 times of that found in a GaAs film (3.5 × 10^{18} cm^{-3}) grown under the same conditions. The much-enhanced n_{max} in $Ga_{1-3x}In_{3x}N_xAs_{1-x}$ films can be explained by considering the N-induced conduction-band modifications. According to the amphoteric native-defect model, the max-

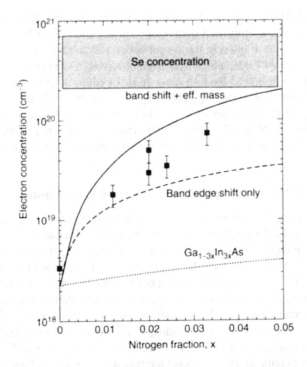

Figure 2.18 Comparison of the measured maximum electron concentration with the calculated values as a function of N fraction in $Ga_{1-3x}In_{3x}N_xAs_{1-x}$. Two different cases of the calculated n_{max} are shown: one includes effects of downward shift of the conduction band only (dashed curve), and the other includes both the band shift and the enhancement of the density of states (solid curve). The calculated n_{max} for samples with no N (i.e., when only the effects from the band-gap lowering produced by In incorporation are considered) are also shown in the figure (dotted curve). The shaded area indicates the range of Se concentration in these samples.

imum free-electron concentration is determined by the Fermi energy, which is constant with respect to the Fermi stabilization energy E_{FS} [71]. Therefore, the downward shift of the conduction-band edge toward E_{FS} and the enhancement of the DOS effective mass in GaInNAs lead to a much larger concentration of uncompensated, electrically active donors for the same location of the Fermi energy relative to E_{FS}. The calculated n_{max} as a function of x for $Ga_{1-3x}In_{3x}N_xAs_{1-x}$ due to the downward shift of the conduction band caused by the level anticrossing only, as well as that including the increase in the effective mass (calculated using Equations 2.14 and 2.15), are shown in Figure 2.18. Comparison of the experimental data with the calculation shows that in order to account for the large enhancement of the doping limits in III-N-V alloys, both the effects of band-gap reduction and the increase in the effective mass have to be taken into account.

Recently, using metalorganic molecular-beam epitaxy, Uesugi and Suemune have successfully grown quaternary GaAsNSe alloys with the

Se concentration up to 15% [72]. Electron concentration as high as $\sim 10^{20}$ cm^{-3} was reported in these GaNAsSe quaternary thin films. Moreover, specific contact resistance on the order of $\sim 10^{-4}$ Ω-cm^2 was achieved for Au/GaNAsSe structure fabricated with these films. This suggests that GaN$_x$As$_{1-x}$ alloys with exceedingly high free-electron concentration can be exploited for fabrication of low-resistivity nonalloyed ohmic contacts.

While Se-doped Ga$_{1-3x}$In$_{3x}$N$_x$As$_{1-x}$ alloys grown by MOVPE have shown enhanced n_{max} in accordance with the BAC model, similar behavior is also observed in S$^+$-implanted GaN$_x$As$_{1-x}$ thin film [73]. Figure 2.19 displays the carrier concentration profiles measured by ECV technique for the S-implanted GaN$_x$As$_{1-x}$ ($x \sim 0.008$) and SI-GaAs samples after RTA. A striking difference in the free-electron concentration n measured in the SI-GaAs and the GaN$_x$As$_{1-x}$ samples is observed. In the S$^+$-implanted SI-GaAs sample, $n \sim 2.5 \times 10^{17}$ cm^{-3} was measured in the bulk of the implanted layer, with a higher $n \sim 5 \times 10^{17}$ cm^{-3} toward the end of the implantation

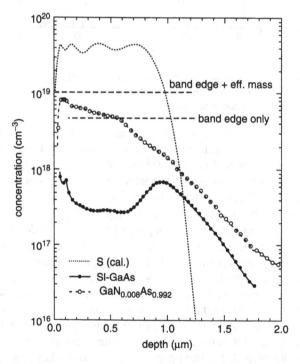

Figure 2.19 Ionized net donor concentration profiles for the GaN$_x$As$_{1-x}$ films and the SI-GaAs standard measured by the electrochemical capacitance-voltage (ECV) technique. The dashed horizontal lines indicate the theoretical free-electron concentrations in Ga$_{1-x}$N$_x$As by considering only the effects of the downward shift of the conduction band (band edge only) and the effects of band-gap reduction and the density-of-states effective-mass enhancement (band edge + effective mass).

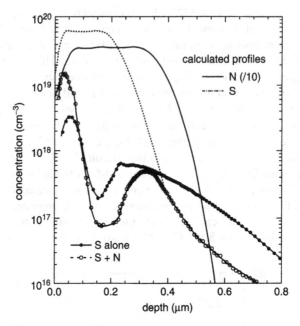

Figure 2.20 The ECV-measured net donor-concentration profiles for the GaAs samples implanted with S alone and S+N after RTA at 945°C for 10 s. The calculated atomic profiles for both the implanted S and N are also shown.

profile. The theoretical n_{max} in GaN_xAs_{1-x} due to the N-induced conduction-band modification within the framework of the BAC model and the amphoteric native-defect model is $\sim 1 \times 10^{19}$ cm^{-3} for the $GaN_{0.008}As_{0.992}$ sample. This value is in a reasonably good agreement with the measured concentration of 6×10^{18} cm^{-3} shown in Figure 2.19.

Attempts were also made to form n-type GaN_xAs_{1-x} thin films with high electron concentration by coimplantation of N and a dopant element in GaAs [74]. Figure 2.20 shows a comparison of the ECV-determined free-electron concentration profiles for the GaAs samples implanted with S alone and coimplanted with S and N (S+N) after RTA at 945°C for 10 s. The calculated, as-implanted S and N atomic distributions are also shown in the figure. The most prominent difference in the electron concentration profiles between the S-only and S+N samples is the much enhanced electron concentration in the S+N sample in a narrow region (~ 500 Å) near the surface.

Considering both the band-gap reduction and the large enhancement of the electron effective mass, the high nmax in the near-surface region of the S+N sample ($\sim 1.5 \times 10^{19}$ cm^{-3}) implies that the N content in this thin, near-surface, dilute-nitride layer is $x = 0.0032$. This value is in good agreement with the calculated N concentration in the surface region ($x \approx 0.003$–0.01). With this N content, the conduction-band edge is shifted downward by 77 meV, and the conduction-band effective mass at the Fermi energy is approximately three times higher than that of GaAs [75].

2.4.2   Mutual Passivation in III-N-V Alloys

In contrast to the observed enhancement of the doping activation of the group VI elements (S, Se), Si and N coimplantation in GaAs only resulted in a highly resistive layer [76]. This asymmetry in the behavior of group VI and IV donors can be explained by an entirely new effect in which an electrically active substitutional group IV donor and an isovalent N atom passivate each other's electronic effects [77]. This mutual passivation occurs in GaN_xAs_{1-x} doped with group IV donors (Si and Ge) through the formation of nearest-neighbor IV_{Ga}-N_{As} pairs.

Figure 2.21 shows the free-electron concentration in MBE-grown, Si-doped $Ga_{0.93}In_{0.07}N_{0.017}As_{0.983}$ and GaAs and MOVPE-grown, Se-doped $Ga_{0.92}In_{0.08}N_{0.024}As_{0.976}$ thin films after RTA for 10 s in the temperature range of 650–950°C. The Si and Se doping levels in these samples are in the range of $2-9 \times 10^{19}$ cm^{-3} and $\sim 2 \times 10^{20}$ cm^{-3}, respectively. For both the GaAs:Si and GaInNAs:Se samples, only slight decreases in electron concentrations, from 1.6×10^{19} to 8×10^{18} cm^{-3} for GaAs:Si and 3×10^{19} to 2×10^{19} cm^{-3} for GaInNAs:Se, are observed as the results of high-temperature RTA. Such a decrease in the electron concentration in GaAs is in agreement with the equilibrium maximum electron concentration (in the range of 10^{18} to 10^{19} cm^{-3}) [78]. The much higher electron concentration

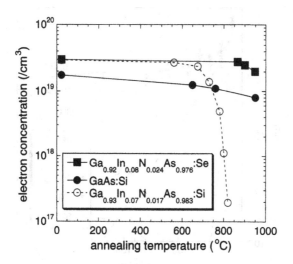

Figure 2.21 Electron concentrations of Si-doped GaAs and $Ga_{0.93}In_{0.07}N_{0.017}As_{0.983}$ as a function of annealing temperature for 10 s. The dependence of the electron concentration on RTA temperature for a MOVPE-grown Se-doped $Ga_{0.92}In_{0.08}N_{0.024}As_{0.976}$ film is also included.

in the Se-doped GaInNAs sample is also consistent with the enhanced donor activation efficiency resulting from the N-induced modification of the conduction-band structure [32].

On the other hand, the free-electron concentration in the GaInNAs:Si sample drops from 1.1×10^{19} cm^{-3} in the as-grown film to 3×10^{17} cm^{-3} after RTA at 950°C for 10 s. In fact, RTA at 950°C for 120 s further reduces the electron concentration to $<10^{15}$ cm^{-3}. The reduced electrical activity of Si donors in GaN$_x$As$_{1-x}$ alloys can be attributed to the formation of nearest-neighbor Si$_{Ga}$-N$_{As}$ pairs. The highly electronegative N atom strongly binds the fourth valence electron of Si, preventing it from acting as a hydrogen donor. Such an explanation suggests that, because of the localized nature of the N-states in GaN$_x$As$_{1-x}$, the passivation is limited to group IV donors that occupy Ga sites. This is supported by the small change in electrical behavior observed in the GaInNAs:Se thin film in which both the N and Se reside in the As sublattice and therefore cannot form nearest-neighbor passivating pairs.

The well-defined onset temperature of about 700°C for the observed reduction of electron concentration in GaInNAs:Si shown in Figure 2.21 roughly corresponds to the annealing condition that allows the Si atoms to diffuse over a length equal to the average distance between randomly distributed Si and N atoms (~7 Å) [77]. The diffusion-controlled passivation process is analyzed in the context of Si diffusion mediated by both neutral Ga vacancies (V_{Ga}^{0}) and triply negatively charged Ga vacancies (V_{Ga}^{3-}) [79]. Figure 2.22 shows the isothermal annealing effects of the normalized

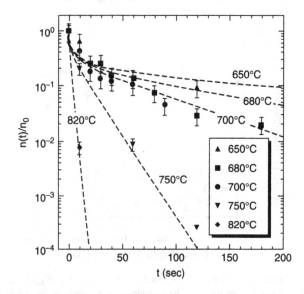

Figure 2.22 Normalized free-electron concentration as a function of annealing time at different annealing temperatures. The dashed curves represent the results from analytical calculations based on Si diffusion via Ga vacancies.

free-carrier concentration of the GaInNAs:Si sample for annealing temperatures in the range of 650 to 820°C. Calculations based on Si diffusion via V_{Ga}^0 and V_{Ga}^{3-} vacancies are shown as dashed lines in the figure. The calculations agree very well with the experimental data. According to the diffusion model, at high annealing temperatures or long annealing time, the Fermi-level-independent, V_{Ga}^0-mediated diffusion becomes increasingly important. This is reflected in the fact that the $\ln[n/n_0] \sim t$ curves approach a linear dependence at high temperatures or long anneal times.

Since isovalent N is responsible for a massive modification of the electronic structure of GaN_xAs_{1-x} alloys, the question arises to what extent the passivation process affects the N-induced modification of the electronic structure of the alloys. PR measurements on the GaInNAs:Si sample show that the band-gap energy increases with increasing RTA temperature. Annealing of the sample at 950°C increases the gap by about 35 meV. If this increase is attributed to deactivation of the N atoms, the concentration of the deactivated N is approximately equal to $0.004 \times 2.2 \times 10^{22}$ cm^{-3} $\approx$ 8×10^{19} cm^{-3}, which is close to the initial total Si concentration in the as-grown sample. This is consistent with the formation of Si_{Ga}-N_{As} pairs being responsible for the mutual passivation of both species. This scenario of the passivation process is further corroborated by PL measurements on the GaInNAs:Si. A strong PL emission peaked at ~0.8 eV is observed when the sample is mutually passivated, indicating the presence of deep states associated with the Si_{Ga}-N_{As} pairs [77].

The general nature of the mutual passivation effect is supported by the investigations of GaN_xAs_{1-x} layers doped with Ge, another group IV donor. Ge-doped GaN_xAs_{1-x} layers were synthesized by sequential implantation of Ge and N ions into GaAs followed by a combination of PLM and RTA [65]. The passivation of the N activity by the Ge atoms is illustrated in the series of PR spectra presented in Figure 2.23. The band-gap energies obtained from the PR spectra are shown in the inset as a function of the duration of 950°C RTA. A fundamental band-gap transition at 1.24 eV is observed for GaAs samples implanted with 2% N alone after PLM-RTA at 950°C for 10 to 120 s, corresponding to a GaN_xAs_{1-x} layer with $x \approx 0.01$. In contrast, the band gap of the samples coimplanted with N and Ge (2% N + 2% Ge) after PLM increases from 1.24 to 1.42 eV (band gap of GaAs) as the RTA duration increases to 60 s, revealing that all N_{As} sites are passivated by Ge. The gradual increase in the band gap of the 2% N + 2% Ge sample as a function of RTA temperature and/or time duration can be attributed to the passivation of N_{As} by Ge_{Ga} through the formation of nearest-neighbor Ge_{Ga}-N_{As} pairs.

Figure 2.24 shows a comparison of the electron concentration of the 2% N + 2% Ge and 2% Ge samples followed by PLM-RTA for 10 s in the temperature range of 650–950°C. The electron concentration of both samples approaches 10^{19} cm^{-3} after PLM. For the 2% Ge sample, thermal

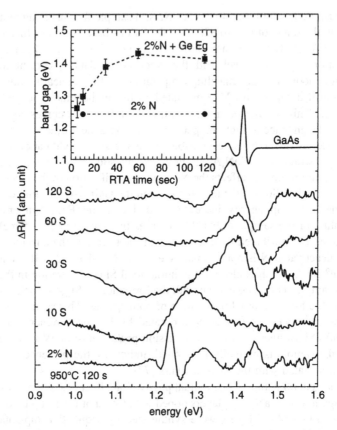

Figure 2.23 PR spectra measured from a series of ion-beam-synthesized, Ge-doped GaN_xAs_{1-x} samples after RTA at 950°C for durations of 5–120 s. The inset shows the band-gap energies determined from the PR measurements.

annealing after PLM drives the system toward equilibrium with an electron concentration of ~1 × 10^{18} cm^{-3}, which is consistent with the amphoteric character of Ge in GaAs [80]. The electron concentration of the 2% N + 2% Ge samples, on the other hand, drops over two orders of magnitude to less than 10^{17} cm^{-3} as the samples are subjected to RTA at temperatures higher than 650°C.

The changes in the band gap and the electrical behavior in the Ge-doped GaN_xAs_{1-x} sample show that the activities of Ge donors and isovalent N mutually passivate each other via the formation of N_{As}-Ge_{Ga} pairs, just as was the case in Si-doped GaN_xAs_{1-x}. These results, together with the mutual passivation of Si and N in GaN_xAs_{1-x}, clearly demonstrate the general nature of this phenomenon. The recently reported restoration of the band-gap energies to the values of the materials without nitrogen by introducing hydrogen into $Ga_{1-y}In_yN_xAs_{1-x}$ and GaN_xAs_{1-x} can be attributed to

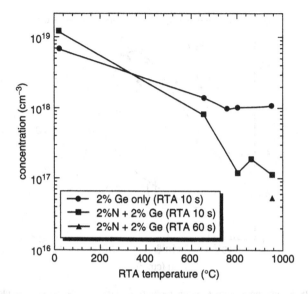

Figure 2.24 Free-electron concentrations of the 2% Ge and 2% N + 2% Ge samples after PLM + RTA at increasing temperature for 10 s obtained by Hall effect measurements. Electron concentration for the 2% N + 2% Ge sample after PLM + RTA at 950°C for 60 s is also shown.

another effect different from what has been discussed above. A full discussion on that topic is given in Chapter 6.

2.4.3 Electron Mobility

It has been widely recognized that the incorporation of small amounts of nitrogen into GaAs leads to a drastic reduction of the electron mobility. The typical mobility of GaN_xAs_{1-x} films ranges from ~10 to a few hundred cm²/Vs [81, 82], which is over an order of magnitude smaller than the electron mobility in GaAs at comparable doping levels. Figure 2.25 shows the change in room-temperature mobility of $Ga_{0.93}In_{0.07}N_{0.017}As_{0.983}$:Si when the electron concentration is reduced by rapid thermal annealing due to Si_{Ga}-N_{As} formation. The mobility shows a nonmonotonic dependence on the electron concentration, with a maximum at $n \sim 5 \times 10^{18}$ cm⁻³. The room-temperature mobility (μ_1) calculated from Equation 2.12 is shown as a short-dashed curve in Figure 2.25. Also shown is the Fermi energy as a function of n calculated from Equation 2.14. At high electron concentrations, when the Fermi energy approaches the original energy level of N localized states in $In_{0.07}Ga_{0.93}As_{0.983}N_{0.017}$ (located at ~0.30 eV above the conduction-band edge of E_M, or 0.54 eV above the conduction-band edge of E_-), the mobility is largely suppressed by the strong hybridization between $|E_N\rangle$ and $|E_M(k)\rangle$. At $n = 2 \times 10^{19}$ cm⁻³, the energy broadening and the scattering lifetime at the Fermi surface are estimated to be 0.25 eV and

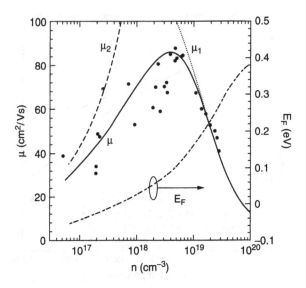

Figure 2.25 Room-temperature electron mobility of $Ga_{0.93}In_{0.07}N_{0.017}As_{0.983}$:Si plotted as a function of electron concentration. The calculated mobilities limited by the conduction-band broadening (μ_1) and by the random field scattering (μ_2) are shown. The calculated Fermi energy is referenced to the bottom of the lowest conduction band (E_-).

3 fs, respectively. The mean free path of free electrons is about 5 Å, which is only a third of the average distance between the randomly distributed N atoms. Therefore, at this electron concentration, the homogeneous broadening resulting from the anticrossing interaction is the dominant scattering mechanism that limits the electron mobility. As is seen in Figure 2.25, at high concentrations, the electron mobility calculated from the BAC model is in quantitative agreement with experimental results. It should be noted that this very good agreement has been obtained without any adjustable parameters.

At lower electron concentrations, the mobility starts to decrease, deviating severely from μ_1. This effect can be attributed to the scattering of the conduction electrons by the random fields caused by the structural and compositional disorder in the alloy. It is well known that, in partially disordered semiconductors, as the Fermi level decreases from the degenerate doping into the nondegenerate doping regime, the conduction electrons experience increasingly strong scattering from the potential fluctuations. As a result, the mobility decreases monotonically with decreasing electron concentration [83, 84]. In the case of $Ga_{1-y}In_yN_xAs_{1-x}$ alloys, the main contribution to the potential fluctuations originates from the random N distribution. An estimate for the electron mobility limited by the random field scattering (μ_2) is shown in Figure 2.25. The solid curve in Figure 2.25 takes into account the contributions of both the level-broadening and random alloy-scattering effects that limit the mobility ($\mu = 1/[\mu_1^{-1} + \mu_2^{-1}]$). This calculated mobility reproduces the nonmonotonic behavior of the mobility measured over two decades of change in electron concentration.

2.5 CONCLUDING REMARKS

The band-anticrossing model has been successfully used not only to explain unusual optical and electronic properties of highly mismatched alloys, but also to predict new effects that have been later experimentally confirmed. Although this chapter is limited to the review of properties of group III-N-V alloys, it should be emphasized that these alloys are only a subgroup of a much broader class of materials whose electronic structure is determined by the anticrossing interaction. Most notably, it has been shown that II-O-VI alloys in which column VI anions are partially replaced with O atoms have properties quite analogous to those of III-N-V alloys [85, 86]. The anticrossing interaction has also been shown to play a crucial role in "moderately" mismatched alloys, such as ZnS_xTe_{1-x} or $ZnSe_xTe_{1-x}$ [87] among group II-VI alloys and GaP_xSb_{1-x} or $GaAs_xSb_{1-x}$ [88, 89] among group III-V alloys.

It has been found recently that the anticrossing interaction is not limited to the conduction band only. For example, it is well known that partial replacement of electronegative Se or S anions in ZnSe or ZnS with more-metallic Te atoms leads to formation of highly localized, donorlike states close to the valence-band edge. It has been shown that for alloylike concentrations of Te, the anticrossing interaction between the Te levels and the extended states of the valence band lead to a drastic modification of the electronic structure [90]. In fact, it has been shown that the interaction is fully responsible for a large band-gap reduction in Se- and S-rich $ZnSe_xTe_{1-x}$ and ZnS_xTe_{1-x} alloys.

Most of the unusual material properties associated with the BAC interaction were discovered or predicted and experimentally confirmed on GaInNAs alloys, the most representative member of the HMA system. Subsequent realization of the general nature of the BAC interactions offers an interesting opportunity to design a large variety of HMAs with desired properties. Improvement of the concentration of electrical active donors demonstrated in GaInNAs alloys suggests that the large BAC-induced shifts of the conduction or valence band could also be used to overcome severe limitations on the n- or p-type doping in some II-VI compounds. It would be very interesting to explore if incorporation of small amounts of O could improve n-type doping of ZnO_xTe_{1-x}, or whether incorporation of Te into ZnS could lead to p-type $ZnTe_xS_{1-x}$. It would also be interesting to investigate whether the mutual passivation and the distinctly different behavior between dopants substituting anion and cation sites are applicable to other HMAs.

One of the many attractive features of the HMAs is that small variations of the composition produce large changes in the values of the material parameters. Consequently, one can envision that the synthesis of HMAs with the combination of ion implantation and pulsed-laser melting could provide efficient means to locally modify the properties of semiconductor materials.

The discovery, more than 10 years ago, of the large band-gap bowing in GaNAs led to very active fundamental and applied research on group III-N-V alloys. However, it was only during the last few years that the significance of this research for the basic understanding of the electronic structure of semiconductor alloys has been fully appreciated. Studies of many of the possible highly mismatched semiconductor alloys are only in their very early stages, and much more effort will be required to fully realize the basic significance and potential for practical applications of these materials.

Acknowledgments

The authors are very grateful to Dr. J.F. Geisz and Prof. C.W. Tu for providing the samples used in this study. Special thanks to J.W. Ager III, E.E. Haller, J. Beeman, M. A. Scarpulla, and O. Dubon for invaluable discussions and technical assistance. This work is supported by the director, Office of Science, Office of Basic Energy Sciences, Division of Materials Sciences and Engineering, of the U.S. Department of Energy under Contract No. DE-AC03-76SF00098.

References

1. Cardona, M., "Optical Properties of the Silver and Cuprous Halides," *Phys. Rev.* 129: 69–78 (1963).
2. Van Vechten, J.A., and Bergstresser, T.K., "Electronic Structures of Semiconductor Alloys," *Phys. Rev. B* 1: 3351–3358 (1970).
3. Richardson, D., "The Composition Dependence of Energy Bands in Mixed Semi-Conductor Systems with Zincblende Structures," *J. Phys. C: Solid State Phys.* 4: L289–291 (1971).
4. Casey, H.C., and Panish, M.B., "Composition Dependence of the $Ga_{1-x}Al_xAs$ Direct and Indirect Energy Gaps," *J. Appl. Phys.* 40: 4910–4912 (1969).
5. Thompson, A.G., Cardona, M., Shaklee, K.L., and Woolley, J.C., "Electroreflectance in the GaAs-GaP Alloys," *Phys. Rev.* 146: 601–610 (1966).
6. Weyers, M., Sato, M., and Ando, H., "Red Shift of Photoluminescence and Absorption in Dilute GaAsN Alloy Layers," *Jpn. J. Appl. Phys.* 31: L853–855 (1992).
7. Bi, W.G., and Tu, C.W., "Bowing Parameter of the Band-Gap Energy of GaN_xAs_{1-x}," *Appl. Phys. Lett.* 70: 1608–1610 (1997).
8. Baillargeon, N., Cheng, K.Y., Hofler, G.F., Pearah, P.J., and Hsieh, K.C., "Luminescence Quenching and the Formation of the $GaP_{1-x}N_x$ in GaP with Increasing Nitrogen Content," *Appl. Phys. Lett.* 60: 2540–2542 (1992).

9. Bi, W.G., and Tu, C.W., "N Incorporation in InP and Band Gap Bowing of InN_xP_{1-x}," *J. Appl. Phys.* 80: 1934–1936 (1996).

10. Harmand, J.C., Ungaro, G., Ramos, J., Rao, E.V.K., Saint-Girons, G., Teissier, R., Le Roux, G., Largeau, L., and Patriarche, G.J., "Investigations on GaAsSbN/GaAs Quantum Wells for 1.3–1.55 μm Emission," *Cryst. Growth* 227–228: 553–557 (2000).

11. Murdin, B.N., Karmal-Saadi, M., Lindsay, A., O'Reilly, E.P., Adams, A.R., Nott, G.J., Crowder, J.G., Pidgeon, C.R., Bradley, I.V., Wells, J.P.R., Burke, T., Johnson, A.D., and Ashley, T., "Auger Recombination in Long-Wavelength Infrared InN_xSb_{1-x} Alloys," *Appl. Phys. Lett.* 78: 1568–1570 (2001).

12. Neugebauer, J., and Van de Walle, C.G., "Electronic Structure and Phase Stability of $GaAs_{1-x}N_x$ Alloys," *Phys. Rev. B* 51: 10568–10571 (1995).

13. Ager III, J.W., and Walukiewicz, W., eds., "III-N-V Semiconductor Alloys," *Semicond. Sci. Technol.* 17: 741–906 (2002).

14. Buyanova, I.A., Chen, W.M., and Monemar, B., "Electronic Properties of Ga(In)NAs Alloys," *MRS Internet J. Nitride Semicond. Res.* 6 (2): 1–19 (2001).

15. Thomas, D.G., Hopfield, J.J., and Frosch, C.J., "Isoelectronic Traps due to Nitrogen in Gallium Phosphide," *Phys. Rev. Lett.* 15: 857–860 (1965).

16. Logan, R.A., White, H.G., and Wiegman, W., "Efficient Green Electroluminescence in Nitrogen-Doped GaP p-n Junctions," *Appl. Phys. Lett.* 13: 139–141 (1968).

17. Hjalmarson, H.P., Vogl, P., Wolford, D.J., and Dow, J.D., "Theory of Substitutional Deep Traps in Covalent Semiconductors," *Phys. Rev. Lett.* 44: 810–813 (1980).

18. Onodera, Y., and Toyozawa, Y., "Persistence and Amalgamation Types in the Electronic Structure of Mixed Crystals," *J. Phys. Soc. Japan* 24: 341 (1968).

19. Mahr, H., "Ultraviolet Absorption of the Mixed System KCl-KBr," *Phys. Rev.* 122: 1464–1468 (1961).

20. Anderson, P.W., "Localized Magnetic States in Metals," *Phys. Rev.* 124: 41–53 (1961).

21. Kocharyan, A.N., "Changes in the Valence of Rare-Earth Semiconductors in the Many-Impurity Anderson Model," *Soc. Phys. Solid State* 28: 6–10 (1986).

22. Ivanov, M.A., and Pogorelov, Y.G., "Electron Properties of Two-Parameter Long-Range Impurity States," *Sov. Phys. JETP* 61: 1033–1039 (1985).

23. Doniach, S., and Sondheimer, E.H., *Green's Functions for Solid State Physicists* (London: Imperial College Press, 1998).

24. Yonezawa, F., and Morigaki, K., "Coherent Potential Approximation," *Suppl. Prog. Theor. Phys.* 53: 1–76 (1973).

25. Elliott, R.J., Krumhansl, J.A., and Leath, P.L., "The Theory and Properties of Randomly Disordered Crystals and Related Physical Systems," *Rev. Mod. Phys.* 46: 465–543 (1974).

26. Shan, W., Walukiewicz, W., Ager III, J.W., Haller, E.E., Geisz, J.F., Friedman, D.J., Olson, J.M., and Kurtz, S.R., "Band Anticrossing in GaInNAs Alloys," *Phys. Rev. Lett.* 82: 1221–1224 (1999).

27. Wu, J., Walukiewicz, W., and Haller, E.E., "Band Structure of Highly Mismatched Semiconductor Alloys: Coherent Potential Approximation," *Phys. Rev. B* 65: 233210/1-4 (2002).

28. Bhat, R., Caneau, C., Salamanca-Riba, L., Bi, W.G., and Tu, C.W., "Growth of GaAsN/GaAs, GaInAsN/GaAs and GaInAsN/GaAs Quantum Wells by Low-Pressure Organometallic Chemical Vapor Deposition," *J. Crystal Growth* 195: 427–437 (1998).

29. Malikova, L., Pollak, F.H., and Bhat, R., "Composition and Temperature Dependence of the Direct Band Gap of $GaAs_{1-x}N_x$ ($0 < x < 0.0232$) Using Contactless Electroreflectance," *J. Electron. Mat.* 27: 484–487 (1998).

30. Keyes, B.M., Geisz, J.F., Dippo, P.C., Reedy, R., Kramer, C., Friedman, D.J., Kurtz, S.R., and Olson, J.M., "Investigation of GaNAs," *AIP Conf. Proc.* 462: 511–515 (1999).

31. Uesugi, K., Marooka, N., and Suemune, I., "Reexamination of N Composition Dependence of Coherently Grown GaNAs Band Gap Energy with High-Resolution X-ray Diffraction Mapping Measurements," *Appl. Phys. Lett.* 74: 1254–1256 (1999).

32. Yu, K.M., Walukiewicz, W., Shan, W., Ager III, J.W., Wu, J., Haller, E.E., Geisz, J.F., Friedman, D.J., and Olson, J.M., "Nitrogen-Induced Increase of the Maximum Electron Concentration in Group III-N-V Alloys," *Phys. Rev. B* 61: R13337ñ13340 (2000).

33. Skierbiszewski, C., Perlin, P., Wisniewski, P., Suski, T., Geisz, J.F., Hingerl, K., Jantsch, W., Mars, D., and Walukiewicz, W., "Band Structure and Optical Properties of $In_yGa_{1-y}As_{1-x}N_x$ Alloys," *Phys. Rev. B* 65: 035207/1-10 (2001).

34. Wu, J., Shan, W., and Walukiewicz, W., "Band Anticrossing in Highly Mismatched III-V Semiconductor Alloys," *Semicond. Sci. Technol.* 17: 860–869 (2002).

35. Perkins, J.D., Masarenhas, A., Geisz, J.F., and Friedman, D.J., Conduction-Band-Resonant Nitrogen-Induced Levels in $GaAs_{1-x}N_x$, *Phys. Rev. B* 64: 121301/1-4 (2001).

36. Wu, J., Walukiewicz, W., Yu, K.M., Ager III, J.W., Haller, E.E., Hong, Y., Xin, H.P., and Tu, C.W., "Band Anticrossing in $GaP_{1-x}N_x$ Alloys," *Phys. Rev. B* 65: R241303/1-4 (2002).

37. Lindsay, A., and O'Reilly, E.P., "Theory of Enhanced Bandgap Non-Parabolicity in GaN_xAs_{1-x} and Related Alloys," *Solid State Comm.* 112: 443–447 (1999).

38. Hofmann, M., Wagner, A., Ellmers, C., Schlichenmeier, S.S., Höhnsdorf, F., Koch, J., Stolz, W., Koch, S.W., Ruhle, W.W., Hader, J., Moloney, J.V., O'Reilly, E.P., Borchert, B., Egorov, A.Y., and Riechert, H., "Gain Spectra of (GaIn)(NAs) Laser Diodes for the 1.3-μm-Wavelength Regime," *Appl. Phys. Lett.* 78: 3009–3011 (2001).

39. Shan, W., Yu, K.M., Walukiewicz, W., Ager III, J.W., Haller, E.E., and Ridgway, M.C., "Reduction of Band-Gap Energy in GaNAs and AlGaNAs Synthesized by N^+ Implantation," *Appl. Phys. Lett.* 75: 1410–1412 (1999).

40. Yu, K.M., Walukiewicz, W., Wu, J., Beeman, J., Ager III, J.W., Haller, E.E., Shan, W., Xin, H.P., Tu, C.W., and Ridgway, M.C., "Synthesis of InN_xP_{1-x} Thin Films by N Ion Implantation," *Appl. Phys. Lett.* 78: 1077–1079 (2001).41.Yu, K.M., Walukiewicz, W., Wu, J., Beeman, J., Ager III, J.W., Haller, E.E., Shan, W., Xin, H.P., Tu, C.W., and Ridgway, M.C., "Formation of Diluted III-V Nitride Thin Films by N Ion Implantation," *J. Appl. Phys.* 90: 2227–2234 (2001).

42. Shan, W., Walukiewicz, W., Ager III, J.W., Haller, E.E., Geisz, J.F., Friedman, D.J., Olson, J.M., and Kurtz, S.R., "Effect of Nitrogen on the Band Structure of GaInNAs Alloys," *J. Appl. Phys.* 86: 2349–2351 (1999).

43. Walukiewicz, W., Shan, W., Ager III, J.W., Chamberlin, D.R., Haller, E.E., Geisz, J.F., Friedman, D.J., Olson, J.M., and Kurtz, S.R., "Nitrogen-Induced Modification of the Electronic Band Structures in Group III-N-V Alloys," *Proc. 195th Mtg. Electrochem. Soc.* 99 (11): 190–200 (1999).

44. Perkins, D.J., Mascarenhas, A., Zhang, Y., Geisz, J.F., Friedman, D.J., Olson, J.M., and Kurtz, S.R., "Nitrogen-Activated Transitions, Level Repulsion, and Band Gap Reduction in $GaAs_{1-x}N_x$ with $x < 0.03$," *Phys. Rev. Lett.* 82: 3312–3315 (1999).

45. Skierbiszewski, C., "Experimental Studies of the Conduction-Band Structure of GaInNAs Alloys," *Semicond. Sci. Technol.* 17: 803–814 (2002).

46. Shan, W., Walukiewicz, W., Yu, K.M., Wu, J., Ager III, J.W., Haller, E.E., Xin, H.P., and Tu, C.W., "Nature of the Fundamental Band Gap in GaN_xP_{1-x} Alloys," *Appl. Phys. Lett.* 76: 3251–3253 (2000).

47. Wu, J., Shan, W., Walukiewicz, W., Yu, K.M., Ager III, J.W., Haller, E.E., Xin, H.P., and Tu, C.W., "Effect of Band Anticrossing on the Optical Transitions in $GaAs_{1-x}N_x$/GaAs Multiple Quantum Wells," *Phys. Rev. B* 64: 085320/1-4 (2001).

48. Shan, W., Fang, X.M., Li, D., Jiang, S., and Shen, S.C., "Photomodulated Transmission Spectroscopy of the Intersubband Transitions in Strained $In_{1-x}Ga_xAs$/GaAs Multiple Quantum Wells under Hydrostatic Pressure," *Phys. Rev. B* 43: 14615–14620 (1991).

49. Jones, E.D., Modine, N.A., Allerman, A.A., Kurtz, S.R., Wright, A.F., Tozer, S.T., and Wei, X., "Optical Properties of InGaAsN: A New 1-eV Bandgap Material System," *SPIE Proc.* 3621: 52–63 (1999).

50. Jones, E.D., Modine, N.A., Allerman, A.A., Kurtz, S.R., Wright, A.F., Tozer, S.T., and Wei, X., "Band Structure of $In_xGa_{1-x}As_{1-y}N_y$ Alloys and Effects of Pressure," *Phys. Rev. B* 60: 4430–4433 (1999).

51. Mattila, T., Wei, S.H., and Zunger, A., "Localization and Anticrossing of Electron Levels in $GaAs_{1-x}N_x$ Alloys," *Phys. Rev. B* 60: R11245–11248 (1999).

52. Kent, P.R.C., and Zunger, A., "Evolution of III-V Nitride Alloy Electronic Structure: The Localized to Delocalized Transition," *Phys. Rev. Lett.* 86: 2613–2616 (2001).

53. Wang, L.W., "Large-Scale Local-Density-Approximation Band Gap-Corrected GaAsN Calculations," *Appl. Phys. Lett.* 78: 1565–1567 (2001).

54. Zhang, Y., Mascarenhas, A., Xin, H.P., and Tu, C.W., "Formation of an Impurity Band and Its Quantum Confinement in Heavily Doped GaAs:N," *Phys. Rev. B* 61: 7479–7482 (2000).

55. Zhang, Y., Mascarenhas, A., Geisz, J.F., Xin, H.P., and Tu, C.W., "Discrete and Continuous Spectrum of Nitrogen-Induced Bound States in Heavily Doped $GaAs_{1-x}N_x$," *Phys. Rev. B* 63: 085205/1-8 (2001).

56. Zhang, Y., Mascarenhas, A., Xin, H.P., and Tu, C.W., "Scaling of Band-Gap Reduction in Heavily Nitrogen Doped GaAs," *Phys. Rev. B* 63: R161303/1-4 (2001).

57. Shan, W., Walukiewicz, W., Yu, K.M., Ager III, J.W., Haller, E.E., Geisz, J.F., Friedman, D.J., Olson, J.M., Kurtz, S.R., Xin, H.P., and Tu, C.W., "Effect of Nitrogen on the Band Structure of III-N-V Alloys," *SPIE Proc.* 3944: 69–79 (2000).

58. Shan, W., Walukiewicz, W., Yu, K.M., Ager III, J.W., Haller, E.E., Geisz, J.F., Friedman, D.J., Olson, J.M., Kurtz, S.R., and Nauka, C., "Effect of Nitrogen on the Electronic Band Structure of Group III-N-V Alloys," *Phys. Rev. B* 62: 4211–4214 (2000).

59. Ves, S., Strossner, K., Kim, K.C., and Cardona, M., "Dependence of the Direct Energy Gap of GaP on Hydrostatic Pressure," *Solid State Comm.* 55: 327–331 (1985).

60. Goni, A.R., Syassen, K., Strossner, K., and Cardona, M., "Effect of Pressure on the Optical Absorption in GaP and $Ga_xIn_{1-x}P$ ($x = 0.36$ and 0.5)," *Phys. Rev. B* 39: 3178–3184 (1989).

61. Martinez, G., "Optical Properties of Semiconductors under Pressure," chap. 4 in *Optical Properties of Solids*, ed. M. Balkanski (Amsterdam: North-Holland, 1980).

62. Madelung, O., ed., *Semiconductors*, Landolt-Börnstein, New Series, group 3, vol. 22, part a (Berlin: Springer-Verlag, 1988).

63. Jasinski, J., Yu, K.M., Walukiewicz, W., Liliental-Weber, Z., and Washburn, J., "Influence of Microstructure on Electrical Properties of Diluted GaN_xAs_{1-x} Formed by Nitrogen Implantation," *Appl. Phys. Lett.* 79: 931–933 (2001).

64. Yu, K.M., Walukiewicz, W., Scarpulla, M.A., Dubon, O.D., Jasinski, J., Liliental-Weber, Z., Wu, J., Beeman, J., Pillai, M.R., and Aziz, M.J., "Synthesis of $GaNxAs_{1-x}$ Thin Films by Pulsed Laser Melting and Rapid Thermal Annealing of N^+-Implanted GaAs," *J. Appl. Phys.* 94: 1043–1049 (2003).

65. Yu, K.M., Walukiewicz, W., Wu, J., Shan, W., Beeman, J., Scarpulla, M.A., Dubon, O.D., Ridgway, M.C., Mars, D.E., and Chamberlin, D.R., "Mutual Passivation of Germanium and Nitrogen in Diluted GaN_xAs_{1-x} Alloys," *Appl. Phys. Lett.* 83:2844–2846 (2003).

66. White, C.W., and Percy, P.S., *Laser and Electron Beam Processing of Materials* (New York: Academic Press, 1980).

67. Williams, J.S., Poate, J.M., and Mayer, J.M., eds., *Laser Annealing of Semiconductors* (New York: Academic Press, 1982), 385.

68. Walukiewicz, W., "Amphoteric Native Defects in Semiconductors," *Appl. Phys. Lett.* 54: 2094–2096 (1989).

69. Walukiewicz, W., "Application of the Amphoteric Native Defect Model to Diffusion and Activation of Shallow Impurities in III-V Semiconductors," *Mat. Res. Soc. Symp. Proc.* 300: 421–432 (1993).

70. Zhang, S.B., Wei, S.H., and Zunger, A., "A Phenomenological Model for Systematization and Prediction of Doping Limits in II-VI and I-III-VI$_2$ Compounds," *J. Appl. Phys.* 83: 3192–3196 (1998).

71. Walukiewicz, W., "Diffusion, Interface Mixing and Schottky Barrier Formation," *Mater. Sci. Forum* 143–147: 519 (1993).

72. Uesugi, K., and Suemune, I., "Highly Conductive GaAsNSe Alloys Grown on GaAs and Their Nonalloyed Ohmic Properties," *Appl. Phys. Lett.* 79: 3284–3286 (2001).

73. Yu, K.M., Walukiewicz, W., Shan, W., Wu, J., Ager III, J.W., Haller, E.E., Geisz, J.F., and Ridgway, M.C., "Nitrogen-Induced Enhancement of the Free Electron Concentration in Sulfur Implanted GaN$_x$As$_{1-x}$," *Appl. Phys. Lett.* 77: 2858–2860 (2000).

74. Yu, K.M., Walukiewicz, W., Shan, W., Wu, J., Beeman, J., Ager III, J.W., and Haller, E.E., "Increased Electrical Activation in the Near-Surface Region of Sulfur and Nitrogen Co-Implanted GaAs," *App. Phys. Lett.* 77: 3607–3609 (2000).

75. Skierbiszewski, C., Perlin, P., Wiśniewski, P., Knap, W., Suski, T., Walukiewicz, W., Shan, W., Yu, K.M., Ager III, J.W., Haller, E.E., Geisz, J.F., and Olson, J.M., "Large, Nitrogen-Induced Increase of the Electron Effective Mass in In$_y$Ga$_{1-y}$N$_x$As$_{1-x}$," *Appl. Phys. Lett.* 76: 2409–2411 (2000).

76. Yu, K.M., "Ion Beam Synthesis and n-Type Doping of Group III-N$_x$-V$_{1-x}$ Alloys," *Semicond. Sci. Technol.* 17: 785–796 (2002).

77. Yu, K.M., Walukiewicz, W., Wu, J., Mars, D.E., Chamberlin, D.R., Scarpulla, M.A., Dubon, O.D., and Geisz, J.F., "Mutual Passivation of Electrically Active and Isoelectronic Impurities: Si Doped GaN$_x$As$_{1-x}$," *Nat. Mater.*, 1, 185–189 (2002).

78. Walukiewicz, W., "Intrinsic Limitations to the Doping of Wide-Gap Semiconductors," *Physica* B302–303: 123–134 (2001).

79. Wu, J., Yu, K.M., Walukiewicz, W., He, G., Haller, E.E., Mars, D.E., and Chamberlin, D.R., "Mutual Passivation Effects in Si-Doped Diluted GaAs$_{1-x}$N$_x$ Alloys," *Phys. Rev. B* B68:195202 (2003)

80. Yeo, Y.K., Ehret, J.E., Pedrotti, F.L., Park, Y.S., and Theis, W.M., "Amphoteric Behavior of Ge Implants in GaAs," *Appl. Phys. Lett.* 35: 197–199 (1979).

81. Geisz, J.F., Friedman, D.J., Olson, J.M., Kurtz, S.R., and Keyes, B.M., "Photocurrent of 1-eV GaInNAs Lattice-Matched to GaAs," *J. Crystl. Growth* 195: 401–408 (1998).

82. Kurtz, S.R., Allerman, A.A., Seager, C.H., Sieg, R.M., and Jones, E.D., "Minority Carrier Diffusion, Defects, and Localization in InGaAsN, with 2% Nitrogen," *Appl. Phys. Lett.* 77: 400–402 (2000).

83. Bonch-Bruevich, V.L., "Interband Optical Transitions in Disordered Semiconductors," *Phys. Stat. Solidi* 42: 35–42 (1970).

84. Zhumatii, P.G., "Intraband Conductivity and Thermopower of Semiconductors with Slowly Varying Gaussian Random Field," *Phys. Stat. Solidi B* 75: 61–72 (1976).

85. Shan, W., Walukiewicz, W., Ager III, J.W., Yu, K.M., Wu, J., Haller, E.E., Nabetani, Y., Mukawa, T., Ito, Y., and Matsumoto, T., "Effect of Oxygen on the Electronic Band Structure in ZnO$_x$Se$_{1-x}$ Alloys," *Appl. Phys. Lett.* 83: 299–301 (2003).

86. Nabetani, Y., Mukawa, T., Ito, Y., and Matsumoto, T., "Epitaxial Growth and Large Band-Gap Bowing in ZnSeO Alloys,î *Appl. Phys. Lett.* 83: 1148ñ1150 (2003).

87. Walukiewicz, W., Shan, W., Yu, K.M., Ager III, J.W., Haller, E.E., Miotkowski, I., Seong, M.J., Alawadhi, H., and Ramdas, A.K., "Interaction of Localized Electronic States with the Conduction Band: Ban Anticrossing in II–VI Semiconductor Ternaries," *Phys. Rev. Lett.* 85: 1552–1555 (2000).

88. Perlin, P., Wisniewski, P., Skierbiszewski, C., Suski, T., Kaminska, E., Subramanya, S.G., Weber, E.R., Mars, D.E., and Walukiewicz, W., "Interband Optical Absorption in Free Standing Layer of G$_{0.96}$In$_{0.04}$As$_{0.99}$N$_{0.01}$," *Appl. Phys. Lett.* 76: 1279–1281 (2000).a

89. Thomas, M.B., Coderre, W.M., and Wooley, J., "Energy Gap Variation in GaA$_x$Sb$_{1-x}$ Alloys," *Phys. Stat. Solidi A* 2: K141 (1970).s

90. Wu, J., Walukiewicz, W., Yu, K.M., Ager III, J.W., Haller, E.E., Miotkowski, I., Ramdas, A.K., Su, C.H., Sou, K., Perera, R.C.C., and Denlinger, J.D., "Origin of the Large Band-Gap Bowing in Highly Mismatched Semiconductor Alloys," *Phys. Rev. B* 67: 035207/1-5 (2003).

CHAPTER 3

Tight-Binding and k·p Theory of Dilute Nitride Alloys

E.P. O'REILLY, A. LINDSAY, and S. FAHY

NMRC University College, Lee Maltings, Prospect Row, Cork, Ireland

3.1 INTRODUCTION

This book reflects the considerable recent interest in and the fascination of $Ga_{1-y}In_yN_xAs_{1-x}$ and related alloys. When a small fraction of arsenic atoms in GaAs is replaced by nitrogen, the energy gap initially decreases rapidly, at about 0.1 eV per percent of N for $x < 0.03$ [1]. This behavior is markedly different to conventional semiconductors, and it is of interest both from a fundamental perspective and also because of its significant potential device

1-591-69019-6/04/$0.00+$1.50
© 2004 by CRC Press LLC

applications. The strong bowing opens the possibility of using GaInNAs to get optical emission on a GaAs substrate at the technologically important wavelengths of 1.3 and 1.55 μm, considerably expanding the capabilities of GaAs for optoelectronics [2–5].

Our understanding of conventional III-V alloys has been built up through a range of approaches. Much progress is based on the use and application of relatively simple models, such as effective-mass theory and the envelope-function method [6–8] to describe electronic states in quantum wells and heterostructures. These simple and well-established models are underpinned and informed by more detailed and fundamental theoretical calculations, as well as by comparison with a wide range of experimental data.

The issue we address in this chapter is the development of appropriate models to describe the electronic structure of ideal dilute nitride alloys; the theoretical analysis of defects is considered later in Chapter 8. The results of band structure calculations using the pseudopotential method have been discussed in Chapter 1 [9–13]. We review here how the tight-binding method [14–17] can also be used to probe and provide insight both into fine details and into trends in the electronic structure. Of equal importance, there is considerable advantage to developing simple models that describe and predict the variation with N composition, x, of such key properties as the energy gap, $E_g(x)$, and the band-edge effective masses, m_c^* and m_v^*. We need to know the applicability and limitations of such models to understand the fundamental electronic properties of this unusual alloy system and also to enable predictive design and analysis of potential optoelectronic devices.

It is well established that when a single N atom replaces an As atom in GaAs, it forms a resonant defect level above the conduction band edge of GaAs [18, 19]. This defect level arises because of the large difference in electronegativity and atomic size between N and As [20–22]. A major break-through was achieved for dilute nitride alloys with the demonstration by Walukiewicz and coworkers (using hydrostatic pressure techniques [23]) that the reduction in energy gap in Ga(In)N$_x$As$_{1-x}$ is due to a band anticrossing (BAC) interaction between the conduction band edge and a band of higher-lying localized nitrogen resonant states. The BAC model is consistent with a wide range of experimental data, but it was not initially supported by a number of theoretical studies, which used the pseudopotential method to carry out a detailed investigation of the lower-lying conduction-band states in ordered GaN$_x$As$_{1-x}$ supercells [9–11]. Experimental studies of ultradilute nitride alloys show a range of resonant defect levels above the conduction band edge due to the formation of N complexes, including, for example, gallium atoms with two N neighbors, giving resonant defect levels close to the low-temperature conduction band edge of GaAs [19, 24, 25]. Similar states are found in empirical pseudopotential [12, 13] and tight-binding [17] studies of N complexes. Such calculations support many aspects of the BAC model, but they also provide additional insight into the role of disorder and nitrogen clustering in GaNAs alloys.

Dilute nitride alloys have many other novel features. Nitrogen is so strongly different from arsenic that adding H can lead to the formation of N-H complexes, which remove the resonant defect levels above the conduction band edge, thereby eliminating the band-gap bowing [26–28], as discussed further in Chapter 5. The electron effective mass increases strongly when even a small amount of N is added to $Ga(In)N_xAs_{1-x}$ [29–31], as predicted by the band anticrossing model [23, 32, 33]. Although the measured increase in some samples is consistent with that predicted by the BAC model, the enhancement observed in other samples is considerably larger than expected.

Overall, adding nitrogen strongly perturbs the band structure of Ga(In)As and related alloys. Alloy scattering is therefore considerably stronger than in conventional III-V semiconductors [33–35]. In addition, composition fluctuations and variations in the N local environment can lead to strong variations in the band-gap and large potential fluctuations in bulk and low-dimensional samples. How then can we derive appropriate theoretical models to describe the electronic structure of Ga(In)NAs?

The approach we take here is to first consider ordered GaNAs, reviewing the insights we have obtained using a carefully parameterized tight-binding method to describe the electronic structure. Using ordered structures, we derive explicitly a two-level BAC model in Section 3.2 to describe the conduction band edge, showing that the concept of localized N resonant states can remain valid even up to $x \sim 0.25$.

We then turn to consider disorder effects in Section 3.3, using the sp^3s^* tight-binding Hamiltonian to investigate how the inevitable disorder in the N distribution in Ga(In)NAs modifies the BAC model and its predictions. We show that even in a disordered alloy, the band-gap bowing (and most other aspects of the electronic structure) can still be described very accurately by the BAC model, but with the BAC model now explicitly treating the random distribution of N resonant states.

Having established the theoretical basis for using the BAC model to describe the extreme band-gap bowing, we then turn in Section 3.4 to show how the conventional eight-band k·p model [6, 8] must be extended to a ten-band model, including two additional spin-degenerate nitrogen resonant bands, to describe the band dispersion in ordered (and disordered) GaN_xAs_{1-x} supercells. This theoretical analysis justifies and underpins the application of such a model to describe the conduction- and valence-band dispersion in bulk GaInNAs and in GaInNAs/GaAs heterostructures [25, 36–38], as reviewed by Klar and Tomic and used by Fehse and Adams in later chapters of this book.

We further consider disorder effects in Section 3.5. It has been widely observed that the energy gap of GaInNAs blueshifts after a rapid annealing process [25, 39–41]. We show that the N nearest-neighbor environment has a strong impact on the N resonance energy, and on the magnitude of its interaction with the conduction band edge. We show that for any given alloy composition, the N resonance energy shifts away from the unperturbed band edge, and the interaction energy decreases as the number of In nearest neighbors increases [42, 43]. The strong blueshift observed upon annealing can then be explained in terms of a rearrangement of the local-neighbor environment, accompanied by an increase in the average number of In neighbors, as discussed further in several later chapters.

We switch direction in Section 3.6 to consider the consequences of the strong band-gap bowing on electron mobility in dilute nitride semiconductors. We show that there is a fundamental connection between the band-gap bowing and the n-type carrier scattering cross-section in the ultradilute limit, and that this imposes general limits on the carrier mobility in such alloys [34, 35]. Within an independent scattering approximation, the carrier mobility is esti-mated to be ~1000 cm^2/V^{-1}s^{-1} for a N atomic concentration of 1%, compa-rable with the highest measured mobility in high-quality GaInNAs samples at these N concentrations, but higher than that found in many samples.

Overall, we conclude that a clear understanding is emerging concerning many aspects of the electronic structure of dilute nitride alloys. Further effort is still needed to provide simple, quantitative descriptions of several areas, including band localization effects and the measured effective masses, particularly in GaN$_x$As$_{1-x}$ at small x. Nevertheless, significant progress has been made, both through the use of the BAC model and more detailed theoretical studies, giving a quantitative description of the elec-tronic structure, and enabling predictive design and analysis of GaInNAs-based heterostructures and optoelectronic devices.

3.2. NITROGEN RESONANT STATES IN ORDERED GaN$_x$As$_{1-x}$ STRUCTURES

The BAC model explains the extreme band-gap bowing observed in In$_y$Ga$_{1-y}$N$_x$As$_{1-x}$ in terms of an interaction between two levels, one at energy E_c associated with the extended conduction band edge (CBE) states of the InGaAs matrix, and the other at energy E_N associated with the localized N impurity states, with the two states linked by a matrix element V_{Nc} describ-ing the interaction between them [23]. The CBE energy of Ga(In)N$_x$As$_{1-x}$, E_-, is then given by the lower eigenvalue of the determinant

$$\begin{vmatrix} E_N & V_{Nc} \\ V_{Nc} & E_c \end{vmatrix} \qquad (3.1)$$

The upper eigenvalue E_+ has been observed in photoreflectance measure-ments [23, 24, 44], as illustrated in Figure 2.5 of the previous chapter,

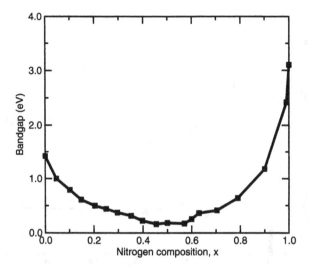

Figure 3.1 Variation of the band-gap energy, E_g, in free-standing GaN_xAs_{1-x}, calculated across the full alloy range ($0 < x < 1$) using a carefully parameterized sp^{3}s* Hamiltonian.)

which showed the measured variation in E_- and E_+ as a function of N composition x in GaN_xAs_{1-x}. However, initial pseudopotential calculations found no direct evidence for the upper state [9–11], although they do confirm its effect on the conduction band edge [12], and it has more recently been identified for relatively low N compositions ($x < {\sim}1\%$) [13].

To investigate the resonant state, and its behavior, we have developed an accurate sp^{3}s* tight-binding (TB) Hamiltonian to describe the electronic structure of $GaInN_xAs_{1-x}$ [16]. This Hamiltonian fully accounts for the observed experimental data, and also gives results in good agreement with pseudopotential calculations [10, 12, 45, 46]. Figure 3.1 shows, for instance, the variation of the band-gap energy across the full alloy range in free-standing GaN_xAs_{1-x}, calculated using the sp^{3}s* Hamiltonian: the observed variation matches well that obtained in the literature [45, 46].

To investigate the resonant state and its behavior, we calculated the electronic structure of ordered GaN_xAs_{1-x} supercells [15, 16]. By comparing the calculated CBE states ψ_{c1} and ψ_{c0} in large supercells ($Ga_{864}N_1As_{863}$ and $Ga_{864}As_{864}$, respectively), we can derive the nitrogen resonant state ψ_{N0} associated with an isolated N atom. In the BAC model, ψ_{c1} is a linear combination of ψ_{c0} and ψ_{N0}, with ψ_{N0} then given by

$$\psi_{N0} = \frac{\psi_{c1} - \alpha\psi_{c0}}{\sqrt{1-\alpha^2}} \qquad (3.2)$$

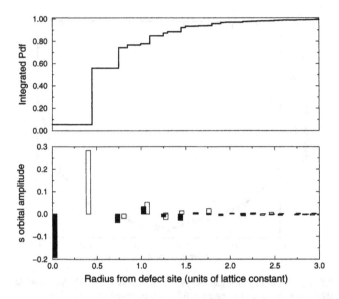

Figure 3.2 (a) Integrated radial probability density function for the resonant state due to an isolated nitrogen at the center of a 1728-atom supercell. (b) Amplitude of the resonant state projected onto the *s*-orbital of each atom as a function of radial distance from the nitrogen site (filled bars: anion sites; empty bars: cation sites).

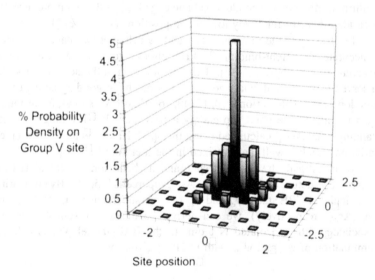

Figure 3.3 Calculated probability density of nitrogen resonant state, $|\psi_{N0}|^2$, projected onto the group V atoms in the [0,0,1] plane of a $Ga_{500}N_1As_{499}$ supercell, with the N atom situated in the center of the plot, at [0,0,0], and the other group V atoms at lattice points $a[m/2,n/2,0]$, $-5 \le m,n \le 5$.

where $\alpha = \langle \psi_{cl} | \psi_{c0} \rangle$. We find that ψ_{N0} is highly localized, with over 50% of its probability density on the N site and the four neighboring Ga atoms (Figures 3.2 and 3.3).

Because the N resonant state is so highly localized in Figures 3.2 and 3.3, it is reasonable to expect that, as we increase the N density, we can associate a similar resonant level with each nitrogen site. To test if this is so, we compared the calculated resonant wavefunctions ψ_N for a series of increasingly smaller unit cells with the resonant state, $\psi_{N(0)}$, predicted by taking a linear combination of resonant wavefunctions from large-unit-cell calculations. We showed that to a very good approximation, we can write the resonant wavefunction ψ_N at the zone center as

$$\psi_N \approx \psi_{N(0)} = \frac{1}{\sqrt{N}} \sum_{n=1}^{N} \psi_{N0,n} \qquad (3.3)$$

where ψ_N is represented by a linear combination of N isolated N states $\psi_{N0,n}$ located on an ordered array of sites $n = 1, \ldots, N$ within the supercell. Figure 3.4(a) shows $|\langle \psi_N | \psi_{N(0)} \rangle|^2$, the modulus squared of the overlap between the predicted and calculated resonant states for simple cubic and face-centered cubic nitrogen arrays. The overlap between the predicted and exact wavefunctions is almost unity for $x < \sim 0.05$, and it remains over 94% for the simple cubic structures, even up to $x = 0.25$ (a $Ga_4N_1As_3$ unit cell). This shows that the nature of the perturbation is indeed related to localized N states, even as far as $x = 0.25$, and, as a consequence, it suggests that we can use a similar representation to accurately describe the N-related states and conduction band edge in disordered GaN_xAs_{1-x} structures. We note that the overlap between the predicted and calculated resonant wavefunctions drops to $\sim 70\%$ for the case of a $Ga_4N_1As_3$ $2 \times 2 \times 1$ face-centered cubic structure [15]. This occurs because there is an infinite chain of gallium atoms, each with two nitrogen nearest neighbors, in this structure. The symmetry is reduced about N-N pairs compared with isolated N states, due primarily to the unsymmetrical nature of the lattice distortion around a N-N pair or other Ga-centered N clusters, as discussed further when we consider disorder in Section 3.3 below.

The filled data points in Figure 3.4b show the calculated resonant energy, E_N, for each structure, found by evaluating $\langle \psi_N | H | \psi_N \rangle$ directly, while the filled data points in Figure 3.4c show the reduction in the conduction band edge energy as a function of N concentration in the ordered structures considered, calculated using the full sp^3s^* model. The calculated values of E_N follow a nonmonotonic trend, with the value of E_N generally being larger in the simple cubic than in the face-centered cubic supercells considered, due to the directional dependence of the resonant-state wave-

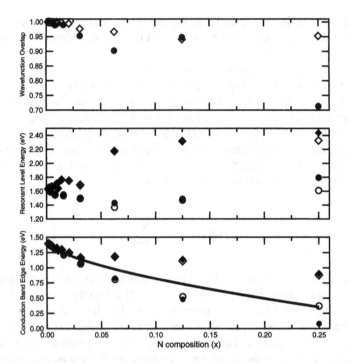

Figure 3.4 (a) Overlap between calculated and predicted resonant wavefunctions in simple cubic (♦) and face-centered cubic (●) supercells. (b) Resonant-state energy, E_N, as a function of composition based on a full calculation (solid data points) and a simplified model (open data points). (c) Conduction band-edge energy, E_c, of GaN_xAs_{1-x}, calculated using the full Hamiltonian (solid data points) and the two-band model of Equation 3.1 (open data points).

function (Figure 3.3, [47]) and of the perturbing potential, ΔV_N, introduced when an As atom is replaced by N.

The open data points in Figure 3.4b show the value of E_N calculated for each structure by directly evaluating $\langle \psi_{N(0)}|H|\psi_{N(0)} \rangle$. These estimated values, $E_{N(0)}$, are in excellent agreement with the values of E_N obtained from the full calculation. Finally, the open data points in Figure 3.4c were obtained by using the estimated values, $E_{N(0)}$ in Equation 3.1, with E_c assumed to vary linearly and V_{Nc} explicitly calculated as $\langle \psi_{N(0)}|\Delta V|\psi_{c0} \rangle$. The very good agreement between the full calculation and this modified two-band model confirms the validity of the two-band model, although we see that the resonant-state energy used, E_N, does depend on local environment, as discussed further below.

3.3   NITROGEN RESONANT STATES IN DISORDERED GaN_xAs_{1-x} STRUCTURES

Overall, Figures 3.2 to 3.4 clearly demonstrate that the conduction band edge in GaN_xAs_{1-x} is being perturbed and pushed downward due to its interaction with a higher-lying localized resonant state, centered on the nitrogen atoms. Why, then, has this state not been identified in previous calculations? To answer this question, and to investigate the role of disorder, we extend the tight-binding and two-level model to disordered GaN_xAs_{1-x} supercells.

We first consider a set of 1000 atom supercells containing up to 15 randomly distributed N atoms. In these supercells we fit the number, but not the distribution, of N-N pairs to the number given statistically, so that each cell contains n isolated N sites and p N-N pairs. For each configuration, we used the GULP molecular relaxation package [48] to calculate the equilibrium positions of all the atoms, using a parameterized valence-force-field model, while using Végard's law to vary the unit cell basis vectors as $a(x) = x\ a_{GaN} + (1 - x)\ a_{GaAs}$. The calculated relaxed bond lengths are in good agreement with those obtained by other authors [46] who used an *ab initio* pseudopotential approach.

In a disordered supercell, we can again try to describe the GaN_xAs_{1-x} conduction band edge by a **L**inear **C**ombination of **I**solated **N**itrogen **R**esonant **S**tates (LCINS) interacting with the unperturbed conduction band edge, ψ_{c0}. For the supercells considered here, we have n resonant basis states, $|\psi_{N0,i}\rangle$, associated with isolated N resonances ($i = 1-n$) and $2p$ resonant basis states associated with the p N-N pairs ($|\psi_{NN+,j}\rangle$ and $|\psi_{NN+,j}\rangle$, $j = 1-p$, which are even and odd, respectively, about the Ga site at the center of the N-N pair). We write the sp^{3}s* Hamiltonian H of the $Ga_{500}N_{n+2p}As_{500-n-2p}$ supercell as

$$H = H_0 + \Delta V_N + \Delta V_{NN} \tag{3.4}$$

where H_0 is the $Ga_{500}As_{500}$ Hamiltonian, ΔV_N is the sum of defect potentials associated with the n isolated N atoms, and ΔV_{NN} is the sum of defect Hamiltonians associated with the p N-N pairs. In extension of the approach for ordered structures, we now determine the GaN_xAs_{1-x} conduction band edge E_- and the N-related conduction-band levels by constructing and solving a $(n + 2p + 1) \times (n + 2p + 1)$ Hamiltonian matrix involving the GaAs conduction-band-edge wavefunction, $|\psi_{c0}\rangle$, and the $n + 2p$ N-related states. We use the sp^{3}s* Hamiltonian to evaluate explicitly each matrix element:

$$H_{\alpha\beta} = \langle \psi_\alpha |H| \psi_\beta \rangle \tag{3.5}$$

where the subscripts α and β each refer to one of the (n+p+1) basis states, $\psi_{N0,i}$, $\psi_{NN\pm j}>$ or $|\psi_{c0}>$. We also evaluate the overlap matrix, S, which has nonzero off-diagonal matrix elements, $S_{\alpha\beta} = \langle \psi_\alpha|\psi_\beta\rangle$, due to the overlap between N resonant states centered on different sites within the supercell.

The eigenvalues ε_α of the LCINS model are then obtained by solving the matrix equation:

$$H \cdot \phi_\alpha = \varepsilon_\alpha S \cdot \phi_\alpha \qquad (3.6)$$

with the eigenstates ϕ_α then given by

$$\phi_\alpha = \sum_i^n u_{N0,i}^\alpha \Psi_{N0,i} + \sum_j^p \left(u_{NN+,j}^\alpha \Psi_{NN+,j} + u_{NN-,j}^\alpha \Psi_{NN-,j} \right) + u_{c0}^\alpha \Psi_{c0} \quad (3.7)$$

Figure 3.5 shows the results of the CBE energy and its Γ_{1c} character calculated using the full tight-binding and LCINS methods for five significantly different fully random 1000-atom $Ga_{500}N_mAs_{500-m}$ supercells containing (i) $m = 5$, (ii) $m = 10$, and (iii) $m = 15$ nitrogen atoms. (Fully random means that we do not fix the number of N-N pairs in this case.) The figures clearly show that both the CBE energy and its fractional Γ_{1c} character are given accurately by the LCINS model, with an excellent correlation in the variation between different random structures. For example, in the

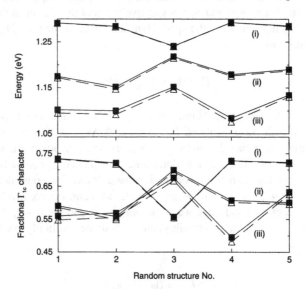

Figure 3.5 Variation in (a) the conduction-band-edge energy, and (b) its fractional Γ_{1c} character between several different random $Ga_{500}N_mAs_{500-m}$ supercell structures calculated using the full tight-binding () and LCINS (Δ) methods for $m = 5$, 10, and 15.

calculations with $m = 15$ N atoms, the maximum variation in CBE energy between different random structures is ~70 meV, while the maximum variation in CBE energy between the LCINS and full tight-binding calculations is only ~7 meV. This excellent correlation confirms the validity of describing the E_- state, and the band-gap bowing, as being due to an interaction between localized N resonant states and the host CBE.

We now turn to a more detailed comparison between the higher-lying conduction states calculated using the LCINS and full tight-binding methods. There are again strong correlations between the two sets of results. The light vertical lines in Figures 3.6a and 3.6b show the distribution of $_{1c}$ character across the conduction-band spectrum of a $Ga_{500}N_{10}As_{490}$ supercell that includes one N-N pair, calculated using the full sp^3s^* tight-binding model. The inverted black lines in Figure 3.6a show the weight of Γ_{1c} character on each of the eigenvalues, ε_α, of the LCINS Hamiltonian of the same structure. The inverted black lines in Figure 3.6b show the projection of the LCINS eigenstates, ϕ_α, weighted by their Γ_{1c} character, $|u_{c0}{}^\alpha|^2$, onto the conduction-band states of the full sp^3s^* calculation.

A comparison of the results in Figures 3.6a and 3.6b shows a remarkable replication of the full Γ_{1c} spectrum by the LCINS model, demonstrating that even the more complex higher-lying interacting states can be described effectively in terms of isolated N and N-N pair states. Figure 3.6 is for a structure with a N concentration of $x = 2\%$.

For other structures considered, the agreement is also excellent, with even closer agreement at smaller x. A detailed analysis of Figure 3.6 shows that the first ten LCINS N levels each correspond to a single sp^3s^* state below the energy of 1.75eV: one at 1.21 eV (the E_- level) with a large Γ_{1c} component; one at 1.44eV (the $E_{NN(+)}$ level) with a small Γ_{1c} component; and eight closely spaced levels in the range 1.50–1.75eV, which have a very weak admixture of Γ_{1c} character and do not contribute to the band-gap bowing. We see from Figure 3.6b(iii), in conjunction with the magnified inset region in Figure 3.6a, that these eight states project onto an equivalent eight levels in the full sp^3s^* calculations.

We identify the eleventh and highest-lying N state in Figure 3.6, at 1.86 eV, with the E_+ level. It has the greatest interaction with the host-matrix conduction-band-edge state, with ~25% Γ_{1c} character, and projects onto a whole spectrum of sp^3s^* states with energy above 1.75eV, as seen in Figure 3.6b(iv). Because of the limited number of basis states in the LCINS model, it cannot replicate the width of a resonance, and the E_+ level is represented by a single state with energy close to the average energy of the resonance and Γ_{1c} character close to the integral sum across the width of the resonance.

Analyzing the eigenvectors of the LCINS calculation, it is then straight-forward to identify the different kinds of N-state combinations and the form of their interaction with the host conduction band edge (CBE). We can identify in Figure 3.6b(i) the N-hybridized CBE, E_-; in Figure 3.6b(ii) a low-lying group of states related to symmetric N-N pair states, $E_{NN(+)}$ (one

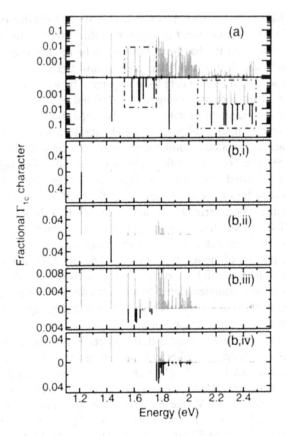

Figure 3.6 (a) GaAs Γ_{1c} character projected onto the conduction-band states of a $Ga_{500}N_{10}As_{490}$ supercell for the full tight-binding (upper) and LCINS (lower) cases. (b) The lower lines are the projection of the (i) first, (ii) second, (iii) third through tenth, and (iv) eleventh LCINS states onto the full tight-binding conduction-band structure (upper, lighter lines).

of which is present in the supercell considered in Figure 3.6); in Figure 3.6b(iii) a complex combination of N states (including the antisymmetric N-N pair state), which interact weakly with the CBE and which lie close to the isolated N-state level; and in Figure 3.6b(iv) a symmetric combination of states on all N atoms — identified as E_+ — that appears as a resonance and interacts strongly with the CBE.

The E_+ state was initially observed at low N concentrations in pseudo-potential calculations, but it proved difficult to track to higher compositions [10]. Using the LCINS model, we can identify why this is so by explicitly projecting the LCINS E_+ eigenvector ϕ_+ onto the eigenstates of a series of full sp^3s^* calculations. Figure 3.7 shows the evolution of the E_+ state as given by the full sp^3s^* Hamiltonian and highlighted using LCINS for various 1000-atom supercell structures containing (i) 0.4%, (ii) 0.8%, (iii)

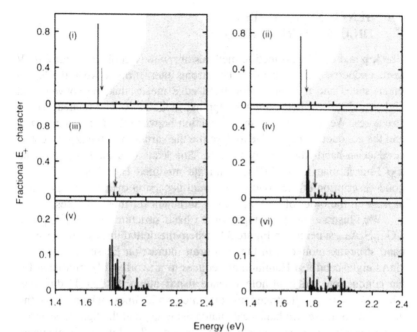

Figure 3.7 Projection of the LCINS E_+ level onto the conduction-band states of disordered 1000-atom GaN$_x$As$_{1-x}$ supercells, with x = (i) 0.4%, (ii) 0.8%, (iii) 1.0%, (iv) 1.4%, (v) 2.0%, and (vi) 3.0%, showing the broadening of the E_+ resonance in the L-related GaAs density of states (shown on a similar scale in [eV-atom]$^{-1}$). The arrow indicates the position of the LCINS E_+ level.

1%, (iv) 1.4%, (v) 2%, and (vi) 3% of randomly distributed N atoms. Figures 3.7(i–iv) show E_+ as a very strongly highlighted feature with a long, quickly decaying tail over higher states. In Figures 3.7(v) and (vi), the E_+ state begins to break up and spread out significantly. This change in the nature of E_+ is consistent with experiment, which shows a relatively strong feature at E_+ for $x <$ ~1% and then becomes weaker and broader with increasing x up to ~3%, where it is hardly distinguishable at all [44]. We note that the strength of the resonance peak depends on where E_+ lies in the density of states of the host system, forming a sharp resonance in the conduction-band Γ-related density of states, which broadens rapidly at higher N composition when the E_+ energy becomes degenerate with the larger L-related density of states.

Overall, we conclude from our detailed analysis of the full band-structure calculations that the conduction band edge in GaN$_x$As$_{1-x}$ is indeed being perturbed and pushed downward due to its interaction with a higher-lying band of localized nitrogen resonant states. We build on this analysis in the next section to derive a ten-band k·p model describing the conduction and valence-band-edge dispersion of GaN$_x$As$_{1-x}$.

3.4   TEN-BAND k·p MODEL FOR DILUTE NITRIDE ALLOYS

The **k·p** and envelope-function methods are widely applied to study III-V semiconductor heterostructures. The strong interaction between the N resonant states and the conduction band edge means that the conventional eight-band **k·p** method cannot be applied to GaInNAs and related heterostructures. We must include the interaction between the N resonant states and the conduction band edge to describe the variation of the (zone-center) conduction-band-edge energy with N. This leads to a modified ten-band **k·p** Hamiltonian for GaInNAs, with the modified Hamiltonian giving a good description of the conduction-band dispersion over an energy range at least on the order of 200 meV [49], sufficient for most analyses.

We illustrate this by comparing the band structure of a $Ga_{32}As_{32}$ and a $Ga_{32}N_1As_{31}$ supercell in Figure 3.8, where the dotted lines show the sp^{3}s* band structure plotted with the spin-orbit interaction E_{so} set to zero. The GaAs eight-band **k·p** Hamiltonian reduces to a two-band Hamiltonian for the conduction and light-hole valence bands along the [0,0,1] direction when $E_{so} = 0$, as illustrated by the thick solid lines, which show the dispersion of these two bands calculated using ψ_{c0} and the light-hole zone-center wavefunction, ψ_{lh0}, as the **k·p** basis states. The **k·p** matrix elements were found by explicitly evaluating $<\psi_{i0}|H(k_z)|\psi_{j0}>$ using the tight-binding Hamiltonian [49].

We must add the nitrogen resonant state ψ_{N0} to the **k·p** Hamiltonian for $Ga_{32}NAs_{31}$. The conduction and light-hole band dispersion are then

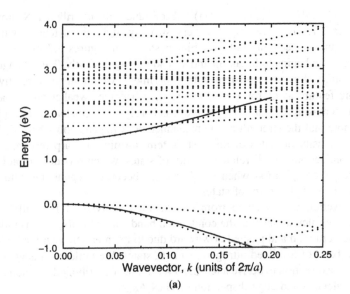

(a)

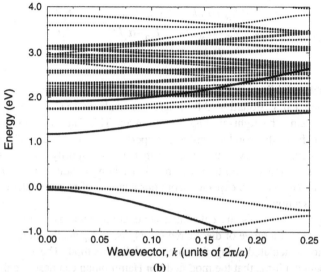

(b)

Figure 3.8 Band structure of (a) a 64-atom unit cell of GaAs, calculated using the tight-binding method (dotted lines) and the two-band **k·p** method (solid lines) and (b) $Ga_{32}N_1As_{31}$, calculated using the tight-binding (dotted) and three-band **k·p** methods (solid lines).

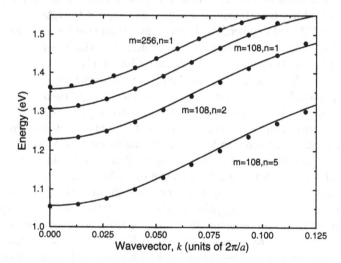

Figure 3.9 Conduction-band dispersion calculated for several different $Ga_mN_nAs_{m-n}$ supercells using the tight-binding (dots) and **k·p** methods (solid lines), with the values of n and m indicated next to each curve. Wave vector in each case is expressed in units of $2\pi/a$, where a is the Végard law average of the GaAs and GaN cubic unit cell sizes.

found by diagonalizing a 3×3 **k·p** model. The most general form of this 3×3 Hamiltonian includes k-dependent diagonal and off-diagonal matrix elements linking the ψ_{N0}, ψ_{c0}, and ψ_{lh0} basis states:

$$H = \begin{vmatrix} E_N + \alpha_N k^2 & V_{Nc} - \alpha_{Nc} k^2 & P_N k \\ V_{Nc} - \alpha_{Nc} k^2 & E_{c0} - \alpha_c k^2 & P_1 k \\ P_N k & P_1 k & E_v + \alpha_v k^2 \end{vmatrix} \qquad (3.8)$$

The thick solid lines in Figure 3.8b show the band structure of $Ga_{32}NAs_{31}$ calculated using Equation 3.8, where we evaluate the matrix elements directly using the tight-binding Hamiltonian. This Hamiltonian gives an excellent fit to the conduction-band dispersion within about 200 meV of the band edge. However, it is notable that the N impurity band in Figure 3.8b does not correspond to a specific higher-lying conduction band in the supercell. This is to be expected from our analysis of resonant states in the previous section.

Figure 3.9 shows the calculated conduction-band dispersion near the zone center for a range of disordered $Ga_mN_nAs_{m-n}$ supercells, along with the dispersion calculated using the three-level $k \cdot p$ method. The overall good agreement confirms that the modified $k \cdot p$ Hamiltonian can describe the conduction- and valence-band dispersion of GaN_xAs_{1-x} for small x. The good agreement obtained here justifies the use of the modified ten-band $k \cdot p$ Hamiltonian, which has by now been widely applied, accounting successfully for the observed interband transition energies in a range of GaN_xAs_{1-x} quantum wells (see [25] and Chapter 5) and also for the variation of the measured gain spectra as a function of energy and carrier density in GaInNAs/GaAs quantum-well (QW) lasers (see [36, 38, 50] and Chapter 12).

Despite these successes, it is important to also note the limitations of the $k \cdot p$ model presented here. The conventional eight-band $k \cdot p$ model breaks down at large wave vector k, typically when $|k| > \sim 0.2 \, k_{max}$, where k_{max} is the maximum wave vector in the first Brillouin zone. The ten-band model as described here can have an even smaller range of validity, as can be seen by considering a Ga_mNAs_{m-1} supercell. The Brillouin zone of this supercell (sc) is m times smaller than that of GaAs, so that $k_{max}(sc) \sim m^{-1/3} k_{max}(GaAs)$, or $x^{1/3} k_{max}(GaAs)$, where $x = 1/m$ is the average N composition. We find $k_{max}(sc) < \approx 0.2 \, k_{max}(GaAs)$ when $x < \sim 0.008$, with the range of validity of the model dropping off further for lower N compositions. This reflects that there are a limited number of N resonant states in the total structure; these can only interact directly with and perturb a fraction on the order of x of the total number of states in the lowest conduction band.

There are far more N-related parameters in the Hamiltonian of Equation 3.8 than can be accurately determined experimentally. The tight-binding calculations provide a good guide as to which are the most important parameters. The dominant terms are those that are independent of k and included in Equation 3.1 to describe the variation of the overall energy gap with N composition. Most implementations of the ten-band Hamiltonian have ignored the k-dependent N-related terms introduced in Equation 3.8. This is justified by a careful analysis of the tight-binding band structure [51].

We note that the **k·p** model introduced in Equation 3.8 appears to ignore any disorder effects. These will be important in actual GaN_xAs_{1-x} alloys, where clustering and other disorder-related effects become important, with a significant influence on the electronic properties [12]. The disorder will introduce a tail of localized states at the band edge [37, 52, 53] and will also lead to a partial breakdown of k-selection. These effects may contribute to some of the anomalously large electron masses observed experimentally, as discussed in Chapter 4. Nevertheless, the ten-band **k·p** model can account for a wide range of data, particularly from experiments such as photoreflectance (PR) and gain measurements, which probe the average optical properties, as discussed in [25] and Chapter 5. The model provides a straightforward framework in which to analyze the electronic properties of GaNAs and related alloys.

Our theoretical analysis shows that E_N, E_c, and V_{Nc} should vary with N composition x in GaN_xAs_{1-x} as follows:

$$E_N = E_{N0} - (\gamma - \kappa)x$$

$$E_c = E_{c0} - (\alpha - \kappa)x$$

$$E_v = E_{v0} + \kappa x$$

$$V_{Nc} = \beta x^{1/2} \tag{3.9}$$

where α, β, and γ determine the variation of the GaN_xAs_{1-x} energy gap with composition, while κ gives the chemical valence-band offset, describing how the valence-band edge of unstrained GaN_xAs_{1-x} shifts on an absolute scale with respect to that of GaAs.

The tight-binding calculations presented earlier provide a good estimate of the magnitude of most of the parameters in Equation 3.9, justifying the use of the ten-band **k·p** Hamiltonian to describe the electronic structure of $Ga(In)N_xAs_{1-x}$/GaAs quantum wells and heterostructures. The ten-band Hamiltonian has now been widely applied, successfully fitting ground- and excited-state interband transition energies in a wide range of experiments, as well as providing the basis to design and analyze Ga(In)NAs-based laser structures.

Further work is, however, still required to confirm the optimum choice of band offset (κ in Equation 3.9), even in a GaN_xAs_{1-x}/GaAs quantum-well structure. We expect from the BAC model, and all evidence supports this expectation, that electrons will be confined in the GaN_xAs_{1-x} layer of such a structure. Whether the holes are also located in the GaN_xAs_{1-x} layer (i.e., the band alignment is Type I) or in the surrounding GaAs layers (i.e., Type II) is not clear *a priori*. Both possible band alignments have been reported in the literature [54–58]. However, the current assembled experimental evidence favors a Type I alignment, although there is still uncertainty as to the best choice of κ. Some of the measurements and analysis being pursued to resolve this key issue are discussed further by Klar and Tomic in Chapter 5.

3.5 LOCAL DISORDER EFFECTS IN GaInNAs

A further source of disorder must be considered when analyzing the quaternary alloy, $Ga_{1-y}In_yN_xAs_{1-x}$. The nature of the isolated resonant state changes with local environment, depending on the average number of In and Ga atoms neighboring each N atom [42, 43]. To investigate this variation, we took a number of 216-atom supercell structures, in which we constrained the central group V site to have a given number, m, of indium nearest neighbors ($m = 0$ to 4) [43], and then placed In atoms at random on the remaining sites to give an overall indium fraction of $y = 0.25$.

We found, for the structures considered, that E_N varies approximately linearly from $E_N = 1.52$ eV for four Ga neighbors to $E_N = 1.75$ eV for four In neighbors. The value of E_N depends weakly on the atomic arrangement in the second shell of group III neighbors, but the calculated effect for a range of supercell calculations indicates that the atomic arrangement in the second shell only shifts E_N by ~±0.021 eV, down by an order of magnitude on the effects of variations in the nearest-neighbor environment. The matrix element linking the N resonant state and the conduction band edge is calculated in the tight-binding method to vary between $2.00\ x^{1/2}$ eV (for a structure where all N have four Ga neighbors) to $1.35\ x^{1/2}$ eV (for a structure where all N have four In neighbors), with the conduction-band-edge energy, E_c, in the 2×2 matrix of Equation 3.1 then varying between 1.10 eV and 1.11 eV for the 216-atom supercells considered, and the band-gap between 0.95 eV and 1.04 eV as the number of In nearest neighbors m increases from 0 to 4. The average interaction in a larger unit cell containing a mixture of nearest-neighbor environments is then given by an appropriately weighted average of the above parameters. This justifies the construction of an effective two-band model to describe the variation of conduction-band-edge energy and energy gap in quaternary $GaInN_xAs_{1-x}$ alloys, but the magnitude of the parameter β in the interaction term $βx^{1/2}$ and the value of E_N will vary, depending on the average N cation nearest-neighbor environment [25]. The experimentally observed blueshift in the energy gap of many GaInNAs samples with mild thermal annealing [25, 39–41] can then be explained by local atomic rearrangements leading to an increase in the average number of In neighbors about each N atom [59–61].

3.6 ALLOY SCATTERING AND MOBILITY IN DILUTE NITRIDE ALLOYS

There has until recently been little progress in developing models to describe the transport and mobility properties of dilute nitride alloys. Even for idealized random alloy crystals, these properties are difficult to analyze precisely because N introduces such a strong perturbation to the band structure of Ga(In)As. This must lead to strong alloy scattering. There is a well-established model [62], based on the Born approximation, to describe the relatively weak alloy scattering that occurs in conventional semiconductor alloys. This model is, however, entirely insufficient for extreme alloys such as GaNAs, underestimating the alloy scattering cross-section by over two orders of magnitude [34, 35]. We describe below how the strong scattering due to N atoms substantially limits the electron mobility in dilute nitride alloys, consistent with the maximum mobility observed experimentally on the order of 1000 cm^2/V^{-1}s^{-1} [63].

We have calculated the scattering cross-section for an isolated N impurity in GaAs using S-matrix theory (distorted Born-wave approach). This was previously applied to successfully describe resonant scattering due to conventional impurities in GaAs [64, 65]. For a sufficiently localized perturbation, ΔV_N, the total scattering cross-section σ for an isolated impurity is given by:

$$\sigma = 4\pi \left(\frac{m^*}{2\pi\hbar^2} \right)^2 |< \psi_{c1} | \Delta V_N | \psi_{c0} >|^2 \, \Omega^2, \qquad (3.10)$$

where m^* is the electron effective mass at the band edge and Ω is the volume of the region in which the wavefunctions are normalized. The state ψ_{c0} is the Γ-point conduction-band Bloch wavefunction (in the absence of the N atom), and ψ_{c1} is the exact band-edge state in the presence of the N atom.

We note that the Born approximation is equivalent to setting $\psi_{c1} = \psi_{c0}$ in the required matrix elements. It is often used in the discussion of conventional alloy and impurity scattering [62], but it is entirely inadequate for the case of N defect scattering in GaAs.

Consider a perfect crystal for which the electron Hamiltonian is H_0 and the conduction-band-edge state has wavefunction ψ_{c0} and energy E_{c0}. When we introduce a single N atom into a large volume Ω of the otherwise perfect lattice, the new Hamiltonian, $H_1 = H_0 + \Delta V_N$, leads to a modified band-edge state ψ_{c1} with energy E_{c1}. We can therefore rewrite the scattering matrix element as:

$$<\psi_{c1}|\Delta V_N|\psi_{c0}> = <\psi_{c1}|H_1 - H_0|\psi_{c0}> = (E_{c1} - E_{c0}) <\psi_{c1} |\psi_{c0}> \quad (3.11)$$

Because $\langle \psi_{c1}|\psi_{c0}\rangle \to 1$ for sufficiently large Ω, we derive that, at low impurity concentrations,

$$\Omega < \psi_{c1} \,|\, \Delta V_N \,|\, \psi_{c0} > = \frac{dE_c}{dn}, \qquad (3.12)$$

where E_c is the conduction-band-edge energy and n is the number of impurities per unit volume. Substituting Equation 3.12 in Equation 3.10, and noting that n is related to the concentration x by $n = 4x/a_0^3$, where a_0 is the GaAs unit cell dimension, the scattering cross-section for an isolated impurity is then given by

$$\sigma = \frac{\pi}{4}\left(\frac{m^*}{2\pi\hbar^2}\right)^2 \left[\frac{dE_c}{dx}\right]^2 a_0^6 . \qquad (3.13)$$

This result is key: it establishes a fundamental connection between the composition-dependence of the conduction-band-edge energy and the n-type carrier scattering cross-section in the ultradilute limit for semiconductor alloys, imposing general limits on the carrier mobility in such alloys.

We can see this by extending the isolated N result of Equation 3.13 to the case of a dilute nitride alloy, GaN_xAs_{1-x}. In an independent scattering model, the mean free path l of carriers depends on the scattering cross-section σ for a single defect and on the number of defects n per unit volume, as $l^{-1} = n\sigma$. Assuming such a classical model and the values of m^* and dE_c/dx at $x = 0$, we estimate for a N content of 1% a mean free path of only 15 nm. This is still more than an order of magnitude larger than the average N separation, suggesting that an independent scattering model should remain appropriate in the dilute random alloy. The mobility μ is related to the mean free path l as $\mu = e\tau/m^*$, with the scattering time $\tau = l/\bar{u}$, where $\bar{u}$ is the mean electron speed. Setting $\bar{u}^2 = 3kT/m^*$, where T is the temperature, we estimate that the mobility μ is given by [34]:

$$\mu^{-1} = \frac{\sqrt{3m^* kT}}{e}\,\pi\left(\frac{m^*}{2\pi\hbar^2}\right)^2 \left[\frac{dE_c}{dx}\right]^2 a_0^3 x . \qquad (3.14)$$

Figure 3.10 shows the estimated variation of the room-temperature electron mobility with x in GaN_xAs_{1-x}, calculated allowing both m^* and dE_c/dx to vary with x based on the two-level model of Equation 3.1. The electron mobility is estimated to be on the order of 1000 $cm^2/V^{-1}s^{-1}$ when $x = 0.01$, of similar magnitude to the highest values observed to date in dilute nitride alloys [63] but larger than that found in many samples, where

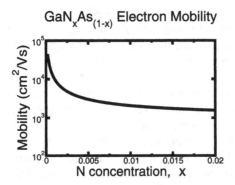

Figure 3.10 Room-temperature variation of alloy-scattering-limited electron mobility, μ, in GaN$_x$As$_{1-x}$, calculated using Equation 3.14.

$\mu \sim 100\text{--}400$ cm^2/V^{-1}s^{-1} [33, 66–70]. We note that factors omitted in the calculation here, including the influence of N-N nearest-neighbor pairs and clusters [12, 17], may contribute to limiting the mobility in actual samples [35]. In addition, film quality and composition fluctuations may also play a role in some samples. The intrinsic alloy-scattering-limited mobility should be larger in GaInNAs samples because of the weaker band-gap bowing observed in indium-containing samples [25].

The intrinsically low electron mobilities in dilute nitride alloys have significant consequences for potential device applications. The low electron mobility, combined with the short nonradiative lifetimes observed to date, limit the electron diffusion lengths and efficiency achievable in GaInNAs-based solar cells. Further efforts may lead to increased nonradiative lifetimes, but we are unlikely to see significant further improvements in the alloy-scattering-limited mobility [63]. The low electron mobility generally observed to date may lead to reduced surface recombination in mesa structures containing GaInNAs quantum wells, which would be of benefit for ultracompact photonic device applications.

3.7 CONCLUSIONS

In summary, this chapter presents an overview of the development of appropriate models to describe the electronic structure of dilute nitride alloys. We have shown how the results of band-structure calculations using the tight-binding method can probe and provide insight both into fine details and also trends in the electronic structure, providing the basis for simpler, k·p-based models to describe the energy states in quantum wells and superlattices.

By applying the tight-binding method to ordered GaN$_x$As$_{1-x}$ supercells, we confirmed the main features of the band anticrossing model, showing that N forms a resonant state above the conduction band edge in Ga(In)As, and that the interaction of the N resonant states with the conduction band edge accounts for the strong band-gap bowing observed in Ga(In)As. The overall achievements and insights gained through the band anticrossing (BAC) model are the chief topic of Chapter 2.

We also considered the consequences of the disorder that is inevitably present in actual Ga(In)NAs samples. We focused on two different aspects of the disorder. First, we showed how the inevitable disorder in the N distribution in Ga(In)NAs, including the formation of N-N pairs and clusters, modifies the electronic structure, demonstrating that even in a disordered alloy the band-gap bowing (and most other aspects of the electronic structure) can still be accurately described using the BAC model. We also showed that the N nearest-neighbor environment in GaInNAs has a strong impact on the N resonance energy and on the magnitude of its interaction with the conduction band edge, leading to a notable increase in the magnitude of the alloy energy gap as the number of In nearest neighbors increases. This effect of environment has been widely observed experimentally, and this is a theme that recurs in Chapter 5, investigating the electronic growth properties of GaInNAs-based heterostructures.

Having established the theoretical basis for using the BAC model in both ordered and disordered GaNAs structures, we then considered its practical application, showing how the conventional eight-band $\mathbf{k \cdot p}$ model must be extended to a ten-band model to describe the electronic structure of Ga(In)NAs/GaAs and related heterostructures, as used in particular by Klar and Tomic (Chapter 5) and by Fehse and Adams (Chapter 12).

The strong band-gap bowing directly affects many of the electronic properties of dilute nitride alloys. We showed in particular that there is a fundamental connection between the composition dependence of the conduction-band-edge energy and the n-type carrier scattering cross-section in dilute nitrides. This imposes general limits on the carrier mobility in such alloys. Within an independent scattering approximation, the carrier mobility in GaN_xAs_{1-x} is estimated to be limited to ~1000 cm²/V⁻¹s⁻¹ for a nitrogen atomic concentration of 1%, of particular consequence for the photovoltaic applications discussed in Chapter 13.

Overall, we conclude that the methods presented here provide a very good basis for the further investigation and analysis of this material, which is fascinating both for its fundamental properties and its potential device applications.

Acknowledgments

We would like to thank many colleagues for useful discussions and collaboration on the electronic structure of GaInNAs. These include Alf Adams, Aleksey Andreev, Stellios Choulis, Robin Fehse, Joerg Hader, Jeff Hosea,

Peter Klar, Stephan Koch, Henning Riechert, Stanko Tomic, and Bernie Weinstein. We are grateful to Science Foundation Ireland for financial support.

References

1. Weyers, M., Sato, M., and Ando, H., "Red Shift of Photoluminescence and Absorption in Dilute GaAsN Alloy Layers," *Jpn. J. Appl. Phys.* 31: L853 (1992).

2. Kondow, M., Kitatani, T., Larson, M.C., Nakahara, K., Uomi, K., and Inoue, H., "Gas-Source MBE of GaInNAs for Long-Wavelength Diode Lasers," *J. Cryst. Growth* 188: 255–259 (1998).

3. Riechert, H., Egorov, A.Y., Livshits, D., Borchert, B., and Illek, S., "InGaAsN/GaAs Heterostructures for Long-Wavelength Light-Emitting Diodes," *Nanotechnology* 11: 201–205 (2000).

4. Jackson, A.W., Naone, R.L., Dalberth, M.J., Smith, J.M., Malone, K.J., Kisker, D.W., Klem, J.F., Choquette, K.D., Serkland, D.K., and Geib, K.M., "OC-48 capable InGaAsN vertical cavity lasers," *Electron. Lett.* 37: 355–356 (2001).

5. Steinle, G., Riechert, H., and Egorov, A.Y., "Monolithic VCSEL with InGaAsN Active Region Emitting at 1.28 μm and CW Output Power Exceeding 500 μW at Room Temperature," *Electron. Lett.* 37: 93–95 (2001).

6. Bastard, G., *Wave Mechanics Applied to Semiconductor Heterostructures* (Paris: Editions de Physique, 1990).

7. Burt, M.G., "Fundamentals of Envelope Function Theory for Electronic States and Photonic Modes in Nanostructures," *J. Phys.: Condensed Matter* 11: R53–R83 (1999).

8. Meney, A.T., Gonul, B., and O'Reilly, E.P., "Evaluation of Various Approximations Used in the Envelope Function Method," *Phys. Rev. B* 50: 10893–10904 (1994).

9. Jones, E.D., Modine, N.A., Allerman, A.A., Kurtz, S.R., Wright, A.F., Tozer, S.T., and Wei, X., "Band Structure of $In_yGa_{1-y}As_{1-x}N_x$ Alloys and Effects of Pressure," *Phys. Rev. B* 60: 4430–4433 (1999).

10. Mattila, T., Wei, S.H., and Zunger, A., "Localization and Anti-Crossing of Electron Levels in $GaAs_{1-x}N_x$ Alloys," *Phys. Rev. B* 60: R11245–R11248 (1999).

11. Zhang, Y., Mascarenhas, A., Xin, H.P., and Tu, C.W., "Scaling of Band-Gap Reduction in Heavily Nitrogen-Doped GaAs," *Phys. Rev. B* 63: 161303 (2001).

12. Kent, P.R.C., and Zunger, A., "Theory of Electronic Structure Evolution in GaAsN and GaAsP Alloys," *Phys. Rev. B* 64: 115208 (2001).

13. Kent, P.R.C., Bellaiche, L., and Zunger, A., "Pseudopotential Theory of Dilute III-V Nitrides," *Semicond. Sci. Technol.* 17: 851–859 (2002).

14. Lindsay, A., and O'Reilly, E.P., "Theory of Enhanced Band-gap Nonparabolicity in GaN_xAs_{1-x} and Related Alloys," *Solid State Commun.* 112: 443–447 (1999).

15. Lindsay, A., and O'Reilly, E.P., "Influence of Nitrogen Resonant States on the Electronic Structure of GaN_xAs_{1-x}," *Solid State Commun.* 118: 313–317 (2001).

16. O'Reilly, E.P., Lindsay, A., Tomic, S., and Kamal-Saadi, M., "Tight-Binding and k.p Models for the Electronic Structure of Ga(In)NAs and Related Alloys," *Semicond. Sci. Technol.* 10: 870–879 (2002).

17. Lindsay, A., and O'Reilly, E.P., "A Tight-Binding Based Analysis of the Band Anti-Crossing in GaN_xAs_{1-x}," *Physica E* 21: 901–906 (2004).

18. Wolford, D.J., Bradley, J.A., Fry, K., and Thompson, J., in *Proceedings 17th International Conference on the Physics of Semiconductors* (New York: Springer, 1994), p. 627.

19. Liu, X., Pistol, M.-E., Samuelson, L., Schwetlick, S., and Seifert, W., "Nitrogen Pair Luminescence in GaAs," *Appl. Phys. Lett.* 56: 1451–1453 (1990).

20. Vogl, P., "Predictions of Deep Impurity Level Energies in Semiconductors," *Adv. Electronics Electron Phys.* 62: 101–159 (1984).

21. Hjalmarson, H.P., Vogl, P., Wolford, D.J., and Dow, J.D., "Theory of Substitutional Deep Traps in Covalent Semiconductors," *Phys. Rev. Lett.* 44: 810–813 (1980).

22. Lindsay, A., and O'Reilly, E.P., "A Universal Model for Trends in A_1-Type Defect States in Zincblende and Diamond Semiconductor Structures," *Physica B* 340:434–439 (2003).

23. Shan, W., Walukiewicz, W., Ager, J.W., Haller, E.E., Geisz, J.F., Friedman, J.M., Olson, J.M., and Kurtz, S.R., "Band Anticrossing in GaInNAs Alloys," *Phys. Rev. Lett.* 82: 1221–1224 (1999).

24. Klar, P.J., Grüning, H., Heimbrodt, W., Koch, J., Höhnsdorf, F., Stolz, W., Vicente, P.M.A., and Camassel, J., "From N Isoelectronic Impurities to N-Induced Bands in the GaN_xAs_{1-x} Alloy," *Appl. Phys. Lett.* 76: 3439–3441 (2000).

25. Klar, P.J., Grüning, H., Heimbrodt, W., Weiser, G., Koch, J., Volz, K., Stolz, W., Koch, S.W., Tomic, S., Choulis, S.A., Hosea, T.J.C., O'Reilly, E.P., Hofmann, M., Hader, J., and Moloney, J.V., "Interband Transitions of Quantum Wells and Device Structures Containing Ga(N,As) and (Ga,In)(N,As)," *Semicond. Sci. Technol.* 17: 830–842 (2002).

26. Polimeni, A., Baldassarri Höger von Högersthal, G., Bissiri, M., Capizzi, M., Fischer, M., Reinhardt, M., and Forchel, A., "Effect of Hydrogen on the Electronic Properties of $In_xGa_{1-x}As_{1-y}N_y$/GaAs Quantum Wells," *Phys. Rev. B* 63: 201304(R) (2001).

27. Baldassarri Höger von Högersthal, G., Bissiri, M., Polimeni, A., Capizzi, M., Fischer, M., Reinhardt, M., and Forchel, A., "Hydrogen-Induced Band-gap Tuning of (InGa)(AsN)/GaAs Single Quantum Wells," *Appl. Phys. Lett.* 78: 3472–3474 (2001).

28. Polimeni, A., Baldassarri Höger von Högersthal, G., Bissiri, M., Capizzi, M., Frova, A., Fischer, M., Reinhardt, M., and Forchel, A., "Role of Hydrogen in III-N-V Compound Semiconductors," *Semicond. Sci. Technol.*, 17: 797–802 (2002).

29. Buyanova, I.A., Pozina, G., Hai, P.N., Chen, W.M., Xin, H.P., and Tu, C.W., "Type I Band Alignment in the GaN_xAs_{1-x}/GaAs Quantum Wells," *Phys. Rev. B* 63: 33303 (2000).

30. Hai, P.N., Chen, W.M., Buyanova, I.A., Xin, H.P., and Tu, C.W., "Direct Determination of Electron Effective Mass in GaNAs/GaAs Quantum Wells," *Appl. Phys. Lett.* 77: 1843–1845 (2000).

31. Geddo, M., Guizzetti, G., Capizzi, M., Polimeni, A., Gollub, D., and Forchel, A., "Photoreflectance Evidence of the N-Induced Increase of the Exciton Binding Energy in an $In_{1-x}Ga_xAs_{1-y}N_y$ Alloy," *Appl. Phys. Lett.* 83: 470–472 (2003).

32. Skierbiszewski, C., Perlin, P., Wisniewski, P., Suski, T., Walukiewicz, W., Shan, W., Ager, J.W., Haller, E.E., Geisz, J.F., Friedman, D.J., Olson, J.M., and Kurtz, S.R., "Effect of Nitrogen-Induced Modification of the Conduction Band Structure on Electron Transport in GaAsN Alloys," *Phys. Stat. Solidi B* 216: 135–139 (1999).

33. Skierbiszewski, C., "Experimental Studies of the Conduction Band Structure of GaInNAs Alloys," *Semicond. Sci. Technol.* 17: 803–814 (2002).

34. Fahy, S., and O'Reilly, E.P., "Intrinsic Limits on Electron Mobility in Dilute Nitride Semiconductors," *Appl. Phys. Lett.* 83: 3731–3733 (2003).

35. Fahy, S., and O'Reilly, E.P., "Theory of Electron Mobility in Dilute Nitride Semiconductors," *Physica E* in press.

36. Hofmann, M., Wagner, A., Ellmers, C., Schlichenmeier, C., Schafer, S., Hohnsdorf, F., Koch, J., Stolz, W., Koch, S.W., Ruhle, W.W., Hader, J., Moloney, J.V., O'Reilly, E.P., Borchert, B., Egorov, A.Y., and Riechert, H., "Gain Spectra of (GaIn)(NAs) Laser Diodes for the 1.3-μm-Wavelength Regime," *Appl. Phys. Lett.* 78: 3009–3011 (2001).

37. Fehse, R., Tomi, S., Adams, A.R., Sweeney, S.J., O'Reilly, E.P., Andreev, A.D., and Riechert, H., "A Quantitative Study of Radiative, Auger and Defect-Related Recombination Processes in 1.3-μm GaInNAs-Based Quantum-Well Lasers," *IEEE J. Selected Top. Quantum Electron.* 8: 801–810 (2002).

38. Tomic, S., O'Reilly, E.P., Fehse, R., Sweeney, S.J., Adams, A.R., Andreev, A.D., Choulis, S.A., Hosea, T.J.C., and Riechert, H., "Theoretical and Experimental Analysis of 1.3-μm InGaAsN/GaAs Lasers," *IEEE J. Selected Top. Quantum Electron.* 9:1228–1238 (2003).

39. Bhat, R., Caneau, C., Salamanca-Riba, L., Bi, W., and Tu, C., "Growth of GaAsN/GaAs, GaInAsN/GaAs and GaInAsN/GaAs Quantum Wells by Low-Pressure Organometallic Chemical Vapor Deposition," *J. Cryst. Growth* 195: 427–437 (1998).

40. Pan, Z., Li, L.H., Zhang, W., Lin, Y.W., Wu, R.H., and Ge, W., "Effect of Rapid Thermal Annealing on GaInNAs/GaAs Quantum Wells Grown by Plasma-Assisted Molecular-Beam Epitaxy," *Appl. Phys. Lett.* 77: 1280–1282 (2000).

41. Kitatani, T., Nakahara, K., Kondow, M., Uomi, K., and Tanaka, T., "Mechanism Analysis of Improved GaInNAs Optical Properties through Thermal Annealing," *J. Cryst. Growth* 209: 345–349 (2000).

42. Kim, K., and Zunger, A., "Spatial Correlations in GaInAsN Alloys and Their Effects on Band-Gap Enhancement and Electron Localization," *Phys. Rev. Lett.* 86: 2609–2612 (2001).

43. Lindsay, A., and O'Reilly, E.P., "Theory of the Electronic Structure of $Ga_{1-y}In_yN_x As_{1-x}$," in *Proceedings 25th Int. Conf. Physics of Semiconductors* (Osaka: Springer, 2001), pp. 63–64.

44. Perkins, J.D., Mascarenhas, A., Zhang, Y., Geisz, J.F., Friedman, D.J., Olson, J.M., and Kurtz, S.R., "Nitrogen-Activated Transitions, Level Repulsion, and Band-gap Reduction in $GaAs_{1-x}N_x$ with $x < 0.03$," *Phys. Rev. Lett.* 82: 3312–3315 (1999).

45. Bellaiche, L., Wei, S.-H., and Zunger, A., "Localization and Percolation in Semiconductor Alloys: GaAsN vs. GaAsP," *Phys. Rev. B* 54: 17568–17576 (1996).

46. Wei, S.-H., and Zunger, A., "Giant Composition-Dependent Optical Bowing Coefficient in GaAsN Alloys," *Phys. Rev. Lett.* 76: 664–667 (1996).

47. O'Reilly, E.P., and Lindsay, A., "Theory of the Electronic Structure of $Ga_{1-y}In_yN_xAs_{1-x}$ and Related Alloys," *High Pressure Res.* 18: 13–20 (2000).

48. Gale, J.D., "GULP — a Computer Program for the Symmetry Adapted Simulation of Solids," *JCS Faraday Trans.* 93: 629–637 (1997).

49. O'Reilly, E.P., and Lindsay, A., "**k.p** Model of Ordered GaN_xAs_{1-x}," *Phys. Stat. Solidi B* 216: 131–134 (1999).

50. Hofmann, M.R., Gerhardt, N., Wagner, A.M., Ellmers, C., Höhnsdorf, F., Koch, J., Stolz, W., Koch, S.W., Rühle, W.W., Hader, J., Moloney, J.V., O'Reilly, E.P., Borchert, B., Egorov, A.Y., Riechert, H., Schneider, H.C., and Chow, W.W., "Emission Dynamics and Optical Gain of 1.3-μm (GaIn)(NAs)/GaAs Lasers," *IEEE J. Selected Top. Quantum Electron.* 38: 213–221 (2002).

51. Lindsay, A., "The Theory of the Electronic Structure of GaN_xAs_{1-x} and $In_{1-y}Ga_yN_x As_{1-x}$" (Ph.D. thesis, University of Surrey, U.K., 2002).

52. Mintairov, A.M., Kosel, T.H., Merz, J.L., Balgnov, P.A., Vlasov, A.S., Ustinov, V.M., and Cook, R.E., "Near-Field Magnetophotoluminescence Spectroscopy of Composition Fluctuations in InGaAsN," *Phys. Rev. Lett.* 87: 277401 (2001).

53. Teubert, J., Klar, P.J., Heimbrodt, W., Volz, K., Stolz, W., Thomas, P., Leibiger, G., and Gottschalch, V., "Enhanced Weak Anderson Localization Phenomena in the Magnetoresistance of n-Type (Ga,In)(N,As)," *Appl. Phys. Lett.* 84:747–749 (2004).

54. Kitatani, T., Kondow, M., Kikawa, T., Yazawa, Y., Okai, M., and Uomi, K., "Analysis of Band Offset in GaNAs/GaAs by X-ray Photoelectron Spectroscopy," *Jpn. J. Appl. Phys. (Part 1)* 38: 5003–5006 (1999).

55. Sun, B.Q., Jiang, D.S., Luo, X.D., Zu, Z.Y., Pan, Z., Li, L.H., and Wu, R.H., "Intersubband Luminescence and Absorption of GaNAs/GaAs Single-Quantum-Well Structures," *Appl. Phys. Lett.* 76: 2862–2864 (2000).

56. Krispin, P., Spruytte, S.G., Harris, J.S., and Ploog, K.H., "Electrical Depth Profile of p-Type GaAs/Ga(As,N)/GaAs Heterostructures Determined by Capacitance-Voltage Measurements," *J. Appl. Phys.* 88: 4153–4158 (2000).

57. Klar, P.J., Grüning, H., Heimbrodt, W., Koch, J., Stolz, W., Vicente, P.M.A., Kamal Saadi, A.M., Lindsay, A., and O'Reilly, E.P., "Pressure and Temperature Dependent Studies of GaN_xAs_{1-x}/GaAs Quantum Well Structures," *Phys. Stat. Solidi B* 223: 163–169 (2001).

58. Bellaiche, L., Wei, S.H., and Zunger, A., "Composition Dependence of Interband Transition Intensities in GaPN, GaAsN, and GaPAs Alloys," *Phys. Rev. B* 56: 10233–10240 (1997).

59. Wagner, J., Geppert, T., Köhler, K., Ganser, P., and Herres, N., "N-Induced Vibrational Modes in GaAsN and GaInAsN Studied by Resonant Raman Scattering," *J. Appl. Phys.* 90: 5027–5031 (2001).

60. Kurtz, S., Webb, J., Gedvilas, L., Friedman, D., Geisz, J., Olson, J., King, R., Joslin, D., and Karam, N., "Structural Changes during Annealing of GaInAsN," *Appl. Phys. Lett.* 78: 748–750 (2001).

61. Kitatani, T., Kondow, M., and Kudo, M., "Transition of Infrared Absorption Peaks in Thermally Annealed GaInNAs," *Jpn. J. Appl. Phys. (Part 2)* 40: L750–752 (2001).

62. Harrison, J., and Hauser, J.R., "Alloy Scattering in Ternary III-V Compounds," *Phys. Rev. B* 13: 5347–5350 (1976).

63. Volz, K., Koch, J., Kunert, B., and Stolz, W., "Doping Behaviour of Si, Te, Zn and Mg in Lattice-Matched (GaIn)(NAs)/GaAs Bulk Films," *J. Cryst. Growth* 248: 451–456 (2003).

64. Sankey, O.F., Dow, J.D., and Hess, K., "Theory of Resonant Scattering in Semiconductors due to Impurity Central-Cell Potentials," *Appl. Phys. Lett.* 41: 664–666 (1982).

65. Fisher, M.A., Adams, A.R., O'Reilly, E.P., and Harris, J.J., "Resonant Electron Scattering due to the Central Cell of Impurities Observed in AlGaAs Using Hydrostatic Pressure," *Phys. Rev. Lett.* 59: 2341–2344 (1987).

66. Geisz, J.F., and Friedman, D.J., "III-N-V Semiconductors for Photovoltaic Applications," *Semicond. Sci. Technol.* 17: 769–777 (2002).

67. Geisz, J.F., Friedman, D.J., Olson, J.M., Kurtz, S.R., and Keyes, B.M., "Photocurrent of 1-eV GaInNAs Lattice-Matched to GaAs," *J. Cryst. Growth* 195: 401 (1998).

68. Hong, Y.G., Tu, C.W., and Ahrenkiel, R.K., "Improving Properties of GaInNAs with a Short Period GaNAs/GaInAs Superlattice," *J. Cryst. Growth* 227–228: 536–540 (2001).

69. Kurtz, S.R., Allerman, A.A., Seager, C.H., Sieg, R.M., and Jones, E.D., "Minority Carrier Diffusion, Defects, and Localization in InGaAsN, with 2% Nitrogen," *Appl. Phys. Lett.* 77: 400–402 (2000).

70. Li, W., Pessa, M., Toivonen, J., and Lipsanen, H., "Doping and Carrier Transport in $Ga_{1-3x}In_{3x}N_xAs_{1-x}$ Alloys," *Phys. Rev. B* 64: 113308 (2001).

CHAPTER 4

Electron Effective Masses of Dilute Nitrides: Experiment

W.M. CHEN and I.A. BUYANOVA

*Department of Physics and Measurement Technology,
Linköping University, Sweden*

1-591-69019-6/04/$0.00+$1.50
© 2004 by CRC Press LLC

4.1   INTRODUCTION

Effective masses of charged carriers are among the most important fundamental parameters in a semiconductor. They are known to play crucial roles in quantum confinement energy, optical dipole matrix elements, band filling, carrier mobility, etc., and thus they play an important role in a wide range of electronic and optoelectronic applications. Electron and hole effective masses are determined by the energy dispersion of their respective conduction and valence bands in the reciprocal lattice. Thus they provide a sensitive means of closely examining the electronic band structure of a semiconductor. In dilute nitrides, the incorporation of N into conventional III-V semiconductors such as GaInAs and-gap has been found to introduce strikingly strong perturbation to the band structure of the host lattices, leading to a number of extraordinary effects, such as a giant band-gap bowing, a strong reduction in the pressure and temperature dependencies of the band-gap energy [1]. It is generally believed that the modification of the electronic band structure by the N-induced perturbation predominantly occurs in the conduction band (CB), which will significantly affect the electron effective mass (denoted as m_e^* values below) near the CB edge. It can, in turn, provide a close inspection of the CB structure and also a critical test of various physical mechanisms proposed to account for the N-induced modification of the band structure.

Due to both fundamental and technological interest, m_e^* in dilute nitrides has been a topic of large research efforts internationally during the last few years. Various experimental and theoretical approaches have been employed to determine m_e^*. In sharp contrast to the agreement on the giant band-gap bowing effect, values of m_e^* reported in the literature scatter over a wide range.

In this chapter we shall provide a brief review of the existing experimental results of m_e^* in dilute nitrides. We shall discuss some general trends as well as discrepancies in the modification of m_e^* by N and In compositions, by electron concentration, and by energy. By pointing out differences and limitations of various experimental approaches employed in determining m_e^*, we hope that possible sources leading to the discrepancies can be identified or hinted at. Our aim is to further stimulate future research efforts in obtaining consistent and reliable information on m_e^* and in gaining a full understanding of its dependence on key compositional, electronic, and structural parameters such as alloy compositions, quantum confinement,

Fermi energy, and strain fields. Such information will provide useful insight into the electronic band structure of dilute nitrides, and will be essential to full exploration of band-gap engineering and optimization of design of device structures.

The chapter is organized as follows. In Section 4.2, we will provide a short description of various experimental approaches that have been employed in retrieving information on m_e^* in dilute nitrides. Advantages and limitations of these approaches will also be briefly discussed. Values of m_e^* reported so far in the literature for various dilute nitrides will be summarized in Section 4.3. Possible sources that have led to reported scattered values and discrepancies will be discussed, taking into account the advantages and limitations of the experimental approaches involved. In Section 4.4, we will compare the experimentally observed values of m_e^* with those expected from theoretical predictions. Conclusions of the existing experimental results and an outlook at future work will be given in Section 4.5.

4.2 EXPERIMENTAL APPROACHES: ADVANTAGES AND LIMITATIONS

In this section we will give a short description of some common experimental approaches that have been employed to directly or indirectly determine m_e^* in dilute nitrides, with the aim of providing the readers with an easy reference. Advantages and limitations of these experimental approaches will also be briefly discussed in an effort to shed light on possible sources that could lead to the reported scattered values and discrepancies to be discussed in Section 4.3.

4.2.1 Optically Detected Cyclotron Resonance (ODCR)

In an external dc magnetic field, a CB electron in a semiconductor will move in a spiral orbit about the direction of the field with a cyclotron orbit radius $r_c = \sqrt{\dfrac{\hbar}{eB}}$ and an angular frequency known as the cyclotron frequency, given by:

$$\omega_c = \frac{eB}{m_e^*} \qquad (4.1)$$

where m_e^* denotes the cyclotron resonance (CR) effective mass of the electron. (A deviation of the CR m_e^* from a bare m_e^* caused by a polaron effect in GaAs [and presumably also GaInGaAs] is estimated to be negligibly small, i.e., about 1% for a bulk material and 3% for a two-dimensional quantum well [QW].) The cyclotron frequency in semiconductors typically falls within the microwave (MW) or far infrared (FIR) range with available magnetic fields. Incident MW or FIR electromagnetic waves with polarization of the electric field at an angle to the dc magnetic field can be absorbed by the electron in the cyclotron motion, with a resonance absorption peak at $\omega = \omega_c$, leading to the so-called cyclotron resonance (CR). Here $\omega = 2\pi\nu$ and ν is the frequency of the MW or FIR electromagnetic waves. In a quantum mechanical description, a strong dc magnetic field can lead to quantization of the cyclotron orbits, transforming the energy band into a series of energy levels known as Landau levels. CR in this description corresponds to a resonant transition between two adjacent Landau levels. From the measured cyclotron frequency ω_c, m_e^* can be unambiguously determined from Equation 4.1. When a damping effect due to collisions is taken into account in a realistic case, however, the cyclotron frequency has to be determined from a CR absorption spectrum of the MW or FIR radiations, which can be calculated from the real part of the conductivity and is given by Equation 4.2 [2]:

$$ P_a = \frac{n_e e^2 \tau}{m_e^*} \cdot \frac{1 + \omega_c^2 \tau^2 + \omega^2 \tau^2}{(1 + \omega_c^2 \tau^2 - \omega^2 \tau^2)^2 + 4\omega^2 \tau^2} \qquad (4.2) $$

when the electric field of the incident MW or FIR electromagnetic waves is linearly polarized perpendicular to the direction of the dc magnetic field, and by Equation 4.3:

$$ P_a = \frac{n_e e^2 \tau}{m_e^*} \cdot \frac{1}{1 + (\omega_c - \omega)^2 \tau^2} \qquad (4.3) $$

when the electric field of the incident MW or FIR electromagnetic waves is circularly polarized perpendicular to the direction of the dc magnetic field. Here n_e is an electron density; τ denotes a momentum scattering time of the electron; and $1/\tau$ measures the number of collisions per unit time. Under the condition of $\omega_c \tau \geq 1$, a CR absorption spectrum exhibits a well-defined resonance peak that determines m_e^*. The CR peak line width, on the other hand, is a critical function of the momentum scattering time.

Unfortunately, sensitivity of conventional CR technique by detecting power loss of the MW and FIR radiations is severely limited. This has prevented it from being used in studies of carrier effective masses in semiconductor thin films and quantum structures. For that, highly sensitive optical detection of CR (ODCR) has been shown successful [3]. Instead of detecting power loss of the MW or FIR radiations, an intensity change of light emissions induced by carrier heating upon CR absorption of the MW or FIR radiations is detected in ODCR [3]. Besides the higher sensitivity, free carriers in ODCR experiments can be generated by optical excitation that does not require high doping of shallow impurities, as in conventional CR. This results in a reduced number of scattering centers and thus a longer momentum scattering time, beneficial to CR. The ODCR intensity should scale with absorption of the MW or FIR radiations as measured in conventional CR. Thus m_e^* and the momentum scattering time can be determined from an analysis of an ODCR spectrum by Equations 4.2 or 4.3, depending on the polarization of the electric field of the applied MW or FIR electromagnetic waves.

CR (and ODCR) is one of the very few experimental techniques that are capable of directly and uniquely determining carrier effective masses in a semiconductor without having to resort to certain assumptions or preknowledge on interdependent parameters. When the condition of $\omega_c \tau \geq 1$ is fulfilled, it provides accurate values of effective masses commonly regarded as the most reliable ones. However, the condition of $\omega_c \tau \geq 1$ for successful CR and ODCR also imposes demanding requirements on quality of materials and experimental conditions. An increase of ω_c requires CR experiments to be carried out at a very high magnetic field for a given effective mass, which is not commonly available. At a high ω_c and high magnetic fields, nonparabolicity of the bands and spin splitting can also complicate interpretation of CR results. If a lower-frequency MW is employed, on the other hand, a long momentum relaxation time τ is required. This demands that CR experiments be performed at a low temperature as well as a high quality of studied materials with high carrier mobility. In dilute nitrides, material quality is known to degrade with increasing N composition, leading to reduced electron mobility. For this reason, the application of the ODCR technique at 95-GHz MW frequency (to be presented in Section 4.3.1.1) is, unfortunately, currently restricted to GaNAs/GaAs quantum wells (QW) with low N compositions (<2%).

4.2.2 Plasma Reflectivity Measurements

The reflectivity of a semi-infinite isotropic medium is given by

$$R = \frac{(n-1)^2 + k^2}{(n+1)^2 + k^2} \tag{4.4}$$

where n is the real part of the refractive index and k is the extinct coefficient. The relation between n and k is described by the complex dielectric function

$$\varepsilon(\omega) = (n - ik)^2 \tag{4.5}$$

The free-electron contribution to the dielectric function of a semiconductor within the Drude model is described by

$$\varepsilon(\omega) = \varepsilon_\infty \left(1 - \frac{\omega_p^2}{\omega(\omega - i\gamma)} \right) \qquad (4.6)$$

where γ is the electron damping constant and [2]

$$\omega_p^2 = \frac{n_e e^2}{m_e^* \varepsilon_\infty \varepsilon_0} \qquad (4.7)$$

where ω_p is the plasma frequency of free electrons; ε_∞ is the high-frequency dielectric constant; ε_0 is the vacuum permittivity; and n_e is the free-electron concentration. For an isotropic CB, n_e is related to the wave vector at the Fermi surface k_F by $n_e(k_F) = \frac{k_F^3}{3\pi^2}$. Since the doping concentration is typically less than 10^{20} cm^{-3}, the plasma frequencies of electrons in semiconductors are usually in the infrared range. Therefore, by obtaining the plasma frequency ω_p from an analysis of infrared reflectivity induced by free electrons, m_e^* can be determined by Equation 4.7, provided that the free-electron concentration is known from independent studies such as Hall measurements.

Advantages of this experimental method include: (1) it is a direct measurement of m_e^* if the electron concentration can be reliably obtained separately, and (2) it provides a convenient means of studying the effective mass as a function of electron concentration (also the wave vector and energy) and thus nonparabolicity of the band dispersion. However, a significantly high doping level is required. Therefore, it cannot be applied to undoped or lightly doped dilute nitrides. It also cannot determine m_e^* near the bottom of the CB. When other contributions to the infrared reflectivity, such as that from interband or phonon absorption, become important, analysis can be rather complicated and perhaps not even straightforward. Difficulties can also arise when electron mobility is very low, such that the reflectivity edge smears.

4.2.3 Magneto-Photoluminescence of Free-to-Bound Transitions

In a strong magnetic field, the energy of the CB electrons in a plane perpendicular to the magnetic field direction becomes quantized into a series of Landau levels described by Equation 4.8.

$$E_n = (n + \frac{1}{2})\hbar\omega_c, \quad n = 0, 1, 2, \ldots \tag{4.8}$$

where ω_c is the cyclotron frequency given in Equation 4.1. Energy of a free-to-bound (F-B) transition, e.g., a transition from the lowest Landau level of the CB to the neutral-carbon acceptor, is given by

$$E_{F-B}(B) = E_c + \frac{1}{2}\hbar\omega_c - E_A \tag{4.9}$$

where E_c denotes the energy of the CB minimum at $B = 0$, and E_A is the energy of the acceptor level. Assuming that the energy shift of E_A in the magnetic fields is negligible as compared with that of the lowest Landau level of the CB, the energy shift of the free-to-bound transition should be obtained by

$$\Delta E_{F-B}(B) = \frac{1}{2}\hbar\omega_c = \frac{e\hbar B}{2m_e^*} \tag{4.10}$$

By analyzing a linear blueshift of the free-to-bound transition in magnetophotoluminescence, m_e^* can be obtained. This approach has a clear advantage in that it is rather simple and straightforward. A drawback is that it requires a well-resolved free-to-bound photoluminescence (PL) peak, which seems to be limited to Ga(In)NAs with a low N composition ($< 0.5\%$).

4.2.4 Magnetooptical Spectroscopy of Diamagnetic Shifts of Excitonic Transitions

In a weak magnetic field regime, when the cyclotron energy $\hbar\omega_c \, (= \frac{\hbar eB}{\mu c})$ of an exciton is lower than the effective exciton Rydberg energy $R^* (= \frac{\mu e^4}{2\hbar^2\varepsilon^2})$, the excitonic transition energy in a semiconductor with a small μ typically exhibits a quadratic dependence on the magnetic field B, often referred to as a diamagnetic shift. Here μ is the exciton reduced mass ($1/\mu = 1/m_e^* + 1/m_h^*$). From first-order perturbation theory, the diamagnetic shift of the 1s ground state of a free exciton can be expressed by [4]

$$\Delta E = D_2 \frac{\varepsilon^2 \hbar^4 B^2}{4\mu^3 c^2 e^2} \quad \text{(cgs unit)} \tag{4.11}$$

Here D_2 is a parameter equal to unity for a purely three-dimensional exciton, and equal to 3/16 for a purely two-dimensional exciton.

Equation 4.11 can therefore be employed to deduce the exciton reduced mass. Unfortunately, the electron and hole effective masses cannot be simultaneously determined from this experimental approach. In order to determine m_e^*, the hole effective mass m_h^* has to be separately determined from an independent study. For dilute nitrides, a procedure has commonly been adopted assuming that the introduction of N predominantly affects the CB, such that m_h^* remains largely unaffected. Nevertheless, effects of quantum confinement and N-induced biaxial strain have to be carefully taken into account in choosing an appropriate value of the valence band (VB) m_h^*.

4.2.5   Photoluminescence (PL) Spectroscopy of Spectral Line Width of Excitonic Transitions

The physical origin of the excitonic line broadening due to compositional disorder lies in the fact that average alloy composition inside the volume occupied by the exciton is different from that inside the volume of another exciton in a different spatial region of the alloy. Using the virtual-crystal approximation (VCA), the exciton PL line shape function $f(E,x)$ is Gaussian, i.e.,

$$f(E, x) = \left[2\pi\sigma_0(x) \right]^{-1/2} \exp\left[-\frac{1}{2}\left(\frac{E - E_g(x)}{\sigma_0(x)} \right)^2 \right] \qquad (4.12)$$

where $E_g(x)$ is the average band-gap energy for the alloy with composition x. From the alloy fluctuation model for PL line widths, σ_0 can be expressed as [5]

$$\sigma_0(x)^2 = K^2 x(1-x)\left(\frac{V_0}{V_{exc}} \right)\left(\frac{\partial E_g}{\partial x} \right)^2 \qquad (4.13)$$

Here V_0 is the smallest volume over which the composition fluctuation can occur, and for the zinc-blende structure, it is equal to $(a_0^3/4)$, a_0 being the lattice constant. $V_{exc} = \frac{4\pi}{3} r_{exc}^3$ is the volume of the exciton, and $r_{exc} = \frac{\hbar^2 \varepsilon}{\mu e^2}$ is the exciton radius. ε is the dielectric constant, and μ is the exciton reduced mass. The term $\frac{\partial E_g}{\partial x}$ is the variation of band-gap energy, e.g., with alloy composition x. Depending on the exact definition of the exciton volume, the value of K in Equation 4.13 varies between 0.32 and 1 [5]. By using $K = 0.4$ and $\varepsilon = 12.5$ for GaAs, the exciton reduced mass in $In_yGa_{1-y}As_{1-x}N_x$ alloys can be determined by the relation [5]

$$\mu = \left(\frac{5.3\sigma(x)}{\sqrt{x(1-x)} \dfrac{\partial E_g}{\partial x}} \right)^{2/3} \tag{4.14}$$

$\sigma(x) = 2\sqrt{2\ln 2}\, \sigma_0(x)$ is the full-width-at-half-maximum PL line width.

To avoid the ill-defined excitonic volume, a quantum-mechanical description was later developed to obtain the line width of an excitonic transition by calculating the mean deviation of its transition energy due to statistical potential fluctuations using first-order perturbation theory. The deduced values were found to be very close to those based on the classical description given above.

Unfortunately, similar to the approach of the excitonic diamagnetic shifts discussed above, the electron and hole effective masses cannot be simultaneously determined using the present experimental approach. It should be pointed out that caution must be exercised when using this approach, which is based on VCA, as the fluctuation of N composition in $In_yGa_{1-y}As_{1-x}N_x$ alloys is found to deviate from that for perfectly random alloys.

4.2.6 Modulation Spectroscopy and Optical Excitation of Band-to-Band Transitions in Quantum-Well Structures

In QW structures, the electron and hole confinement energies as well as their subband structures are sensitive functions of the electron and hole masses. These physical properties are commonly retrieved from band-to-band optical transitions between the CB and the VB by using photoluminescence excitation (PLE), photovoltaic measurements, or a variant of modulation spectroscopy. In the latter, instead of measuring optical reflectance (or transmittance), the spectral response modified by a repetitive perturbation (e.g., light, an electric field, a heat pulse, or stress) is evaluated. This gives rise to derivativelike spectral features in the photon energy region corresponding to interband transitions or critical points of the band structures.

To determine m_e^* from this experimental approach, solid knowledge on some other parameters that are also important for interband transitions is required. For example, values of conduction- and valence-band offsets at the heterointerfaces, the hole effective mass, strain field, and exact QW width should be separately obtained or self-consistently fitted or assumed. Depending on the choices of these parameters, a large uncertainty can arise in some cases, as will be illustrated in Section 4.3.1 in connection with the presentation of experimental results. Typically appropriate theoretical treatments by, for example, a modified k·p model, as described in Chapters 3 and 5, are required to obtain a self-consistent fitting of experimental results.

4.2.7   Magnetotransport

The electron density-of-states effective mass at the Fermi level can be calculated from solutions to the Boltzmann transport equation using the relaxation time approximation [6],

$$
m_e^* = \left(\frac{3}{\left| R_H \right| e\pi} \right)^{2/3} \frac{e\hbar^2}{k_B^2 T} \left(\alpha - \frac{Q}{\left| R_H \right| \sigma} \right) \tag{4.15}
$$

where σ is the conductivity, α is the Seebeck coefficient, Q is the Nernst coefficient, R_H is the Hall coefficient, and k_B is the Boltzmann's constant. The equation is valid in a weak-field regime ($\mu_e B \ll 1$), where μ_e denotes the electron mobility. By independently measuring the conductivity, Hall, Seebeck, and Nernst coefficients of a single sample (the so-called method of four coefficients), m_e^* can be deduced according to Equation 4.15. The accuracy of m_e^* obtained by this method is determined by the reliability and accuracy of the four coefficients from electrical measurements.

Another magnetotransport technique that has been employed in studies of dilute nitrides is the magnetotunneling spectroscopy technique [7]. It explores the effect that electron tunneling into the CB can give rise to resonant peaks in the current-voltage characteristics. By varying a magnetic field applied perpendicular to the current direction, one can tune an electron to tunnel into a given k state. Since the applied voltage tunes the energy, one can map out the form of the energy-momentum dispersion curve of the studied CB. Though a heavy electron mass for the E_+ CB of GaAsN is indicated from the magnetotunneling results, it has not been possible to extract its exact value because a precise correlation between the measured voltage value and the energy is unknown [7].

4.3   EXPERIMENTALLY DETERMINED CONDUCTION-BAND ELECTRON EFFECTIVE MASSES IN DILUTE NITRIDES

In this section, existing experimental results of m_e^* in various dilute nitrides will be summarized. See Tables 4.1 to 4.3. Our present understanding of

Table 4.1 Experimentally Determined Values of the Electron Effective Mass m_e^* in GaN$_x$As$_{1-x}$

N (%)	n_e (10^{19} cm^{-3})	QW or "bulk"	m_e^* (m_0)	μ (m_0)	T (K)	Experimental Method
1.2	undoped	7-nm QW	0.12	—	4.6	ODCR
2.0			0.19	—		Hai et al. [8]
0.043	undoped	epilayer	0.074	—	30	Magneto-PL of free-to-bound
0.095			0.13	—		transitions
0.21			0.13	—		Masia et al. [9]
0.5			0.14	—		
1.2	undoped	7-nm QW	0.098	—	RT	Photoreflectance of interband
1.6			0.102	—		transitions
2.0			0.104	—		Wu et al. [10]
2.8			0.107	—		
1.1	undoped	100-nm	—	0.063	LT	Magneto-PL of exciton
1.3		epilayer	—	0.090		diamagnetic shifts
1.7			—	0.083		Wang et al. [11]
0.9	undoped	7-nm QW	0.56	—	RT	Electroreflectance of interband
1.2			0.475	—		transitions
1.6			0.431	—		Zhang et al. [12]
2.0			0.37	—		
2.8			0.225	—		
4.5			0.165	—		
0	0.499	0.96 μm	0.084	—	RT	Thermomagnetic transport
0.01	0.756	2.0 μm	0.07	—		Young et al. [13]
0.1	0.506		0.061	—		
0.13	0.642		0.036	—		
0.4	0.548		0.029	—		

Note: m_0 is the bare electron mass in vacuum; n_e denotes the electron concentration; RT and LT denote the measurement temperature at room temperature and low temperature, respectively; μ denotes the exciton-reduced mass, defined by $1/\mu = [(1/m_e^*) + (1/m_h^*)]$, where m_h^* is the hole effective mass. In cases of QW structures, m_e^*, m_h^*, and μ refer to those in the QW plane.

the effects of alloy compositions and band nonparabolicity will be presented. Possible sources that have led to reported scattered values and discrepancies will also be discussed, taking into account the advantages and limitations of the experimental approaches involved.

Table 4.2 Experimentally Determined Values of the Electron Effective Mass m_e^* in $Ga_{1-y}In_yN_xAs_{1-x}$

In (%)	N (%)	n_e (10^{19} cm^{-3})	QW or "bulk"	m_e^* (m_0)	μ (m_0)	T (K)	Experimental Method
3	1	0.02	1–3 μm	0.08	—	RT	Plasma
3	1	0.97		0.16	—		reflectivity
3	1	1.9		0.20	—		Skierbiszews
3	1	2.2		0.26	—		ki et al.
5	1	2.2		0.26	—		[14–16]
5	1.9	4.0		0.38	—		
5	2.1	1.7		0.17	—		
5	2.2	2.0		0.18	—		
8	3.3	3.7		0.24	—		
8	3.3	6.0		0.4	—		
3	1	undoped	epilayer	—	0.10	4	Magneto-PL of
6	2			—	0.13		exciton line width Jones et al. [17]
7	2	undoped	>200-nm epilayer	—	0.13	2	Magneto-PL of exciton diamagnetic shifts Jones et al. [18]
14.3	0	undoped	8.6-nm QW	0.062	—	RT	Photoreflectance
14.3	1			0.08	—		of interband
14.3	1.4			0.09	—		transitions
14.3	1.7			0.085	—		Héroux et al. [19]
23	0	undoped	6.6-nm QW	0.060	—	RT	Transmission of
23	0.5			0.088	—		interband
23	1			0.090	—		transitions Duboz et al. [20]

Table 4.2 Experimentally Determined Values of the Electron Effective Mass $m_e{}^*$ in $Ga_{1-y}In_yN_xAs_{1-x}$ (Continued)

In (%)	N (%)	n_e (10^{19} cm^{-3})	QW or "bulk"	$m_e{}^*$ (m_0)	μ (m_0)	T (K)	Experimental Method
25	0	undoped	6-nm QW	0.065	0.042	150–180	Magneto-PL of
25	0.7		6-nm QW	0.092	0.052		exciton
32	0		6.4-nm QW	0.061	0.039		diamagnetic
32	1.1		6.0-nm QW	0.086	0.049		shifts
34	0		7-nm QW	0.057	0.038		Baldassarri Höger von
34	2.7		7-nm QW	0.072	0.045		Högersthal et
38	0		8.0-nm QW	0.055	0.038		al. [21]
38	5.2		8.2-nm QW	0.08	0.048		
30	0	undoped	6.8-nm QW	0.055	—	RT	Photovoltaic of
30	0.35			0.073	—		interband
30	0.7			0.084	—		transitions
30	1			0.094	—		Pan et al. [22]
38	1.5	undoped	4-nm QW	$m_e{}^*$(x=0) +0.03	—	5	Polarized PLE of interband transitions Hetterich et al. [23]

Note: m_0 is the bare electron mass in vacuum; n_e denotes the electron concentration; RT denotes room temperature during the measurement; μ denotes the exciton-reduced mass, defined by $1/\mu = [(1/m_e{}^*) + (1/m_h{}^*)]$, where $m_h{}^*$ is the hole effective mass. In cases of QW structures, $m_e{}^*$, $m_h{}^*$, and μ refer to those in the QW plane.

4.3.1   GaInNAs

In the past years, GaInNAs has generated greater interest due to its existing and potential applications for efficient optoelectronic devices operating in the near-infrared spectral range, especially 1.3–1.55-μm lasers for fiber-optic communications. This is evident from a large number of published experimental studies of this alloy system, including studies of $m_e{}^*$ (see Tables 4.1 and 4.2) [8–23]. In the following sections, we will provide a brief account of the effects of N and In compositions, electron concentration, and electron energy on $m_e{}^*$ in GaInNAs.

4.3.1.1   Effects of N composition

Effects of N composition on $m_e{}^*$ in GaInNAs have been studied by a number of experimental techniques, including all those described in Section 4.2. In most cases, several sets of samples were studied and compared, each set

Table 4.3 Experimentally Determined Values of the Electron Effective Mass, m_e^*, in InN_xAs_{1-x}, InN_xSb_{1-x}, and GaN_xP_{1-x}

	N (%)	n_e (10^{19} cm^{-3})	QW or "bulk"	m_e^* (m_0)	T (K)	Experimental Method
InNAs	0	0.00264	—	0.024	RT	Plasma reflectivity
	0.2	0.191	2465 nm	0.063		Hung et al. [25]
	2.2	0.28	2670 nm	0.058		
	2.4	0.324	2460 nm	0.068		
	3	1.69	2580 nm	0.326		
InNSb	0	—	2400 nm	0.033	RT	Cyclotron resonance
	1.1	0.0387		0.044		Murdin et al. [26]
GaNP	2.5	undoped	QW	0.9	10	Photoluminescence Xin et al. [27]

Note: m_0 is the bare electron mass in vacuum; n_e denotes the electron concentration; RT denotes room temperature during the measurement.

with a fixed In composition. The experimental results obtained so far are summarized in Figure 4.1, plotted as a function of N composition. Except for two cases, to be discussed below, all experimental results point to a clear trend that m_e^* increases with increasing N composition.

As a representative case of the experimental observation, we will here present experimental results of m_e^* in 7-nm-wide GaNAs QWs obtained from the ODCR studies by Hai et al. [8]. Figure 4.2 shows typical ODCR spectra obtained at 4.6 K and 94.9 GHz, for $x = 0$, 1.2, and 2.0%. The observed shift of the rather well-defined electron CR peak at different N compositions clearly indicates a corresponding change of m_e^*. By fitting Equation 4.2 to the experimental ODCR spectra following the procedure described in Section 4.2.1, accurate and reliable values of m_e^* can be obtained without having to resort to any assumptions or input parameters. A strong increase of m_e^* was observed with increasing N composition, nearly three times from $x = 0\%$ to $x = 2.0\%$. The ODCR studies also revealed that the momentum scattering time τ shortened with increasing N composition (see the inset in Figure 4.2), a clear sign of decreasing electron mobility that was largely expected for this alloy and was experimentally confirmed.

Another commonly employed experimental approach is magneto-PL of exciton diamagnetic shifts. It was first explored by Jones et al. to obtain m_e^* in GaInNAs [18]. Figure 4.3 shows typical diamagnetic shifts in applied magnetic fields seen in PL spectra of excitonic transitions in GaInNAs from a more recent work by Baldassarri Höger von Högersthal et al. [21]. It is

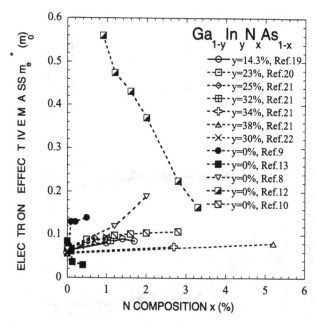

Figure 4.1 A summary of reported experimentally determined values of the electron effective mass in $Ga_{1-y}In_yN_xAs_{1-x}$, plotted as a function of N composition. Except for the filled circles and filled squares that are from thick epilayers, all data are from QWs with a well width within the range of 6 to 8.6 nm. The lines are guides for the eye.

very clear that the diamagnetic shifts can be strongly suppressed by the incorporation of N in the material, as a result of increasing exciton-reduced mass. As the magnetic field strength used spans across from weak to intermediate field regimes, an analysis of the exciton diamagnetic shifts over the entire field range in this particular case becomes somewhat more complicated, and a numerical method was employed for the purpose [21]. Provided m_h^* is known, m_e^* can be deduced from the diamagnetic shifts as described in Section 4.2.4. As seen from Figure 4.4, where the results discussed here are highlighted by the filled symbols and the solid lines, m_e^* is shown to increase with increasing N composition.

In sharp contrast to the trend of increasing m_e^* with increasing N composition, supported by the vast majority of the reported experimental studies, two reports claimed an opposite trend [12, 13]. One was based on electroreflectance measurements of band-to-band transitions from 7-nm-wide GaNAs QWs [12]. A huge jump of m_e^*, from $m_e^* = 0.067\ m_0$ at $x = 0\%$ to $m_e^* = 0.56\ m_0$ at $x = 0.9\%$ was reported, followed by a steady decline with increasing N composition over the entire range ($0.9\% < x < 3.3\%$) of N composition studied in that work (see Table 4.1 and Figure 4.1). This conclusion was later challenged by two separate studies of similar QW

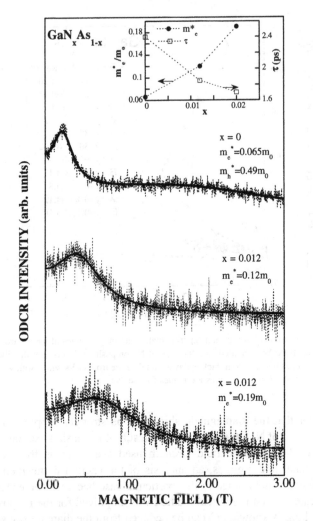

Figure 4.2 Electron ODCR spectra from the 7-nm GaN$_x$As$_{1-x}$ QWs for x = 0, 1.2, and 2%, obtained at 4.6 K and 94.9 GHz. The direction of the applied dc magnetic field is perpendicular to the QW planes. The solid lines are simulated ODCR spectra, with the specified parameters obtained from a best fit of Equation 4.2 to the experimental results. The inset illustrates the dependencies of the electron effective mass m_e^* and momentum scattering time τ on N composition, determined from the ODCR.

structures, one by photoreflectance measurements of band-to-band transitions [10], and one by the direct method ODCR [8]. Both studies confirmed the trend of increasing m_e^* with increasing N composition. The claim made by Zhang et al. [12] is now commonly believed to be most likely erroneous, possibly due to misassignments of the observed optical transitions. The other study by thermomagnetic transport [13], which predicted a decrease

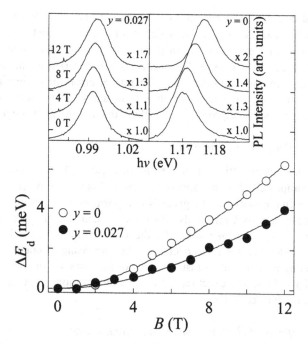

Figure 4.3 Diamagnetic shift, ΔE_d, measured at T = 100 K in $In_{0.32}Ga_{0.68}As_{1-y}N_y$ QW for the specified N-composition y vs. magnetic field B. The inset shows the PL spectra of these two samples at representative magnetic field values. Multiplication factors are given. (From Ref. [21].)

of m_e^* with increasing N composition, also seems to be questionable. To our knowledge, this is the only case that m_e^* in GaNAs is reported to fall well below that of the N-free GaAs, despite a moderate level of electron concentration that is generally believed to cause an increase of m_e^* due to a band nonparabolicity. Experimental evidence for the band nonparabolicity in GaInNAs has been well documented, and this will be discussed further in Section 4.3.1.3. Taking into account that many input parameters are required in the thermomagnetic transport studies, it is quite conceivable that the results from the "method of four coefficients" could be inaccurate.

If we can discard the two likely erroneous reports [12, 13], a consistent trend of increasing m_e^* with increasing N composition can be obtained. The exact values of m_e^*, however, scatter over a wide range, and a quantitative agreement has not been reached so far. It should be noted that the divergence among the data from GaNAs seems to be stronger than that from GaInNAs. One possible reason for the more consistent findings in GaInNAs could be due to the fact that in this alloy, with a large In composition, both quantum confinement and the compressive strain in the QW plane favor the heavy-hole (HH) state to be at the topmost of the VB, such that the ordering between the HH and light-hole (LH) states is secured and insensitive to structural design or experimental conditions. In GaNAs QW, on the contrary, the HH-LH ordering can vary, depending on the competition between the counteracting effects of the in-plane tensile strain and quantum confinement. It can, for example, be sensitive to small variations in structural design and experimental conditions. Therefore, errors or uncertainties in values of m_e^* may be introduced during an analysis of results from an indirect experimental approach such as those that require knowledge of m_h^*.

4.3.1.2   Effects of In composition

From a close examination of the results in Figure 4.4, some correlations with In composition become apparent. First, m_e^* is shown to decrease with increasing In composition at a given N composition, which is expected, as the N-free host alloy GaInAs also displays the same dependence. Second, apart from the results for $y = 14.3\%$, the slope of the N-induced increase of m_e^*, i.e., $\partial m_e^*/\partial x$, seems to decrease with increasing In composition. It is interesting to note that a drastic change of the slope occurs between $y = 32\%$ and $y = 34\%$. The exact reason for this observation is not understood and requires further study.

4.3.1.3   Effects of electron concentration and energy

Effects of electron concentration on m_e^* were studied by Skierbiszewski et al. [14, 15] from free-electron plasma reflectivity measurements of a set of thick n-type GaInNAs epilayers designed to be lattice-matched to a GaAs substrate. The results are summarized in Figure 4.5. A consistent increase of m_e^* was observed with increasing electron concentration, indicating a strong nonparabolicity of the CB.

By a careful analysis of the interband transitions of undoped GaNAs QWs as a function of hydrostatic pressure studied by photomodulated reflectance, Klar et al. [24] showed that m_e^* strongly increases with increasing electron energy above the bulk CB edge due to a strong nonparabolicity of the CB of GaNAs QWs (Figure 4.6).

4.3.2   Other Dilute Nitrides

Recently, the electron effective masses in less-known InNAs [25] and InNSb [26] were determined from plasma reflectivity measurements and CR (Table 4.3). Again, a general trend of increasing m_e^* with increasing N composition or increasing electron concentration was observed. The experimental findings were in both cases interpreted as a consequence of a strong anticrossing between the extended CB state and the localized resonant state of N based on a band anticrossing (BAC) model. In GaNP with 2.5% N composition, a large CB $m_e^* \approx 0.9\ m_0$ was estimated from an analysis of the quantum confinement energies by using an infinite-barrier model and various well widths [27].

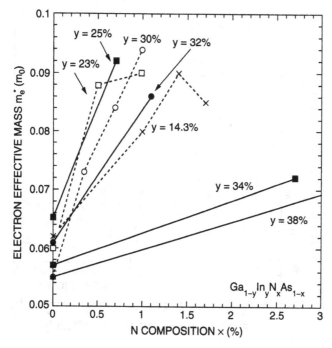

Figure 4.4 A zoom-in of Figure 4.1 for $x < 3\%$ depicting the reported experimentally determined values of the electron effective mass in $Ga_{1-y}In_yN_xAs_{1-x}$. All data are from QWs with a well width within the range of 6 to 8.6 nm. The lines are guides for the eye. Data represented by the filled symbols are from the study given in Ref. [21]. Data represented by crosses, open squares, and open circles are from Refs. [19, 20, 22], respectively.

4.4   COMPARISON WITH THEORETICAL PREDICTIONS

Several theoretical treatments have been employed in the past years to explain the observed unusual properties of dilute nitrides, based on the BAC model [28, 29], tight-binding and **k·p** models [30, 31], and the pseudopotential theory [32, 33]. For details of these theoretical treatments, the readers are referred to Chapters 1–3.

Within the BAC model, the interaction of the CB edge with the dispersionless N level results in a characteristic level anticrossing that leads to a splitting of the CB into two highly nonparabolic subbands, $E_-(k)$ and $E_+(k)$. In first-order perturbation theory, the hybridized lowest CB $E_-(k)$ is given by [28, 29]

$$E_-(k) = \frac{1}{2}\left[\left(E_{M\Gamma}(k) + E_N\right) - \sqrt{\left(E_{M\Gamma}(k) - E_N\right)^2 + 4xC_{NM}^2}\right] \quad (4.16)$$

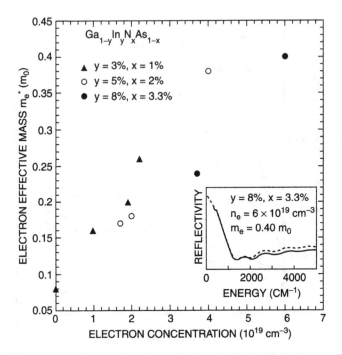

Figure 4.5 A summary of reported experimentally determined values of the electron effective mass in n-type $Ga_{1-y}In_yN_xAs_{1-x}$, plotted as a function of electron concentration. The lines are guides for the eye. The inset shows an infrared reflectivity spectrum for $In_{0.08}Ga_{0.92}N_{0.033}As_{0.967}$ (courtesy of Skierbiszewski). Solid and dashed lines represent experimental and simulated spectra, respectively.

Here $E_{M\Gamma}(k)$ is the dispersion relation at the Γ CB edge; E_N is the energy of the localized state derived from the substitutional N atoms; x is the mole fraction of substitutional N; and C_{NM} is the coupling parameter depending on the semiconductor matrix. The density-of-states electron effective mass of the corresponding CB can be obtained from Equation 4.16 as

$$m_e^*(E) = m^*\left[1 + \left(\frac{C_{MN}x^{1/2}}{E_N - E}\right)^2\right]\tag{4.17}$$

where m^* is the electron effective mass of the N-free semiconductor matrix with $x = 0$.

Equation 4.17 can be employed to predict the electron effective masses in dilute nitrides. The predicted values for GaNAs are shown in Figure 4.7 by using the parameters C_{NM} and E_N determined from a best fit of Equation 4.16 to the experimentally determined band-gap energies. It clearly shows that the m_e^* increases with increasing N composition, in agreement with

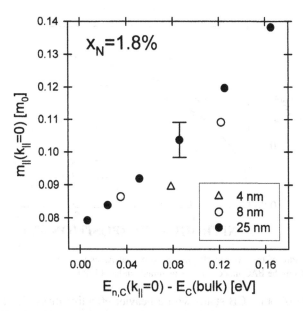

Figure 4.6 In-plane electron effective mass of the nth CB subband as a function of energy difference between the subband minimum $E_{n,c}(k = 0)$ and bulk conduction band edge E_c(bulk) for all the enhh$m$ transitions of the GaN$_{0.018}$As$_{0.982}$/GaAs quantum wells with well widths of 4, 8, and 25 nm. enhhm denotes the interband transition between the mth confined heavy-hole state and the nth confined electron state in Type I quantum-well structures. (From Ref. [24].)

the experimentally observed trend reported in the literature (see Figures 4.1 and 4.4, and Tables 4.1 and 4.2). On the other hand, the rate of m_e^* increase is predicted to slow down significantly when $x > 1\%$. As only very few experimental studies were reported for high N compositions ($x > 1$), it is difficult to confirm whether this is in agreement with experiments. The BAC model can also provide a rather satisfactory description for the dependence of m_e^* on energy and hydrostatic pressure [28, 29].

A modified ten-band **k·p** model was also developed [30, 31], including the two spin-degenerate N states, and has expectedly provided similar predictions as that from the BAC model for Ga(In)NAs with low N compositions, as shown in Figure 4.7. An added value of this model can be found in analyses of subband structures of the conduction and valence bands in quantum structures by properly taking into account the additional interactions between the band states and strong band nonparabolicity.

The empirical pseudopotential method and large supercells [32, 33] can provide a qualitative explanation for most of the experimentally observed physical properties of dilute nitrides, by considering strong multiband Γ-L-X coupling effects induced by N due to breaking of the translational symmetry of the host lattice. The coupling was also taken as an argument for the observed increase in m_e^* of the CB edge, by admixing

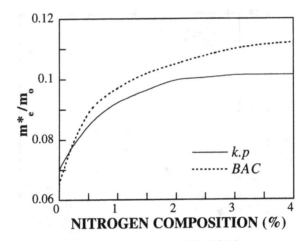

Figure 4.7 Plot of the electron effective mass of GaNAs as a function of N composition, calculated from the BAC model and the modified ten-band **k·P** theory.

characters of other CB states with a heavier effective mass. Effect of the amalgamation on m_e^* needs to be closely examined experimentally.

4.5 CONCLUSIONS AND OUTLOOK

The m_e^* in dilute nitrides has been found by various experimental techniques to possess a general trend to increase with increasing N composition, with energy above the bulk CB edge, and with electron concentration and hydrostatic pressure. The exact values of m_e^*, however, differ to various degrees between different experimental approaches and experimental conditions, particularly in GaNAs. This is believed to arise from the complication that the effects of the tensile strain and quantum confinement are counteracting in GaNAs, leaving a large uncertainty on values and even ordering of the heavy-hole and light-hole states that many indirect experimental methods rely on as crucial input of analyses. This is less a problem in compressively strained GaInNAs with a high In composition, explaining the better agreement on the general trends of m_e^* obtained in this quaternary alloy system using different experimental techniques. The observed increase of m_e^* in dilute nitrides has theoretically been attributed to a strong N-induced perturbation on the host conduction states by a mutual repulsion between the CB-edge state and the dispersionless localized N states or a strong multiband Γ-L-X coupling that brings to the CB edge the character of other CB states of heavier masses. The observed increase of m_e^* with increasing electron concentration and energy has provided compelling evidence for a strong nonparabolicity of the CB in GaInNAs.

Future strong efforts are required to reach quantitative agreement on the values of m_e^* in dilute nitrides, which are at present rather scattered and poorly understood. Strategies should be designed to obtain a reliable documentation of the values as a function of various key compositional, electronic, and structural parameters and experimental conditions, and more importantly to gain a complete understanding of the underpinning physical processes so that accurate predictions of the physical properties and optimized designs of device structures can be made.

References

1. Buyanova, I.A., Chen, W.M., and Monemar, B., "Electronic Properties of Ga(In)NAs Alloys," *MRS Internet J. Nitride Semicond. Res.* 6: 1–19 (2001).
2. Seeger, K., *Semiconductor Physics: An Introduction* (Springer-Verlag, Berlin, 1997).
3. Godlewski, M., Chen, W.M., and Monemar, B., "Optical Detection of Cyclotron Resonance for Characterization of Recombination Processes in Semiconductors," *Crit. Rev. Solid State Mater. Sci.* 19: 241–301 (1994).
4. Rogers, D.C., Singleton, J., Nicholas, R.J., Foxon, C.T., and Woodbridge, K., "Magneto-Optics in GaAs-$Ga_{1-x}Al_xAs$ Quantum Wells," *Phys. Rev. B* 34: 4002–4009 (1986).
5. Jones, E.D., Allerman, A.A., Kurtz, S.R., Modine, N.A., Bajaj, K.K., Tozer, S.W., and Wei, X., "Photoluminescence-Linewidth-Derived Reduced Exciton Mass for $In_yGa_{1-y}As_{1-x}N_x$ Alloys," *Phys. Rev. B* 62: 7144–7149 (2000).
6. Young, D.L., Geisz, J.F., and Coutts, T.J., "Nitrogen-Induced Decrease of the Electron Effective Mass in $GaAs_{1-x}N_x$ Thin Films Measured by Thermomagnetic Transport Phenomena," *Appl. Phys. Lett.* 82: 1236–1238 (2003).
7. Endicott, J., Patanè, A., Ibáñez, J., Eaves, L., Bissiri, M., Hopkinson, M., Airey, R., and Hill, G., "Magnetotunneling Spectroscopy of Dilute Ga(AsN) Quantum Wells," *Phys. Rev. Lett.* 91: 126802 (2003).
8. Hai, P.N., Chen, W.M., Buyanova, I.A., Xin, H.P., and Tu, C.W., "Direct Determination of Electron Effective Mass in GaNAs/GaAs Quantum Wells," *Appl. Phys. Lett.* 77: 1843–1845 (2000).
9. Masia, F., Polimeni, A., Baldassarri Höger von Högersthal, G., Bissiri, M., Capizzi, M., Klar, P.J., and Stolz, W., "Early Manifestation of Localization Effects in Diluted Ga(AsN)," *Appl. Phys. Lett.* 82: 4474–4476 (2003).
10. Wu, J., Shan, W., Walukiewicz, W., Yu, K.M., Ager III, J.W., Haller, E.E., Xin, H.P., and Tu, C.W., "Effect of Band Anticrossing on the Optical Transitions in $GaAs_{1-x}N_x$/GaAs Multiple Quantum Wells," *Phys. Rev. B* 64: 085320 (2001).
11. Wang, Y.J., Wei, X., Zhang, Y., Mascarenhas, A., Xin, H.P., Hong, Y.G., and Tu, C.W., "Evolution of the Electron Localization in a Nonconventional Alloy System $GaAs_{1-x}N_x$ Probed by High-Magnetic-Field Photoluminescence," *Appl. Phys. Lett.* 82: 4453–4455 (2003).
12. Zhang, Y., Mascarenhas, A., Xin, H.P., and Tu, C.W., "Formation of an Impurity Band and Its Quantum Confinement in Heavily Doped GaAs:N," *Phys. Rev. B* 61: 7479–7482 (2000).
13. Young, D.L., Geisz, J.F., and Coutts, T.J., "Nitrogen-Induced Decrease of the Electron Effective Mass in $GaAs_{1-x}Nx$ Thin Films Measured by Thermomagnetic Transport Phenomena," *Appl. Phys. Lett.* 82: 1236–1238 (2003).
14. Skierbiszewski, C., Perlin, P., Wisniewski, P., Knap, W., Suski, T., Walukiewicz, W., Shan, W., Yu, K.M., Ager, J.W., Haller, E.E., Geisz, J.F., and Olson, J.M., "Large, Nitrogen-Induced Increase of the Electron Effective Mass in $In_yGa_{1-y}N_xAs_{1-x}$," *Appl. Phys. Lett.* 76: 2409–2411 (2000).

15. Skierbiszewski, C., Perlin, P., Wisniewski, P., Suski, T., Geisz, J.F., Hingerl, K., Jantsch, W., Mars, D.E., and Walukiewicz, W., "Band Structure and Optical Properties of In$_y$ Ga$_{1-y}$N$_x$As$_{1-x}$ Alloys," *Phys. Rev. B* 65: 035207/1-10 (2001).

16. Skierbiszewski, C., Lepkowski, S.P., Perlin, P., Suski, T., Jantsch, W., and Geisz, J.F., "Effective Mass and Conduction Band Dispersion of GaAsN/GaAs Quantum Wells," *Physica E* 13: 1078–1081 (2003).

17. Jones, E.D., Allerman, A.A., Kurtz, S.R., Modine, N.A., Bajaj, K.K., Tozer, S.W., and Wei, X., "Photoluminescence-Linewidth-Derived Reduced Exciton Mass for In$_y$Ga$_{1-y}$As$_{1-x}$N$_x$ Alloys," *Phys. Rev. B* 62: 7144–7149 (2000).

18. Jones, E.D., Modine, N.A., Allerman, A.A., Fritz, I.J., Kurtz, S.R., Wright, A.F., Tozer, S.W., and Wei, X., "Optical Properties of InGaAsN: a New 1eV Bandgap Material System," in *PIE Conference Proceedings*, ed. E.F. Schubert, I.T. Ferguson, and H.W. Yao (Bellingham, WA: International Society for Optical Engineering, 1999), pp. 52–63.

19. Héroux, J.B., Yang, X., and Wang, W.I., "Photoreflectance Spectroscopy of Strained (In)GaAsN/GaAs Multiple Quantum Wells," *J. Appl. Phys.* 92: 4361–4366 (2002).

20. Duboz, J.-Y., Gupta, J.A., Byloss, M., Aers, G.C., Liu, H.C., and Wasilewski, Z.R., "Intersubband Transitions in InGaNAs/GaAs Quantum Wells," *Appl. Phys. Lett.* 81: 1836–1838 (2002).

21. Baldassarri Höger von Högersthal, G., Polimeni, A., Masia, F., Bissiri, M., Capizzi, M., Gollub, D., Fischer, M., and Forchel, A., "Magnetophotoluminescence Studies of (InGa)(AsN)/GaAs Heterostructures," *Phys. Rev. B* 67: 233304/1-4 (2003).

22. Pan, Z., Li, L.H., Lin, Y.W., Sun, B.Q., Jiang, D.S., and Ge, W.K., "Conduction Band Offset and Electron Effective Mass in GaInNAs/GaAs Quantum-Well Structures with Low Nitrogen Concentration," *Appl. Phys. Lett.* 78: 2217–2219 (2001).

23. Hetterich, M., Dawson, M.D., Egorov, A., Bernklau, D., and Riechert, H., "Electronic States and Band Alignment in GaInNAs/GaAs Quantum-Well Structures with Low Nitrogen Content," *Appl. Phys. Lett.* 76: 1030–1032 (2000).

24. Klar, P.J., Grüning, H., Heimbrodt, W., Koch, J., Stolz, W., Tomi, S., and O'Reilly, E.P., "Monitoring the Non-Parabolicity of the Conduction Band in GaN$_{0.018}$As$_{0.982}$/GaAs Quantum Wells," *Solid State Electron.* 47: 437–441 (2003).

25. Hung, W.K., Cho, K.S., Chern, M.Y., Chen, Y.F., Shih, D.K., Lin, H.H., Lu, C.C., and Yang, T.R., "Nitrogen-Induced Enhancement of the Electron Effective Mass in InN$_x$As$_{1-x}$," *Appl. Phys. Lett.* 80: 796–798 (2002).

26. Murdin, B.N., Adams, A.R., Murzyn, P., Pidgeon, C.R., Bradley, I.V., Wells, J.-P.R., Matsuda, Y.H., Miura, N., Burke, T., and Johnson, A.D., "Band Anticrossing in Dilute InN$_x$Sb$_{1-x}$," *Appl. Phys. Lett.* 81: 256–258 (2002).

27. Xin, H.P., and Tu, C.W., "Photoluminescence Properties of GaNP/GaP Multiple Quantum Wells Grown by Gas Source Molecular Beam Epitaxy," *Appl. Phys. Lett.* 77: 2180–2182 (2000).

28. Shan, W., Walukiewicz, W., Yu, K.M., Ager III, J.W., Haller, E.E., Geisz, J.F., Friedman, D.J., Olson, J.M., Kurtz, S.R., Xin, H.P., and Tu, C.W., "Band Anticrossing in III-N-V Alloys," *Phys. Stat. Solidi B* 223: 75–85 (2001).

29. Wu, J., Dhan, W., and Walukiewicz, W., "Band Anticrossing in Highly Mismatched III-V Semiconductor Alloys," *Semicond. Sci. Technol.* 17: 860–869 (2002).

30. Lindsay, A., and O'Reilly, E.P., "Theory of Enhanced Bandgap Non-Parabolicity in GaNxAs$_{1-x}$ and Related Alloys," *Solid State Commun.* 112: 443–447 (2002).

31. O'Reilly, E.P., Lindsay, A., Tomi, S., and Kamal-Daadi, M., "Tight-Binding and k·P Models for the Electronic Structure of Ga(In)NAs and Related Alloys," *Semicond. Sci. Technol.* 17: 870–879 (2002).

32. Mattila, T., Wei, S.-H., and Zunger, A., "Localization and Anticrossing of Electron Levels in GaAs$_{1-x}$Nx Alloys," *Phys. Rev. B* 60: R11245–11248 (1999).

33. Kent, P.R.C., and Zunger, A., "Theory of Electronic Structure Evolution in GaAsN and GaPN Alloys," *Phys. Rev. B* 64: 115208 (2001).

CHAPTER 5

Electronic Properties of (Ga,In)(N,As)-Based Heterostructures

PETER J. KLAR and STANKO TOMIĆ

Philipps-University, Marburg, Germany
CCLRC Daresbury Laboratory, United Kingdom

5.1   INTRODUCTION

This chapter presents an overview of the present knowledge of the electronic states of (Ga,In)(N,As)/Ga(N,As) quantum-well (QW) structures. The unusual band-formation process, when incorporating small amounts of N into (Ga,In)As and GaAs semiconductor hosts, results in a rather complex band structure of bulk (Ga,In)(N,As) and Ga(N,As) [1–10]. Appropriate theoretical concepts for describing the band structure of these materials are currently developed [11–26]. Furthermore, there exists a structural metastability of the quaternary alloy $Ga_{1-y}In_yN_xAs_{1-x}$ that arises due to subtle microscopic effects when both In and N are present. As one result, the band-gap of $Ga_{1-y}In_yN_xAs_{1-x}$ is not simply a function of composition y and x, but also depends strongly on the material history, e.g., growth conditions as well as annealing conditions. Therefore, to understand this class of semiconductor materials, it is essential to correlate structural and electronic investigations.

The unusual band structure of (Ga,In)(N,As), together with the aspect of metastability, has a number of implications for studies of (Ga,In)(N,As)-containing heterostructures. As in bulk, structural and electronic investigations need to be combined. Great care has to be taken in comparing results published in the literature on QW structures of nominally similar composition. New $\mathbf{k}\cdot\mathbf{P}$-type models need to be developed to describe the electronic states of such heterostructures.

This chapter is divided into two sections. In the first section, it is shown how the metastability of (Ga,In)(N,As) manifests itself in the electronic properties of corresponding heterostructures grown by molecular-beam epitaxy (MBE) as well as by metalorganic vapor-phase epitaxy (MOVPE). It is demonstrated that by careful procedures, the alloy can be driven into a stable configuration. It is essential that such a stable configuration exists and can be reliably attained if this class of semiconductor alloys is to be employed in devices. The second section provides an overview of the studies into the electronic states of such structures. The weight of the discussion will be on Ga(N,As)/GaAs heterostructures. The reasons are the following. On the one hand, metastability is not a major issue for ternary Ga(N,As). Therefore, to a first approximation, the electronic band structure of bulk GaN_xAs_{1-x} only depends on N-composition x, as demonstrated by the excellent agreement of the various optical spectroscopic studies [1–8]. On the other hand, the problems arising in describing the electronic states of heterostructures containing dilute-N ternary layers are essentially the same as for the technologically relevant quaternary layers. The theoretical apparatus developed and tested for describing Ga(N,As)/GaAs QWs can be easily transferred to (Ga,In)(N,As)/GaAs QWs by simply adjusting the set of material parameters.

5.2   METASTABILITY OF QUATERNARY (Ga,In)(N,As) SAMPLES

5.2.1   Thermal Annealing of MBE and MOVPE Samples

Thermal annealing is commonly used to improve the photoluminescence (PL) properties of $Ga_{1-y}In_yN_xAs_{1-x}$ structures grown by MOVPE or MBE [27–36]. A common feature after annealing, apart from an increased PL efficiency due to the removal of defects, is a strong blueshift of the band-gap of quaternary samples, sometimes up to about 100 meV. Usually the blueshift for quaternary (Ga,In)(N,As) samples is much bigger than for ternary Ga(N,As) samples. There are several mechanisms that might contribute to the blueshift of the band gap. The magnitude of the various contributions strongly depends on the growth conditions (e.g., MBE or MOVPE, choice of precursors, N and In concentrations, etc.) as well as the annealing conditions (annealing time, ambient conditions, annealing temperature, etc.). Possible contributions are intermixing of the barrier/well interface [27, 29, 32], redistribution of In [35], or N inhomogeneities [30] and rearrangement of the local N environment [34].

5.2.2   Local N-Environment and Fine Structure of the Band Gap

In the following, we will focus on the rearrangement of the local N environment. In many samples grown by MOVPE or MBE, there are indications of changes of the local N environment from a configuration with four Ga nearest neighbors (nn) to configurations with more In nns [34, 37, 38]. This change of the local N environment can be detected by changes of the local vibrational modes of N in (Ga,In)(N,As) in Raman or infrared-absorption spectra [36–39]. Very convincing evidence for such changes is obtained by Raman spectroscopy close to the E_+ resonance with excitation energies of 2.18 eV [36] and 1.92 eV [37] or by infrared absorption [38, 39]. Recently, these findings were also confirmed by X-ray absorption spectroscopy [40]. It is worth mentioning that under certain growth and annealing conditions, there do not seem to be significant changes of the nn environment of N in (Ga,In)(N,As), i.e., the four-Ga environment dominates even after annealing [41].

To reach a better understanding of the effect of the nitrogen nn environment on the band structure of $Ga_{1-y}In_yN_xAs_{1-x}$, full sp^3s^* tight-binding supercell calculations have been performed, in which the central group V site was constrained to have a given number m (= 0 to 4) of In nns. In atoms were placed at random on the remaining sites to give the desired overall y. The Hamiltonian was written as $H_1 = H_0 + \Delta H$, where H_0 is the Hamiltonian for (Ga,In)As, and ΔH is the change due to N incorporation. Pairs of supercells H_0 and H_1 were defined by placing As and N, respectively, onto the central group V site. Calculating separately the wavefunctions ψ_{c0} and ψ_{c1} for H_0 and H_1, one can derive a nitrogen resonant level wavefunction $\psi_N \propto (\psi_{c1} - \langle \psi_{c0}|\psi_{c1}\rangle \psi_{c0})$. This allows one to relate the supercell calculations to the simple level-repulsion model by $V_{Nc} = \langle \psi_N |H_1|\psi_{c0}\rangle$, $E_N = \langle \psi_N|H_1|\psi_N\rangle$, and $E_c = \langle \psi_{c0}|H_1|\psi_{c0}\rangle$. Details of the model are given in Refs. [17, 25] and Chapter 3 of this volume. These results are corroborated by other theoretical models [23].

Figure 5.1a shows results of 216-atom supercell calculations (corresponding to $x \approx 1\%$). For each y, the E_N for the five nn configurations are equally spaced with the value for four-Ga nns being always about 220 meV lower than that of four-In nns. Such strong dependence of the energy of isolated, strongly localized impurities on the nn environment is common [42]. Calculations, where the atomic arrangement in the second group III shell and higher shells was altered, only shift E_N by ±20 meV. This agrees with experiments on Ga(As,P):N, where a broadening of 30 meV of the localized N state is observed due to disorder on the group V sublattice [43]. The derived matrix element V_{Nc} linking the N resonant state and the conduction band edge varies between about 2.00 eV$\cdot x^{1/2}$ for four-Ga nns to about 1.35 eV$\cdot x^{1/2}$ for four-In nns, i.e., the strength of the perturbation of the crystal decreases with increasing number of In nns. The large differences of the five nn environments of N are also reflected in the derived conduction-band-edge energies E_-. For each y at $x = 1\%$, the E_- values for the five nn configurations are evenly spread over an energy range of about 80 meV below the corresponding unperturbed E_c of (Ga,In)As, with that for four-Ga nns being lowest in energy. The influence of the different nn configurations on the valence-band edge is small in comparison with that on the conduction band edge. The magnitude of the splitting E_- (4Ga) $- E_-$ (4In) increases strongly with x for all y in Figure 5.1b. The three curves were calculated varying the number of the atoms in the supercell between 1728 ($x \approx 0.1\%$) and 64 ($x \approx 3.1\%$). The inset of Figure 5.1b depicts the probability density of the band-edge wavefunction ψ_{c1}. About 40% of it is localized around the N impurity and its four nns. Changing the nn configuration of N modifies this localized band-edge wavefunction strongly. Such strong impuritylike band-gap behavior has also been reported for Ga(N,As) [20] and (Ga,In)N [26]. It explains the strong effect of the N-impurity environment on band-gap structure of (Ga,In)(N,As).

Next, we want to discuss experimental results concerning the influence of the rearrangement of the nn environment of N on the band structure. The samples studied were $Ga_{1-y}In_yN_xAs_{1-x}$/GaAs QW structures ($y \leq 35\%$, $x \leq$ 3%, well width ≈ 8 nm) and $Ga_{1-y}In_yN_xAs_{1-x}$ layers ($y \leq 15\%$, $x \leq 3\%$) grown by MOVPE. Pieces of the samples were annealed at various temperatures at a constant tertiarybutyl arsine (TBAs) partial pressure of 0.0272 mbar for 1 h followed by 25 min either under vacuum (unstabilized) or at the same TBAs partial pressure (As-stabilized). For the annealed samples, no indications for

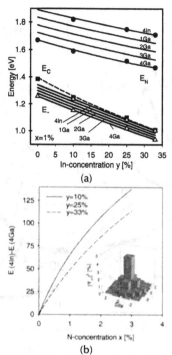

Figure 5.1 (Supercell calculation results of the conduction-band structure of $Ga_{1-y}In_yN_xAs_{1-x}$ for various nn environments of the N center: (a) y dependence, constant x; (b) x dependence, constant y. Inset: probability density of the conduction-band-edge wavefunction around the N-center.

a loss of nitrogen in the layers, interdiffusion in the QW structures, or phase separation were found by secondary-ion mass spectroscopy (SIMS), transmission electron microscopy (TEM), or X-ray diffraction (XRD).

Figure 5.2 depicts details of room-temperature photomodulated reflectance (PR) spectra of as-grown QW samples and of corresponding samples annealed unstabilized at 625°C. For the as-grown $Ga_{1-y}In_yN_xAs_{1-x}$/GaAs QWs with $x \approx 2\%$ (Figure 5.2a), at energies below 1 eV, a double-peak structure, corresponding to the lowest QW state, develops on increasing y from 10% to 35%. After annealing (Figure 5.2b), the peak structure below 1.0 eV has changed: The low-energy signal of the double-peak structure has disappeared for $y \geq 0.2$, whereas its high-energy signal has increased in oscillator strength. The energy separation between the two signals of about 30 to 40 meV is, approximately, independent of y. The disappearance of the low-energy peak in PR is also reflected in PL, where a shift of the PL band to higher energies occurs. The situation is completely different for the ternary QWs, as depicted in Figures 5.2c and 5.2d. Neither the $GaN_{0.015}As_{0.985}$/GaAs QW nor the $Ga_{0.8}In_{0.2}As$/GaAs QW showed a blueshift or a dramatic change of the PR spectra after annealing. This is also con-

PETER J. KLAR and STANKO TOMIĆ

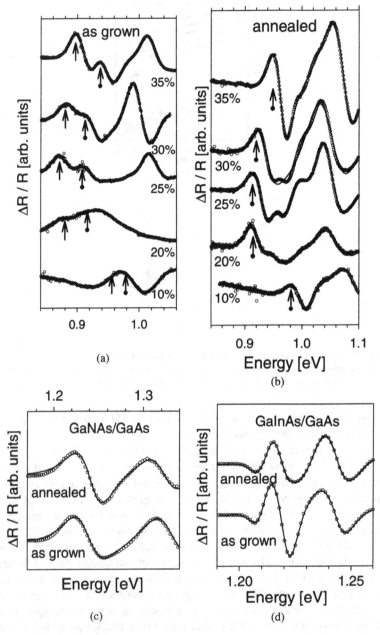

Figure 5.2 (a): (b) x ᵃ 2%, various y; (c) x =1.5%, y = 0%; (d) x = 0%, y = 20%." by "(a) x ᵃ 2%, various y, as grown; (b) x ᵃ 2%, various y, annealed; (c) x =1.5%, y = 0%, as grown and annealed; (d) x = 0%, y = 20%, as grown and annealed.

firmed by the room-temperature PL spectra, which do not show a blueshift after thermal treatment.

The inset of Figure 5.3a shows the dependence of the room-temperature PL of a $Ga_{0.7}In_{0.3}N_{0.01}As_{0.99}$/GaAs QW structure recorded after As-stabilized annealing. Apart from the increase of the PL intensity, which is mainly due to a reduction of nonradiative recombination centers during annealing, an unusual blueshift of the PL maximum together with significant changes of the PL line shape are observed. The PL spectra consist of overlapping bands separated by about 30 meV, whose relative intensities are changed such that the maximum of the PL is shifted to higher energies. At room temperature, these PL bands must originate from the band-band transitions, which is confirmed by the negligible Stokes shifts between the PL and the PR signals. Therefore, the shift of the maximum of the room-temperature PL is also a measure for the redistribution of the combined density of states at the band-gap in (Ga,In)(N,As).

The dependence of the blueshift of the PL maximum obtained from the spectra shown in the inset is depicted as curve 1 in Figure 5.3a. Below a critical annealing temperature T_c of about 600°C, the band-gap remains stable, but for higher temperatures, a blueshift is observed that reaches about 70 meV for $T > 700$°C. Curve 2 was recorded using pieces of the same $Ga_{0.7}In_{0.3}N_{0.01}As_{0.99}$/GaAs QW structure as before, but with unstabilized annealing conditions. The T_c of curve 2 is smaller than that of curve 1, and the blueshift saturates already at lower annealing temperatures.

Figure 5.3b is a plot of blueshifts measured by PL and PR at 300 K for various $Ga_{1-y}In_yN_xAs_{1-x}$/GaAs QWs and $Ga_{1-y}In_yN_xAs_{1-x}$ epitaxial layers with $1\% < x < 2\%$ after unstabilized annealing at 625°C as a function of y. At $y \approx 7\%$, a steplike increase of the energy shift is observed. After a plateau at about 30 meV, the blueshift increases to about 60 meV as y increases.

The double-peak structure for the lowest transition in the $Ga_{1-x}In_xN_yAs_{1-y}$ QWs in the PR spectra (Figure 5.2a) as well as the PL line-shape changes (Figure 5.3a) can be interpreted as a genuine proof of the existence of local band gaps arising from different nn configurations of the N-impurity. The dependence of the annealing effect on As vapor pressure (Figure 5.3a) leads to the conclusion that the change of N environment is enhanced by As vacancies. Thus, the reconfiguration process takes place on the group V sublattice. It probably arises due to a hopping of N ions to more favorable In-rich environments via an As vacancy. Combining the results of Figure 5.3b with ion statistics shows that this hopping process is of very short range and mainly takes place between neighboring sites. The N-hopping process is driven by the competition between chemical bonding and local strain contributions to the site energies of the different N environments. During the growth process, chemical bonding aspects dominate at the surface, and these favor Ga–N bonds instead of In–N bonds [44].

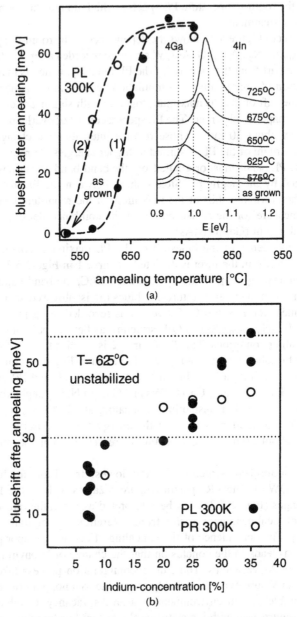

Figure 5.3a (a) Photoluminescence blueshift of a $Ga_{0.7}In_{0.3}N_{0.01}As_{0.99}$/GaAs QW vs. annealing temperature: (1) annealed As-stabilized and (2) unstabilized; inset: room-temperature photoluminescence corresponding to (1). (b) Blueshift after unstabilized annealing at 625°C of $Ga_{1-y}In_yN_xAs_{1-x}$/GaAs QWs ($x \approx 2\%$) vs. y.

This surface state is frozen in during the nonequilibrium growth process. In contrast to the surface, In-rich nn configurations of N are favored in bulk at equilibrium due to the dominance of local strain effects. Therefore, the frozen nonequilibrium bulk state can be transformed into the equilibrium bulk state by annealing under appropriate conditions.

The discrete character of the set of five bands comprising the actual band-gap and the x-dependence of the energy separation between the bands is verified by PR at low temperatures. Figure 5.4a depicts PR spectra of four pieces of a $Ga_{0.7}In_{0.3}N_{0.01}As_{0.99}$/GaAs QW as grown and annealed As-stabilized at 625°C, 650°C, and 725°C. The e1hh1 and e2hh2 transitions dominate the spectra; forbidden hh transitions do not contribute significantly. Light-hole transitions do not need to be considered, as they are at higher energies due to strain and confinement. The e1hh1 signal shows a distinct fine structure corresponding to the conduction-edge states of different nitrogen environments, e.g., four-Ga and three-Ga nn environments for the as-grown sample. Throughout the series, the energy positions of the five bands do hardly change, but with increasing annealing temperature, the oscillator strength of the signals is redistributed such that the In-rich environments become more dominant. This redistribution is also reflected in the line shape of the e2hh2 signal,* which consists of overlapping contributions of the five environments. However, these are smeared out so that individual signals can no longer be distinguished, similar to the corresponding room-temperature PL (inset Figure 5.3a). The conduction band edge is characterized by wavefunctions that are strongly localized near the N centers, leading to large energy separations between the corresponding conduction band edges. The unique phenomenon of inherent order in the frame of disorder is that — despite the alloy disorder — five discrete band gaps corresponding to the five ligand fields exist and are resolvable by modulation spectroscopy.

Figure 5.4b shows PR spectra of three as-grown $Ga_{0.7}In_{0.3}N_xAs_{1-x}$/GaAs QWs with different x ($\approx 0.5\%$, 1%, and 2%). In all spectra, the e1hh1 signal shows a fine structure originating from the four-Ga and three-Ga nn environments. With increasing x, all the QW states shift to lower energies, as expected from the increasing level-repulsion effect, but the energy splitting between the two conduction-band-edge states (due to four-Ga and three-Ga nn environments) also increases from about 20 meV for $x \approx 0.5\%$ to about 40 meV for $x \approx 2\%$. The corresponding total splittings E_- (four Ga)-E_- (four In) agree reasonably well with the calculated values (Figure 5.1b) considering an enhancement at lower temperatures as the band-edge E_c approaches the N-level E_N.

---

* By e$n$hh$m$, we denote the interband transition between the mth confined heavy-hole state and nth confined electron state in Type I quantum-well structures. Similarly, we use e$n$lh$m$ for the corresponding light-hole-related transitions.

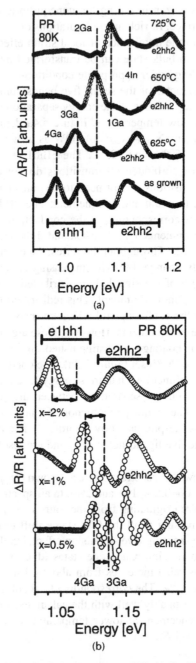

Figure 5.4 (a) Photomodulated reflectance spectra at 8 K: (a) $Ga_{0.7}In_{0.3}N_{0.01}As_{0.99}$/GaAs QW as grown and annealed As-stabilized at 625°C, 650°C, and 750°C; (b) $Ga_{0.7}In_{0.3}N_xAs_{1-x}$/GaAs QW with x = 0.5%, 1%, and 2%.

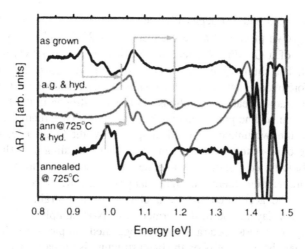

Figure 5.5 Photomodulated reflectance spectra at 300 K: $Ga_{0.7}In_{0.3}N_{0.01}As_{0.99}$/GaAs QW as grown and annealed at 725°C before (black) and after (gray) hydrogenation treatment.

Further confirmation that the fine structure of the e1hh1 transition observed in these MOVPE-grown $Ga_{0.7}In_{0.3}N_xAs_{1-x}$/GaAs QW structures arises due to different nitrogen nn environments is given when hydrogenating the samples. It is well established by hydrogenation experiments of Ga(N,As)- or (Ga,In)(N,As)-based structures that hydrogen almost completely neutralizes the effect of N on the band structure of the GaAs and (Ga,In)As host [45–49]. For more details, see also Chapter 6.

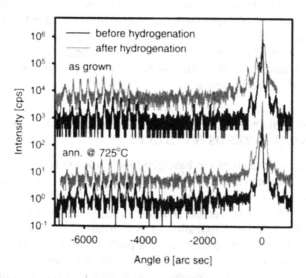

Figure 5.6 X-ray diffraction traces at 300 K: $Ga_{0.7}In_{0.3}N_{0.01}As_{0.99}$/GaAs QW as grown and annealed at 725°C before (black) and after (gray) hydrogenation treatment.

Figures 5.5 and 5.6 depict PR spectra and XRD traces, respectively, of an as-grown piece and a piece annealed at 725°C under As-stabilized conditions of a $Ga_{0.7}In_{0.3}N_{0.01}As_{0.99}$/GaAs before and after hydrogenation treatment. The similarity of the XRD traces of the as-grown and annealed QWs before hydrogenation underlines again, as discussed before, that under these moderate annealing conditions, no loss of nitrogen in the layers, no interdiffusion, and no phase separation occur in the QW structures. The PR spectra of the two unhydrogenated samples show the familiar features for the QW ground-state transition e1hh1, i.e., after annealing the e1hh1 transition is blueshifted significantly and the double-peak structure disappears. After hydrogenation, there is a slight change of the XRD traces observable, which can be accounted for by a change of the strain state of the (Ga,In)(N,As) layers due to the formation of N-H complexes [49]. More interesting are the PR spectra of the hydrogenated samples. After hydrogenation, when the effect of N on the band structure is virtually switched off, the ground-state transitions of both specimens are further blueshifted such that both transition energies are approximately the same, corresponding to that of the ground-state transition of an N-free QW. This result confirms again that the fine structure of the lowest QW transition arises due to different local N environment and that the blueshift due to annealing in these samples is almost entirely caused by a rearrangement of the local N environment, i.e., interdiffusion of Ga and In, loss of N, etc. play a minor role.

In summary, the band-gap of $Ga_{1-y}In_yN_xAs_{1-x}$ depends on the local N environment. Therefore, it is not simply a function of composition y and x, but also depends strongly on the material history, i.e., growth conditions as well as annealing conditions. This is a strong manifestation of the metastability of this quaternary alloy system. However, controlled annealing of the alloy leads to a stable situation that is likely to be the same for MBE- and MOVPE-grown samples.

5.3 INTERBAND TRANSITIONS OF (Ga,In)(N,As)-BASED HETEROSTRUCTURES

The immense redshift of the band-gap [1–8], when only a small fraction of N is incorporated into GaAs or (Ga,In)As, allows one to reach the telecommunication wavelengths of 1.3 or even 1.55 μm. But the N incorporation leads to considerable additional changes of the band structure, such as the N-induced formation of new bands [7–10] and the strong nonparabolicity of the conduction band predicted by theory [17, 50] and confirmed by various experiments [51–58]. All these modifications of the band structure have to be accounted for when designing new (Ga,In)(N,As)-based laser devices. Therefore, it is essential to develop useful theoretical descriptions of QW structures containing (Ga,In)(N,As). In particular, the unusual increase of the effective mass with decreasing energy gap observed in Ga(N,As)/GaAs QWs cannot be described by the commonly used eight-band k·P Hamiltonian for a direct-gap zinc-blende III-V semiconductor. However, an appropriate k·P-like description of the electronic states of Ga(N,As)/GaAs and (Ga,In)(N,As)/GaAs QWs can be obtained by using a ten-band Hamiltonian based on the band anticrossing effect [25, 59, 59a]. It consists of the conventionally used eight bands and two additional spin-degenerate N-related levels that are coupled to the two Γ_6 conduction-band levels. In many cases, the results of the ten-band k·P Hamiltonian can be approximated by a simpler analytical model for describing the conduction band states of such QWs [60]. A justification of the k·P approach is given by the tight-binding modeling of the (Ga,In)(N,As) band structure described in detail in Chapter 3.

5.3.1   Ga(N,As)/GaAs QWs

5.3.1.1   Conduction-band structure and band alignment

An interesting implication of the band-structure modification due to N in Ga(N,As) and (Ga,In)(N,As) is the strong nonparabolicity of the conduction band predicted by theory [17, 50]. In comparison with GaAs, the electron effective mass is expected to increase by about 50% at $k = 0$ for $x \approx 1\%$, with a further mass increase for $k \neq 0$. There are, to date, several experimental studies of this interesting effect. Skierbiszewski et al. [51] observed a strong increase of the electron effective mass at the Fermi level in (Ga,In)(N,As):Se with increasing free-electron concentration up to a value of 0.4 m_0 for $n = 6 \times 10^{19}$ cm^{-3}. Hai et al. [52, 53] showed by cyclotron resonance on 7-nm GaN$_x$As$_{1-x}$/GaAs QWs that the electron effective mass at the conduction band edge in GaN$_x$As$_{1-x}$ increases with x, whereas the hole effective masses are similar to those of GaAs. They reported values of 0.12 m_0 and 0.19 m_0 for $x = 1.2\%$ and 2.0%, respectively. Wu et al. [54] pointed out that the confined states of Ga(N,As)/GaAs QWs can only be correctly described when the N-induced changes of the electron effective mass are taken into account. Baldassari Höger von Högersthal et al. [55] reported a value of 0.15 m_0 for $x = 1.6\%$ determined by magnetophotoluminescence. A sharp increase of the electron effective mass was observed for N contents below 0.5% [56]. Similar experiments were performed by Wang et al. [58]. The electron effective mass is usually reported to strongly increase with increasing N content, apart from a report of transport experiments where the opposite trend was claimed [57]. The experimental data scatter considerably. This is partly a result of different assumptions underlying the models used for extracting the actual effective-mass values from the experimental data obtained by different experimental techniques. In addition, as discussed below in detail, effective-mass results obtained on bulk samples and quantum wells

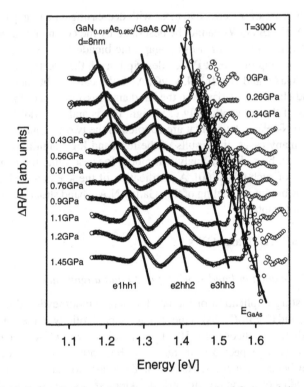

Figure 5.7 (a) Series of photomodulated reflectance spectra of 8-nm and 4-nm GaN$_{0.018}$ As$_{0.982}$/GaAs QWs obtained under hydrostatic pressure at 300 K. The solid curves are fits to the experimental data (open circles). The straight lines are a guide for the eye and indicate the pressure shifts of the e$n$hh$n$ QW transitions and the GaAs barrier.

of the same x are not necessarily comparable. The effective-mass issue is discussed from a different perspective in Chapter 4.

A further complication arises because the conduction band of GaN$_x$As$_{1-x}$ is strongly nonparabolic. This can also be understood qualitatively in the framework of the level-repulsion model. The closer the conduction-band states of the host are to the N level, the stronger is the level repulsion. Consequently, this means for QWs that the effective masses of the electron subbands in the conduction band must depend on N content, well width, and confinement energy. This manifests itself in the hydrostatic-pressure dependence of the interband transitions e$n$hh$n$ of Ga(N,As)/GaAs QWs, as can be seen in Figure 5.7. The figure shows series of PR spectra obtained under hydrostatic pressure at 300 K for two GaN$_{0.018}$As$_{0.982}$/GaAs QWs of width 8 nm and 4 nm. In the first series of spectra in Figure 5.7a, three signals can be clearly detected at all pressures; a fourth one can be discerned in the spectra for pressures exceeding 0.7 GPa. The signal at the highest energy originates from the GaAs barrier. The signals energetically

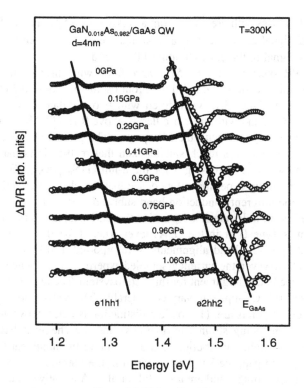

Figure 5.7 (b) (continued)

below the barrier signal correspond to interband transitions between confined QW states. They are assigned to the three allowed transitions e$n$hh$n$.
No signals arising from interband transitions e$n$lh$n$ between light-hole
subbands and electron subbands can be discerned in the spectra of this
sample. However, this is typical of Ga(N,As)/GaAs QWs when the corresponding transition energies are bigger than that of e1hh1. The e1lh1
transition is observed only in the PR spectra if this transition is lowest in
energy. This is the case for wider quantum wells, where bandshifts due to
tensile strain in the Ga(N,As) layer dominate quantum-confinement effects
for the lowest energy transitions [61, 62]. In the second series of spectra
in Figure 5.7b, the situation is very similar. Two signals can be clearly seen
at ambient pressure: the GaAs signal at 1.42 eV and the QW transition
e1hh1 at approximately 1.25 eV. Both signals are shifting to higher energies
with increasing pressure. The line shape of the GaAs-barrier signal is
changing dramatically in the pressure range between 0.15 GPa and 0.29
GPa. This is due to an interference between the GaAs barrier signal and
the signal of the e2hh2 QW transition. A second electron state becomes
confined in the QW in this pressure range due to the increasing interaction
between the N resonant state and the conduction band edge as the value of

$E_N - E_c$ decreases with pressure. From a pressure of 0.5 GPa up to 1.1 GPa, the e2hh2 signal can be seen as a weak feature on the low-energy side of the GaAs signal in the corresponding PR spectra.

The PR spectra in Figure 5.7a were fitted with three oscillators for pressures of less than 0.7 GPa and with four oscillators above 0.7 GPa. The PR spectra in Figure 5.7b were fitted with two oscillators for pressures below 0.15 GPa and with three oscillators above 0.15 GPa. In both cases, the fits agree very well with the PR spectra. The pressure dependence of the e$n$hh$n$ transitions is much smaller than that of the GaAs barrier and even decreases with increasing n for the 8-nm GaN$_{0.018}$As$_{0.982}$/GaAs QW. The latter is strong experimental evidence that the electron effective mass varies for the different conduction-band subbands [61–63].

Whether the holes are also located in the Ga(N,As) layers (i.e. the band alignment is Type I) or in the GaAs layers (i.e., Type II) is not *a priori* clear. Both possible band alignments were reported in the literature [52, 61, 64–67]. However, the current assembled experimental evidence strongly favors a Type I band alignment for this heterosystem. The reported valence-band offsets (VBO) appear to vary considerably. The reasons are manifold. Two important points are: (1) no clear distinction is made between the net VBO ratio (including strain contributions) and the chemical VBO ratio (corrected for strain contributions), and (2) due to the nonlinearity of the GaN$_x$As$_{1-x}$ band gap, the VBO ratios given in the literature are only valid for the corresponding x and are not valid for all x. A tentative value of 30% ± 15% for the chemical VBO ratio in GaN$_{0.018}$As$_{0.982}$/GaAs QW structures was proposed from studying the light-hole and heavy-hole splitting [61]. Krispin et al. [66, 67] reported a net VBO of only 5% for GaN$_{0.03}$As$_{0.97}$/GaAs QW structures determined by capacitance-voltage measurements.

It is important to note that changes of both the VBO and the electron effective mass affect the interband transition energies. However, assuming that the heavy-hole effective masses in Ga(N,As) and GaAs are equal (which seems a reasonable approximation as the N incorporation mainly affects the CB structure of the alloy) allows a separation of the effects. The confinement situation for light hole and heavy hole in the QW only depends on the strain situation, the VBO, and the hole effective masses. As the QW layers are under tensile strain, the degeneracy of light-hole and heavy-hole bands at $k = 0$ is lifted, such that the e1lh1 transition is lowest in energy for wide QWs. With decreasing well width, the e1lh1 transition energy approaches and finally crosses the e1hh1 energy because confinement effects become more significant and are greater for the light hole than for the heavy hole due to its lighter effective mass. Because both transitions involve the first electron state, the well width, where the crossing occurs, is independent of the electron effective mass. Therefore, the VBO value can be determined from the crossing point of the e1hh1 and e1lh1 transition energies. The value of the VBO of 30% ± 5% for GaN$_x$As$_{1-x}$ of $x = 1.8\%$

was determined this way by Klar et al. [61]. After fixing the VBO, the electron effective mass then plays the key role in determining the magnitude of the transition energies enhhn. In terms of the level-repulsion model, the resulting electron effective mass of a conduction-band subband at $k_{\parallel} = 0$ will depend on the energy difference of the subband minimum $E_{c,n}$ and the energy E_N of the N level. In the **k·P** model described below, the coupling parameter between the N level and the conduction band, and their energy separation, determine the degree of level repulsion and hence the magnitude of the effective mass and the degree of nonparabolicity. A rigorous test of the model parameters relating conduction band and N levels are hydrostatic-pressure experiments [9, 61, 62], because applying hydrostatic pressure allows one to continuously tune the degree of level repulsion as the N level and conduction band edge approach each other under pressure. Therefore, fitting the well-width dependence and the pressure dependence using the same set of material parameters for all samples [59, 68], a VBO of 30% [61] and N-related model parameters adjusted within the boundaries of the tight-binding predictions [17, 25, 50] must yield a good theoretical description of the dispersions of the electron subbands of the GaN$_{0.018}$As$_{0.982}$/GaAs QWs.

5.3.1.2   Theoretical modeling

5.3.1.2.1   Ten-Band k·P Model

The **k·P** and envelope-function methods are widely applied to study III-V semiconductor heterostructures [69]. In this section, we extend the well-established eight-band **k·P** Hamiltonian [70] by introducing two additional spin-degenerate N-related states. This extension of the conventional eight-band **k·P** method is necessary (a) to account for the strong interaction between the N resonant states and the conduction band edge in (Ga,In)(N,As) and related heterostructures and (b) to describe the variation of the (zone-center) conduction-band-edge energy with N [17, 25]. The modified ten-band **k·P** Hamiltonian for (Ga,In)(N,As) is given by [59a]:

$$H = \begin{pmatrix} E_n & V_{Nc} & -\sqrt{3}T_{N+} & \sqrt{2}U_N & -U_N & 0 & 0 & 0 & -T_{N-} & -\sqrt{2}T_{N-} \\ & E_{CB} & -\sqrt{3}T_{+} & \sqrt{2}U & -U & 0 & 0 & 0 & -T_{-} & -\sqrt{2}T_{N-} \\ & & E_{HH} & \sqrt{2}S & -S & 0 & 0 & 0 & -R & -\sqrt{2}R \\ & & & E_{LH} & Q & T_{N+}^{*} & T_{+}^{*} & R & 0 & \sqrt{3}S \\ & & & & E_{SO} & \sqrt{2}T_{N+}^{*} & \sqrt{2}T_{+}^{*} & \sqrt{2}R & -\sqrt{3}S & 0 \\ & & & & & E_{N} & V_{Nc} & -\sqrt{3}T_{N-} & \sqrt{2}U_N & -U_N \\ & & & & & & E_{CB} & -\sqrt{3}T_{-} & \sqrt{2}U & -U \\ & & & & & & & E_{HH} & \sqrt{2}S^{*} & -S^{*} \\ & & & & & & & & E_{LH} & Q \\ & & & & & & & & & E_{SO} \end{pmatrix}$$

$$(5.1)$$

where only the upper triangular block is shown because the matrix is Hermitian. The matrix elements are

$$E_{CB} = E_{c0} + \frac{\hbar^2}{2m_0} s_c (k_\parallel^2 + k_z^2) + \delta E_{CB}^{hy}$$

$$E_{HH} = E_{v0} - \frac{\hbar^2}{2m_0} (\gamma_1 + \gamma_2) k_\parallel^2 - \frac{\hbar^2}{2m_0} (\gamma_1 - 2\gamma_2) k_z^2 + \delta E_{VB}^{hy} - \varsigma$$

$$E_{LH} = E_{v0} - \frac{\hbar^2}{2m_0} (\gamma_1 - \gamma_2) k_\parallel^2 - \frac{\hbar^2}{2m_0} (\gamma_1 + 2\gamma_2) k_z^2 + \delta E_{VB}^{hy} + \varsigma$$

$$E_{SO} = E_{v0} - \Delta_{so} - \frac{\hbar^2}{2m_0} \gamma_1 (k_\parallel^2 + k_z^2) + \delta E_{VB}^{hy}$$

$$T_\pm = \frac{1}{\sqrt{6}} P(k_x \pm ik_y)$$

$$U = \frac{1}{\sqrt{3}} Pk_z$$

$$S = \sqrt{\frac{3}{2}} \frac{\hbar^2}{m_0} \gamma_3 k_z (k_x \pm ik_y)$$

$$R = \frac{\sqrt{3}}{2} \frac{\hbar^2}{2m_0} [(\gamma_2 + \gamma_3)(k_x - ik_y)^2 - (\gamma_3 - \gamma_2)(k_x + ik_y)^2]$$

$$Q = -\frac{1}{\sqrt{2}} \frac{\hbar^2}{m_0} \gamma_2 k_\parallel^2 + \sqrt{2} \frac{\hbar^2}{m_0} \gamma_2 k_z^2 - \sqrt{2}\varsigma \qquad (5.2)$$

and

$$s_c = \frac{1}{m_c^*} - \frac{E_p}{3} \left[\frac{2}{E_g} + \frac{1}{E_g + \Delta_{so}} \right] \qquad (5.3)$$

where the subscripts N, CB, HH, LH, and SO stand for nitrogen resonant, conduction, heavy-hole, light-hole, and spin-orbit split-off bands, respectively. Full details about 10 band Hamiltonian basis functions are given elsewhere [59a[. Because of the difference in Ga(N,As) and GaAs lattice constants, Ga(N,As) QWs are generally under biaxial tensile strain. The nitrogen resonant band is labeled by N, with

$$E_N = E_{N0} + \delta E_N^{hy} \qquad (5.4)$$

while

$$V_{Nc} = -\beta\sqrt{x} \tag{5.5}$$

describes the coupling between the N states and the conduction band edge, including its dependence on x. $\gamma_1 = \gamma_1^L - E_p/(3E_g)$ and $\gamma_{2,3}^L = \gamma_{2,3} - E_p/(6E_g)$ are modified Luttinger-Kohn parameters; Δ_{so} is the magnitude of the spin-orbit splitting at $k_\parallel = 0$; and $P = \sqrt{2m_0 E_p / \hbar^2}$ is the Kane matrix element for the conduction band related to the Kane energy E_p. $\delta E_b^{hy} = 2a_b(1 - c_{12}/c_{11})\varepsilon_{xx}$ and $\varsigma = -b_{ax}(1 + 2c_{12}/c_{11})\varepsilon_{xx}$ describe the influence of the hydrostatic and shear strain components on the band structure, where a_b are hydrostatic deformation potentials (with index b = N, CB, and VB for nitrogen, conduction, and valence bands, respectively); c_{11} and c_{12} are elastic constants; b_{ax} is the axial deformation potential; and ε_{xx} is the strain in the layer plane. In the QW case, k_z is no longer good quantum number, and it should be replaced by its operator $k_z \mapsto -i\frac{\partial}{\partial z}$. There are far more N-related parameters in the Hamiltonian of Equation 5.1 than can be accurately determined experimentally. Symmetry arguments indicate that the matrix elements $T_{N\pm}$ and U_N should vary linearly with k, coupling the N level to the valence band. The sp^{3}s* tight-binding calculations provide a guide as to which are the most important parameters (see O'Reilly et al. [25] and Chapter 3). We find that the dominant terms are those that are independent of k, which are included in Equation 5.1 to describe the variation of the overall energy gap with N composition. Consequently, we set $T_{N\pm}$ and U_N to be zero.

As pointed out above, the band-offset ratios in (Ga,In)(N,As)/GaAs QW structures are still under debate. Although a Type II band alignment was originally proposed for GaAs$_{1-x}$N$_x$ heterojunctions, based on band offsets calculated using the dielectric model [71], more recent first-principles calculations [72] as well as experiments [52, 61] strongly support a Type I alignment.

We have fitted the input parameters for our band-structure calculations to photoreflectance (PR) spectra from extensive measurements on Ga(N,As)/GaAs heterostructures [63, 68]. For band-structure calculations, we assume that the energy of unstrained nitrogen-level E_{N0}, conduction E_{c0}, and valence E_{v0} band-edge energies at the Γ point all vary linearly with N composition x as:

$$E_{N0} = E_{N0} - (\gamma - \kappa)x$$

$$E_{c0} = E_{c0} - (\alpha - \kappa)x$$

$$E_{v0} = E_{v0} + \kappa x \qquad (5.6)$$

where α and γ determine the variation of the GaN_xAs_{1-x} energy gap with x, while κ gives the chemical valence-band offset, describing how the valence-band edge of unstrained GaN_xAs_{1-x} shifts on an absolute energy scale with respect to that of GaAs. The band-edge energies are otherwise assumed to have the same variation with built-in hydrostatic strain and applied hydrostatic pressure as for GaAs.

Having determined the band-energy dispersion by solving the ten-band Hamiltonian (Equation 5.1) along k_z or $k_{\parallel}$ direction in the case of bulk material or along $k_{\parallel}$ in the QW case, the band-edge effective mass along each particular direction can be found by differentiation of the $E(k)$ curves at the zone center ($k = 0$) as

$$\frac{1}{m^*_{\perp(\parallel)}} = \frac{1}{\hbar^2 k_{\perp(\parallel)}} \left| \frac{\partial E(k_{\perp(\parallel)})}{\partial k_{\perp(\parallel)}} \right| \qquad (5.7)$$

and similarly for each particular confined subband in the case of QWs.

5.3.1.2.2 Analytical Model For deriving the analytical model, it is sufficient to account for the band anticrossing effect of the N levels with the conduction-band states and treat the valence band separately [60]. The conduction-band structure of the well and the barrier layers can both be described by the following band-anticrossing Hamiltonian:

$$H(x) = \begin{pmatrix} E_N & V_{Nc} \\ V_{Nc} & E_{CB} + bk^2 \end{pmatrix} \qquad (5.8)$$

where

$$E_N = E_{N0} - (\gamma - \kappa)x + 2a_N(1 - c_{12}/c_{11})\varepsilon_{xx} + (\partial E_N/\partial p)p$$

and

$$E_{CB} = E_{c0} - (\alpha - \kappa)x + 2a_c(1 - c_{12}/c_{11})\varepsilon_{xx} + (\partial E_{CB}/\partial p)p$$

are the energies of the N level and the conduction band edge. As before, the linear shifts $(\alpha - \kappa)x$ and $(\gamma - \kappa)x$ account for "normal" alloying effects, whereas the interaction-matrix element $V_{Nc} = \beta x^{1/2}$ reflects the "unusual" band formation due to the level repulsion. The latter is the cause of the effective-mass increase. The term bk^2 describes the conduction-band dispersion, i.e., $b_{w(b)} = \hbar^2/2m^*_{cw(b)}$, where m^*_{cw} is the effective mass of the host material and m^*_{cb} is the effective mass of the barrier material. In the case of N-free barriers the description still holds, as N-level and conduction band edge are decoupled, i.e., $V_{Nc} = 0$. The summand $2a_{N(c)}(1 - c_{12}/c_{11})\varepsilon_{xx}$ accounts for the shift of the nitrogen (conduction) band due to the biaxial strain in the layers. The summands $(\partial E_c/\partial p)p$ and $(\partial E_N/\partial p)p$ describe the shifts of the conduction band edge and the N level under hydrostatic pressure p. The corresponding eigenvalue problem $[H - EI]\psi = 0$, where I is the unit matrix and Ψ is the two-component wavefunction of the mixed nitrogen-host conduction-band state, can be solved for all layers. Applying continuity conditions at the well–barrier interfaces for the two-component wavefunction and its first derivative allows one to derive analytical expressions for the energies and eigenfunctions of the subbands of the conduction-band QW, very similar to those for the conventional one-band effective-mass model. The eigenenergies of a QW of width d can be determined, to a first approximation, by solving the following equations:

$$\frac{k_{z-}}{m^*_{cw}} \tan\left(k_{z-} \frac{d}{2} \right) = \frac{\kappa_{z-}}{m^*_{cb}} \tag{5.9}$$

with the relationship between energy E and wave vectors k_{z-} in the well and κ_{z-} in the barrier determined by solving the 2×2 Hamiltonian of Equation 5.8:

$$k^2_{z-} = \frac{1}{b_w}\left[\frac{V^2_{Nc}}{E_N - E} - (E_{cw} - E) \right] \text{ and } \kappa^2_{z-} = \frac{1}{b_b}(E_{cb} - E) \tag{5.10}$$

A similar expression to Equation 5.9 can also be used for odd states, with $\tan(k_{z-}d/2)$ replaced by $-\cot(k_{z-}d/2)$. The two-band model of Equation 5.8 can give electron ground- and excited-state confinement energies in good agreement with more complete calculations using a ten-band **k·P** Hamiltonian, as shown in Figure 5.8.

We turn now to the use of the two-band model to describe the electron effective mass in bulk GaN_xAs_{1-x} and within the plane of $GaN_xAs_{1-x}/GaAs$ QW structures. The mixing between the N level and the conduction band edge reduces the band dispersion in GaN_xAs_{1-x} relative to the uncoupled bands, with the inverse band-edge mass, m^{*-1}_{c-}, given in the two-band band anticrossing Hamiltonian of Equation 5.8 by:

$$\frac{1}{m^*_{c-}(k_{z-})} = |A_{c-}|^2 \frac{1}{m^*_{cw}} \tag{5.11}$$

where m_{cw}^* was introduced in Equation 5.8. The term $|A_{c-}|^2$ is the x-dependent squared amplitude of the conduction-band character of the lower eigenvalue of Equation 5.8,

$$E_- = [(E_N - E_{cw}) - \sqrt{(E_n - E_{cw})^2 + 4V_{Nc}^2},$$

and is given by:

$$|A_{c-}|^2 = \frac{1}{2}\left[1 + \frac{(E_N - E_{cw}) - b_w k_{z-}^2}{\sqrt{[(E_N - E_{cw}) - b_w k_{z-}^2]^2 + 4V_{Nc}^2}}\right] \qquad (5.12)$$

with $|B_{N-}|^2 = 1 - |A_{c-}|^2$ corresponding to the x-dependent squared amplitude of the nitrogen character of the E_- band-edge state. It can be seen that the conduction character of the lower bulk band-edge state decreases dramatically between $x = 0\%$ and 1%. It more or less saturates at $|A_{c-}|^2 = 0.55$ for higher x at room temperature. This mixing is primarily responsible for the observed enhancement of the bulk effective mass in Ga(N,As). The enhancement will depend both on the N content and on the confinement energy in Ga(N,As)/GaAs QWs.

The perpendicular or in-plane band-edge effective mass, $m_\perp^*$ or $m_\parallel^*$, for the ith confined QW state is given in the conventional one-band model by [73]:

$$\frac{1}{m_{\perp(\parallel)i}^*} = \frac{P_i^{(w)}}{m_{\perp(\parallel)i}^{(w)*}} + \frac{P_i^{(b)}}{m_{\perp(\parallel)i}^{(b)*}} \qquad (5.13)$$

where $P_i^{(w)}$ and $P_i^{(b)} = 1 - P_i^{(w)}$ are the probabilities of finding the ith confined state in the well and in the barrier, respectively, and the superscripts w and b denote well and barrier regions. The calculated wavefunction penetration and in-plane mass have a markedly different behavior in the two-band model depicted in Equation 5.8. Integrating the two-dimensional wavefunction ψ over the QW width yields the probabilities:
For even states:

$$P^{(w)} = \frac{\sin(k_{z-}d)/2k_{z-} + d/2}{|A_{c-}|^2 \cos^2(k_{z-}d/2)/\kappa_{z-} + \sin(k_{z-}d)/2k_{z-} + d/2} \qquad (5.14a)$$

For odd states:

$$P^{(w)} = \frac{-\sin(k_{z-}d)/2k_{z-} + d/2}{|A_{c-}|^2 \sin^2(k_{z-}d/2)/\kappa_{z-} - \sin(k_{z-}d)/2k_{z-} + d/2} \quad (5.14b)$$

where it can be shown using first-order perturbation theory that the perpendicular or in-plane band-edge effective mass is given by:

$$\frac{1}{m^*_{\perp(\parallel)i}} = |A_{c-}|^2 \frac{P_i^{(w)}}{m_{\perp(\parallel)i}^{(w)*}} + \frac{P_i^{(b)}}{m_{\perp(\parallel)i}^{(b)*}} \quad (5.15)$$

This leads to a markedly different behavior compared with Equation 5.13. First, there is significantly less wavefunction penetration into the barrier, because wavefunction matching only occurs with the conduction-band component of the well wavefunction. In addition, as the confinement energy increases, the relative magnitude of the nitrogen-related component (first term on right of Equation 5.15) becomes progressively more important. This tends both to reduce the wavefunction penetration into the barrier, and also to increase the average effective mass, $m^*_\perp$ or $m^*_\parallel$, within the well. Hence, the zone-center effective mass tends to increase with decreasing well width and also with increasing confinement energy (increasing i) for a fixed well width.

The corresponding heavy- and light-hole subband energies can be calculated by using a simple textbook one-dimensional quantum-well model, assuming that the unstrained valence-band edge shifts as $E_v = \kappa x$ in unstrained GaN_xAs_{1-x}, and assuming parabolic bands for heavy and light holes with effective masses $m_{hh} = (\gamma_1 - 2\gamma_2)^{-1}$ and $m_{lh} = (\gamma_1 + 2\gamma_2)^{-1}$. The biaxial strain in the layer leads to additional shifts of ζ for the heavy-hole band and of $-(\Delta_{so} + \varsigma)/2 + \sqrt{(9\varsigma^2 - 2\varsigma\Delta_{so} + \Delta_{so}^2)/4}$ (due to interaction of the light-hole and the spin-orbit split-off bands) for the light-hole band. Combining the eigenenergies of the conduction-band and valence-band subbands yields the transition energies.

5.3.1.3 Comparison of experiment and theory

We have tested the model by comparing calculated and experimental values of the transition energies of 21 GaN_xAs_{1-x}/GaAs QWs with well widths between 2 and 25 nm and 1% < x < 4% grown by MBE as well as by MOVPE. Figure 5.8 is a comparison of the transition energies of a series of $GaN_{0.018}As_{0.982}$/GaAs QWs of different well width determined by PR with those calculated using the ten-band k·P model and the analytical model. It is worth noting that the experimental values for the allowed transitions enhhn obtained from the MBE and MOVPE samples form a consistent data set. This underlines again that the Ga(N,As) system does not show signs of metastability. The theoretical results of the ten-band k·P

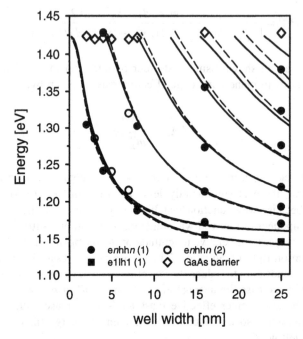

Figure 5.8 Comparison of transition energies of $GaN_{0.018}As_{0.982}$/GaAs QWs of different width extracted from photomodulated reflectance spectra with those calculated using the ten-band k·P model (solid lines) and the analytical model (dashed lines) (ambient pressure and $T = 300$ K).

calculation are in good agreement with all allowed transition energies e$n$hh$n$ throughout the series. The analytical model using the material parameters deduced from the ten-band k·P Hamiltonian gives excellent agreement with the full calculation and with experiment up to the e3hh3 transition. The differences between the models increase for higher QW transitions. This is to be expected, as the two-band Hamiltonian of Equation 5.8 underestimates the conduction-band nonparabolicity of Ga(N,As) and hence overestimates slightly the confinement energy of the higher-lying conduction states, ei ($i > 3$).

Figure 5.9 shows the hydrostatic-pressure dependence of the transition energies of 9-nm $GaN_{0.013}As_{0.987}$/GaAs, 8-nm $GaN_{0.018}As_{0.982}$/GaAs, and 7-nm $GaN_{0.024}As_{0.976}$/GaAs QWs extracted from PR spectra, compared with those calculated in the framework of the ten-band k·P model and the analytical model. The most striking result in Figure 5.9 is that the pressure coefficients of the e$n$hh$n$ transitions decrease with increasing n for all three samples, although the pressure coefficient of the GaAs is bigger than that of bulk Ga(N,As). This directly reflects the strong nonparabolic dispersion of the conduction band in Ga(N,As), which is explained by an increasing effective electron mass due to an increased level repulsion as the N level is energetically approached. The resulting nonparabolic effect is much greater than in GaAs/(Al,Ga)As QWs [74]. Again the agreement between

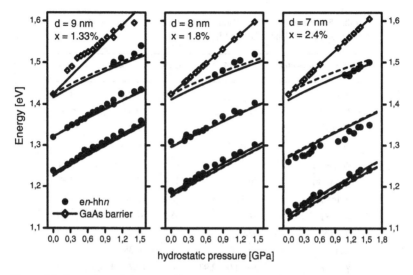

Figure 5.9 Comparison of the hydrostatic pressure dependence of the transition energies of 9-nm $GaN_{0.013}As_{0.987}$/GaAs, 8-nm $GaN_{0.018}As_{0.982}$/GaAs, and 7-nm $GaN_{0.024}As_{0.976}$/GaAs QWs extracted from photomodulated reflectance spectra with those calculated in the framework of the ten-band $k \cdot P$ model (solid lines) and the analytical model (dashed lines) ($T = 300$ K).

experiment and both theories is very good. Similar quality of the fits is obtained for all other samples using the same set of material parameters given in Tables 5.1 and 5.2.

The values of the parameters α, β, and γ are close to the values obtained from full sp^3s^* tight-binding calculations [17, 25, 50], but they have been adjusted here to the experimentally observed composition dependence of the E_- and the E_+ transition energies.

The fits shown in Figures 5.8 and 5.9 agree very well with the experimental data. This justifies extraction of the effective-mass values $m_\parallel$ ($k_\parallel = 0$) for the electron subbands from parabolic fits of the minima of the corresponding theoretical in-plane dispersions. The extracted $m_\parallel$ ($k_\parallel = 0$) values are plotted in Figure 5.10 as a function of energy difference of subband minimum $E_{c,n}$ ($k_\parallel = 0$) and bulk conduction band edge E_c. With increasing energy difference, the effective mass increases considerably, with all samples showing the same trend. This qualitatively confirms the predictions of the tight-binding calculations [17, 25, 50]. Our derived values are also in good agreement with those of Wu et al. [54], who give a value of about 0.11 m_0 for the effective mass of GaN_xAs_{1-x} for $1.2\% < x < 2.8\%$ determined at 300 K (in a similar fashion as in this work) from QWs with well width of less than 10 nm. The results of Hai et al. [52, 53] determined at 5 K by optically detected cyclotron resonance on 7-nm GaN_xAs_{1-x}/GaAs QWs of $x = 1.2\%$ and 2.0% yielded values of 0.12 m_0 and 0.19 m_0, respectively. The values determined at low temperatures are in general greater than those

Table 5.1 Relevant Material Parameters of the Binary Compounds GaAs, InAs, and Zinc-Blende GaN and InN

Material parameter	GaAs	GaN	InAs	InN
a_0 [Å]	5.6533	4.50	6.0853	4.98
E_g [eV]	1.424	—	0.355	—
E_p [eV]	25.7	—	22.2	—
m_0^*	0.0665	—	0.024	—
c_{11} [GPa]	122.1	293	83.3	182
c_{12} [GPa]	56.6	159	45.3	125
a_c [eV]	−7.17	−2.2	−5.08	−2.5
a_v [eV]	−1.16	−5.2	−1.0	−0.5
b_{ax} [eV]	−2.0	−2.2	−1.8	−1.3
Δ_0 [eV]	0.341	—	0.38	—
$\gamma_1, \gamma_2, \gamma_3$	6.98, 2.06, 2.93	—	19.7, 8.4, 9.3	—

Sources: Choulis, S.A. et al. *Phys. Rev. B* 66: 165321/01-09 (2002); Kim, K. et al. *Phys. Rev. B* 53: 16310–16236 (1996); Kim, K. et al., *Phys. Rev. B* 56: 7018–7021 (1997); Vurgaftman, I. et al., *J. Appl. Phys.* 89: 5815–5875 (2001).

Table 5.2 N-Related Model Parameters for Ga(N,As)

	Host		N-level
α [eV]	1.55	$E_{N0} - E_{c0}$ [eV]	0.23
κ [eV]	3.00	γ [eV]	3.00
—	—	β [eV]	2.45

determined at 300 K. A probable reason is the different temperatures at which the effective mass was determined. With decreasing temperature, the CB edge approaches the N level, and thus level-repulsion effects increase, leading to an increasing effective electron mass.

5.3.2   (Ga,In)(N,As)/Ga(N,As) QWs

5.3.2.1   Conduction-band structure and band alignment

There are presently only a few studies of the interband transitions of (Ga,In)(N,As)/Ga(N,As) QWs [59, 78–82]. These mainly focus on Ga_{1-y}

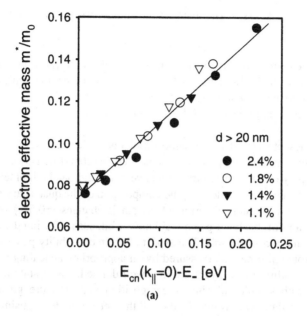

(a)

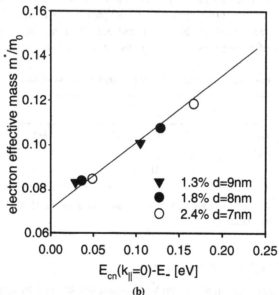

(b)

Figure 5.10 Effective mass values $m_\parallel$ ($k_\parallel = 0$) for the electron subbands from parabolic fits of the minima of the corresponding theoretical in-plane dispersions. The extracted $m_\parallel$ ($k_\parallel = 0$) values are expressed as a function of energy difference of subband minimum $E_{c,n}$ ($k_\parallel = 0$) and bulk conduction band edge E_c.

In$_y$N$_x$As$_{1-x}$/GaAs QWs. The main results of these works can be summarized as follows:

1. The electron effective mass is increased in (Ga,In)(N,As)/GaAs compared with (Ga,In)As/GaAs QWs [78–82] and increases with increasing conduction-band subband index n [59, 82].
2. The band alignment is of Type I [59, 78–82].

It becomes obvious that the same problems as for Ga(N,As)/GaAs are encountered in the analysis of the observed interband transitions. Again, the interplay of varying electron effective mass and band alignment needs to be accounted for. In addition, the composition of the quaternary material needs to be accurately determined, as small changes (±0.5%) of the N content can shift the band-gap by ±50 meV while also affecting the electron effective mass considerably. The aspect of the metastability poses an additional problem; it has to be assured by an appropriate annealing treatment that the (Ga,In)(N,As) layers are in a well-defined, stable state and comparable with samples of similar composition and structure grown by a different method. Unfortunately, the full theoretical modeling using the ten band **k·P**-model has so far only been applied in the analysis of as-grown (Ga,In)(N,As)/GaAs QWs [59]. The model parameters used and derived in this analysis are listed in Tables 5.1 and 5.3. It cannot be considered an accurate set of parameters yet and needs further improvement.

Table 5.3 N-Related Model Parameters for Ga$_{0.7}$In$_{0.3}$N$_x$As$_{1-x}$

	Host		N-level
α [eV]	1.55	$E_{N0} - E_{c0}$ [eV]	0.485
κ [eV]	3.5	γ [eV]	3.5
—	—	β [eV]	1.675

Source: From Choulis, S.A. et al., *Phys. Rev. B* 66: 165321/01–09 (2002).

5.3.2.2 Type I–Type II transition

Additional information about the VBO might be obtained by analyzing the Type I to Type II transition occurring in (Ga,In)(N,As)/GaN$_x$As$_{1-x}$ QWs when x is increased or when an external electric field is applied. Studies along this line have been reported for various II-VI and III-V QW systems [83–87] where the band alignment of individual QW structures has been tuned by applying external electric or magnetic fields.

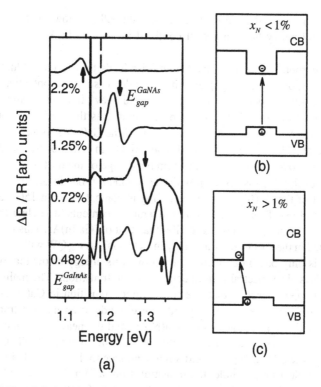

Figure 5.11a (a) Series of photomodulated reflectance spectra of $Ga_{0.77}In_{0.23}As/GaN_xAs_{1-x}$ QWs with x ranging from 0.48% to 2.2% ($T = 300$ K). (b) Type I band alignment for $x < 1$%. (c) Type II band alignment for $x > 1$%.

PR spectra of four $Ga_{0.77}In_{0.23}As/GaN_xAs_{1-x}$ QWs with x ranging from 0.48% to 2.2% are shown in Figure 5.11a. The large N-induced redshift of the Ga(N,As) barrier with increasing x can be seen clearly. Arrows indicate the energetic position of the Ga(N,As) band gap. The solid line represents the band-gap of $Ga_{0.77}In_{0.23}As$ pseudomorphically grown on (100) GaAs. The dashed line indicates the position of the lowest transition in the spectrum of the sample with $x = 0.48$%. The PR spectrum of this particular sample is typical for a QW with a Type I band alignment (Figure 5.11b). The lowest QW transition e1hh1 at 1.19 eV is blueshifted with respect to the band-gap of the corresponding strained (Ga,In)As bulk. Furthermore, a second QW transition can be clearly distinguished at 1.26 eV, which corresponds to the e2hh2 transition. Both heavy-hole-related transitions are comparable with the Ga(N,As) barrier signal in terms of oscillator strength. Light-hole-related transitions are not observed in the spectrum. Due to the compressive strain in the (Ga,In)As QW layer and due to tensile strain in the Ga(N,As) layer, the confining potential for the light holes is very shallow and shifted to higher energies by about 100 meV. The observation of the two strong heavy-hole transitions e1hh1 and e2hh2 shows that both electrons and heavy holes are well confined in the (Ga,In)As layer for $x = 0.48\%$.

Incorporating more N into the barrier leads to a further reduction of the corresponding band-gap and thus to a decrease of the potential height of the confining potentials for electrons and heavy holes. It is anticipated that this reduction of the band-gap mainly affects the confining potential of the electrons, as the band-gap reduction of the Ga(N,As) is a conduction-band effect. It can be understood qualitatively in terms of the band anti-crossing model, where the conduction band edge of the host is repulsed by the N level, which is resonant with the conduction-band states [8, 9]. The redshift of the Ga(N,As) with increasing N content is about 150 meV per percent N [8]. Therefore, even at moderate N contents, it is clear that the Ga(N,As) band-gap becomes smaller than that of (Ga,In)As. This is clearly reflected in the spectrum of the $Ga_{0.77}In_{0.23}As/GaN_xAs_{1-x}$ QW with $x = 2.2\%$. There is only one broad feature visible with its energy below that of bulk (Ga,In)As. The spectral feature comprises two transitions. The main signal at the higher energy is that due to the direct transition in Ga(N,As), and the shoulder on the low-energy side corresponds to an indirect transition between electrons located in the Ga(N,As) and the heavy holes located in the (Ga,In)As layer, as schematically depicted in Figure 5.11c. This situation is typical for a Type II band alignment. Due to the Coulomb attraction between electron and hole, the transition from a Type I to a Type II band alignment is less abrupt, as suggested by Figures 5.11b and 5.11c. Therefore, it is difficult to determine the actual N concentration where a flat-band situation in the conduction band arises and the transition takes place. This is also reflected by the PR spectra of the $Ga_{0.77}In_{0.23}As/GaN_xAs_{1-x}$ QWs with intermediate N contents, i.e., $x = 0.72\%$ and 1.25%. In the case of the $Ga_{0.77}In_{0.23}As/GaN_{0.0072}As_{0.9928}$ QW, the e1hh1 is already strongly redshifted with respect to that for $x = 0.48\%$, and its oscillator strength is decreased considerably. In addition, there is no longer a e2hh2 transition visible in the spectrum. These observations are strong indications that the confining potential for the electrons (though still located in the (Ga,In)As layer) has become very shallow. When x is increased to 1.25%, the Ga(N,As) band-gap is still greater than that of the (Ga,In)As. However, typical for a Type II situation, there is no clear feature of a e1hh1 transition at energies between those of bulk (Ga,In)As and Ga(N,As). This suggests that the Type I to Type II transition in the $Ga_{0.77}In_{0.23}As/GaN_xAs_{1-x}$ QWs occurs for $0.72\% < x < 1.25\%$.

Figure 5.12 is a plot of the transition energies (symbols) in $Ga_{0.77}In_{0.23}As/GaN_xAs_{1-x}$ QWs as a function of N-content x. The solid lines are calculated transition energies assuming constant (i.e., independent of x) confining potentials for the heavy holes, with heights corresponding to 20%, 30%, and 40% of the band-gap difference of 260 meV between GaAs ($x = 0\%$)

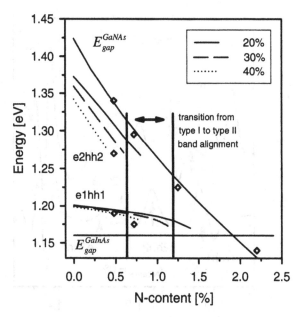

Figure 5.12 Plot of the transition energies (symbols) in $Ga_{0.77}In_{0.23}As/GaN_xAs_{1-x}$ QWs vs. N-content x. The solid lines are calculated assuming constant confining potentials for the heavy holes with heights corresponding to 20%, 30%, and 40% of the band-gap difference between GaAs and strained $Ga_{0.77}In_{0.23}As$.

and strained $Ga_{0.77}In_{0.23}As$. For simplicity, constant electron effective masses of $0.1\ m_0$ and $0.05\ m_0$ (where m_0 is free-electron mass) were assumed for the electrons in the Ga(N,As) and the (Ga,In)As, respectively. This is only a crude approximation, in particular for the GaN_xAs_{1-x}, where the electron effective mass is known to vary strongly with x. We used a heavy-hole effective mass of $0.35\ m_0$. The variation of the GaN_xAs_{1-x} band-gap with x was described by a band anticrossing model [8]. Thus, the calculated curves in Figure 5.12 should only be considered as a rough estimate of dependence of the transition energies on the VBO in the vicinity of the Type I to Type II transition. Nevertheless, already these simple calculations show some typical features that are sensitive to the VBO, such as the sudden increase of the redshift of the e1hh1 transition or the energetic position of the e2hh2 for x where the confinement situation is still Type I. The pair of solid lines in Figure 5.12 indicates the range of x where the Type I to Type II transition takes place.

In the same fashion, the band alignment in $Ga_{0.7}In_{0.3}N_{0.005}As_{0.995}/GaN_xAs_{1-x}$ QWs of different x can be analyzed. Again for $x = 0\%$, the band alignment is of Type I. As for the ternary QWs, the confinement potential for the electrons is strongly reduced with increasing N content until a Type I to Type II transition of the band alignment occurs. Of course, due to the

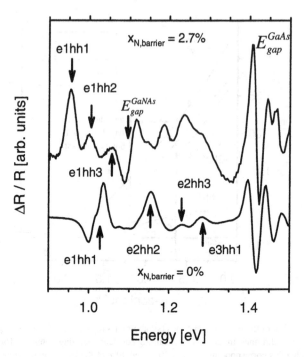

Figure 5.13 Photomodulated reflectance spectra of two $Ga_{0.7}In_{0.3}N_{0.005}As_{0.995}/GaN_xAs_{1-x}$ QWs with $x = 2.7\%$ (top) and $x = 0\%$ (bottom) ($T = 300$ K; QW signals and barrier signals indicated by arrows).

higher In content of 30% and the incorporated small amount of N of about 0.5% in the (Ga,In)(N,As) layer, the height of the electron-confining potential for $x = 0\%$ will be considerably greater than for the (Ga,In)As with an In content of only 23%, discussed above. Therefore, it is anticipated that the Type I to Type II transition for the quaternary QW layers takes place at a considerably higher N-content x in the GaN_xAs_{1-x} than for the ternary QW layers. This is confirmed by the experimental results depicted in Figures 5.13 and 5.14. Figure 5.13 depicts two PR spectra of $Ga_{0.7}In_{0.3}N_{0.005}$ $As_{0.995}/GaN_xAs_{1-x}$ QWs with $x = 0\%$ and $x = 2.7\%$. The spectrum of the sample with N-free barriers shows, at about 1.0 eV, the lowest QW transition e1hh1. At higher energies below the GaAs barrier signal at 1.42 eV, further QW transitions are visible and assigned to e2hh2, e2hh3, and e3hh1. Again no light-hole-related transitions are observed. In the spectrum of the sample with $x = 2.7\%$, the barrier signal shifted to about 1.1 eV. In this case, the signal at 1.42 eV is due to the GaAs substrate. In comparison with the $x = 0\%$ sample, the e1hh1 transition is shifted to lower energies, and the e2hh2 has vanished, whereas the forbidden e1hh2 and e1hh3 transitions are still visible in the spectrum. Thus, the confinement situation for this sample is also Type I. However, the disappearance of the e2hh2 and the strong redshift

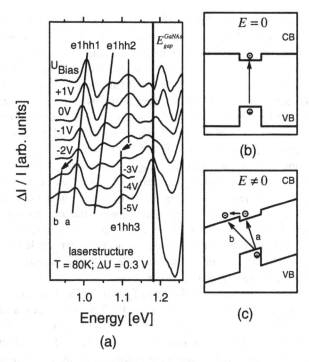

Figure 5.14a (a) Electromodulated absorption spectra of a laser structure with $Ga_{0.7}In_{0.3}N_{0.005}As_{0.995}/GaN_{0.024}As_{0.976}$ QWs for different bias voltages ($T = 80$ K; QW signals indicated by solid lines). (b) Band alignment for zero bias. (c) Band alignment for an applied bias.

of the e1hh1 indicate the vicinity of the Type I to Type II transition for $x =$ 2.7%. Estimating the magnitude of the redshift of the e1hh1 transition when changing the band-gap of the barrier going from GaAs to $GaN_{0.027}As_{0.973}$ suggests again that the confinement situation mainly changes in the conduction band.

To further investigate the Type I to Type II transition in such quaternary QWs, we performed electron absorption (EA) experiments on a laser structure with $Ga_{0.7}In_{0.3}N_{0.005}As_{0.995}/GaN_{0.024}As_{0.976}$ QWs in the active region. In the EA experiment, a dc bias voltage can be applied in addition to the modulation voltage. This results in a dc electric field across the active region of the laser structure, which allows one to tilt the band structure such that an indirect transition between the electrons in the Ga(N,As) and the heavy holes located in the (Ga,In)(N,As) occurs (as shown schematically in Figures 5.14b and 5.14c). This is commonly referred to as an electric-field-induced Type I to Type II transition. The spectra in Figure 5.14a were recorded at different dc bias voltages varying from +1 V (forward bias) to −5 V (reverse bias) by using a modulation voltage of ±0.3 V. The behavior of the QW signals with increasing reverse bias is very interesting. Up to −2 V, the three QW transi-

tions show the typical quantum-confined Stark-shift to lower energies indicative for a Type I band alignment. At a reverse bias of -3 V, the e1hh3
transition reveals an abrupt shift to lower energies. In the EA spectra
obtained at reverse bias of -4 V and -5 V, an additional sideband suddenly
appears on the low-energy side of the e1hh1 signal. Such changes in
modulated reflection or absorption spectra of QWs are typical signatures
of a Type I to Type II transition [86]. In this case, the sideband appearing
on the low-energy side of the e1hh1 transition is the indirect transition of
the first heavy-hole state confined in the $Ga_{0.70}In_{0.30}N_{0.005}As_{0.995}$ to the conduction band edge of the $GaN_{0.024}As_{0.976}$ barrier. A rough estimate of the
height of the conduction-band potential can be obtained by assuming that
the entire bias voltage drops across the active region of the device (about
450 nm). The corresponding energy shift between the left and right barrier
(see Figure 5.14c) of an 8-nm QW is about 70 meV for a reverse bias of
-4 V. Assuming that the indirect transition occurs when the left top of the
barrier is level with the right bottom of the well, this directly yields the
height of the confining potential at zero bias.

The results suggest that the Type I to Type II transition in $Ga_{0.7}In_{0.3}N_{0.005}$
$As_{0.995}/GaN_xAs_{1-x}$ QWs takes place for $x \approx 3\%$. This is further corroborated
by Figure 5.15, where transition energies (symbols) in $Ga_{0.7}$ $In_{0.3}N_{0.005}$
$As_{0.995}/GaN_xAs_{1-x}$ QWs vs. N content x are plotted. The solid lines are
calculated assuming constant confining potentials for the heavy holes, with
heights corresponding to 20%, 30%, and 40% of the gap difference between
GaAs and strained $Ga_{0.7}In_{0.3}N_{0.005}As_{0.995}$. The Ga(N,As) parameters used are
the same as for Figure 5.12. For (Ga,In)(N,As), we assumed an electron
effective mass of 0.07 m_0, a heavy-hole effective mass of 0.35 m_0, and a
bulk band-gap of 0.93 eV.

Both $Ga_{0.77}In_{0.23}As/GaN_xAs_{1-x}$ and $Ga_{0.7}In_{0.3}N_{0.005}As_{0.995}/GaN_xAs_{1-x}$
QWs show a Type I to Type II transition of the band alignment with
increasing x at about 1% and 3%, respectively. More sophisticated calculations based on a ten-band $\mathbf{k \cdot P}$ model (described above), where N-related
changes of the electron effective mass, strain, and conduction-band structure
are properly accounted for, should yield valuable information about the
VBOs in these structures.

5.4 SUMMARY AND CONCLUSIONS

This chapter gave an overview of the electronic structure of (Ga,In)(N,As)-
based QW structures. The first part of the chapter dealt with aspects of
metastability. Heterostructures containing quaternary (Ga,In)(N,As), in
contrast to structures based on ternary Ga(N,As), are metastable. Therefore,
the structure of the electronic states depends on the sample history. Nevertheless, it appears that by thermal annealing, a stable electronic configuration might be achieved that seems reproducible and, within limits, inde-

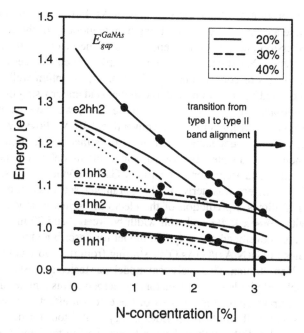

Figure 5.15 Plot of the transition energies (symbols) in $Ga_{0.7}In_{0.3}N_{0.005}As_{0.995}/GaN_xAs_{1-x}$ QWs vs. N-content x. The solid lines are calculated assuming constant confining potentials for the heavy holes.

pendent of the growth method, i.e., MBE or MOVPE. However, further studies are required to confirm this. Of course, the existence of such a stable electronic configuration is the fundamental prerequisite for any theoretical description or parameterization of the electronic band structure of these alloys. This prerequisite can be considered fulfilled for heterostructures based on ternary Ga(N,As), such as Ga(N,As)/GaAs QWs and (Ga,In) As/Ga(N,As) QWs. Therefore, the second part of this chapter focused mainly on studies of the electronic states in GaN_xAs_{1-x}/GaAs heterostructures and suitable models for describing them.

Because the band-formation process and the band structure are, in principle, the same for the ternary and the quaternary alloys, the developed models are also applicable to quaternary (Ga,In)(N,As)-based QWs. Although the microscopic mechanisms behind the unusual band formation of bulk Ga(N,As) and (Ga,In)(N,As) are not entirely understood to date and are still controversially discussed, it is commonly accepted that a parameterization of their band structures on the basis of the band anticrossing model (or level-repulsion model) introduced by Shan et al. [9] is adequate. In its simplicity, the band anticrossing model already accounts for a large number of experimental observations in Ga(N,As) and (Ga,In)(N,As) arising due to the modification of the conduction-band structure by incorporating N into GaAs or (Ga,In)As. In the context of quantum wells, these are, in particular, the strong nonparabolicity of the conduction-band dispersion and the strong dependence of the electron effective mass on N-content. This suggests that existing $\mathbf{k \cdot P}$ models for describing the electronic states of conventional III-V-containing quantum-well structures can be modified by combining them with the band anticrossing model. This is the approach we followed. Along these lines, we have extended the existing eight-band $\mathbf{k \cdot P}$ model for conventional III-V alloys by including two additional spin-degenerate N-related states that couple directly to the Γ_6 conduction-band states. By comparison of the model with experimental data for interband transitions of various Ga(N,As)/GaAs QWs, a set of material parameters was derived. This parameter set, in conjunction with the model, allows a prediction of the electronic states of any GaN_xAs_{1-x}/ GaAs with $1\% < x < 4\%$ and well width between 2 and 20 nm.

We also developed an analytical model for describing the conduction-band states of Ga(N,As)/GaAs QWs starting from the two-band band anti-crossing Hamiltonian. This analytical model was tested against the ten-band $\mathbf{k \cdot P}$ model and yields very satisfactory agreement. This simple model also accounts implicitly for the changes of the electron effective mass with N content x and well width. It provides a very useful tool for describing N-containing III-V QWs without having to use the full ten-band $\mathbf{k \cdot P}$ formalism. Because of the problems arising due to the metastability of the quaternary (Ga,In)(N,As) alloy, it is not yet possible to obtain a parameter set of quality comparable with that for Ga(N,As), which can be employed for modeling electronic states in $Ga_{1-y}In_yN_xAs_{1-x}$/GaAs QWs using the ten-band $\mathbf{k \cdot P}$ model or the corresponding analytical model. In order to determine such parameter sets for describing N-containing III-V QW structures, it is essential to obtain complementary information about the band alignment and the electron effective masses, as the latter depend strongly on the N content and the confinement situation. Additional information about the band alignment in these heterostructures might be obtained by studying the Type I to Type II transition in (Ga,In)As/GaN_xAs_{1-x} or (Ga,In)(N,As)/ GaN_xAs_{1-x} QWs with increasing x.

In conclusion, our description of the electronic states in (Ga,In)(N,As)-based QWs is still rather incomplete. Further experimental work and comparison between experiment and theory are required. However, a suitable theoretical apparatus for describing the unusual band structure of these heterostructures is being developed and successfully tested on Ga(N,As)/GaAs QWs. In this case, the agreement between theory and experiment is excellent and already allows the prediction of the properties of QWs over a wide range of sample parameters. This leads one to hope that this approach could be applicable to heterostructures of this whole class of nonamalgamation-type alloys.

Acknowledgments

The authors would like to thank their colleagues and collaborators in Germany, England, Ireland, and Sweden who contributed to this work. We are also grateful for funding by the DFG (Germany) and EPSRC (U.K.).

References

1. Makimoto, T., Saito, H., Nishida, T., and Kobayashi, N., "Excitonic Luminescence and Absorption in Dilute $GaAs_{1-x}N_x$ alloy ($x < 0.3\%$)," *Appl. Phys. Lett.* 70: 2984–2986 (1997).

2. Weyers, M., Sato, M., and Ando, H., "Red Shift of Photoluminescence and Absorption in Dilute GaAsN Alloy Layers," *Jpn. J. Appl. Phys.* 31: L853–L855 (1992).

3. Kondow, M., Uomi, K., Hosomi, K., and Mozume, T., "Gas-Source Molecular Beam Epitaxy of GaN_xAs_{1-x} Using a N Radical as the N Source," *Jpn. J. Appl. Phys.* 33: L1056–L1058 (1994).

4. Pozina, G., Ivanov, I., Monemar, B., Thordson, J.V., and Andersson, T.G., "Properties of Molecular-Beam Epitaxy-Grown GaNAs from Optical Spectroscopy," *J. Appl. Phys.* 84: 3830–3835 (1998).

5. Bi, W.G., and Tu, C.W., "Bowing Parameter of the Band-Gap Energy of GaN_xAs_{1-x}," *Appl. Phys. Lett.* 70: 1608–1610 (1997).

6. Grüning, H., Chen, L., Hartmann, T., Klar, P.J., Heimbrodt, W., Höhnsdorf, F., Koch, J., and Stolz, W., "Optical Spectroscopic Studies of N-Related Bands in Ga(N,As)," *Phys. Stat. Solidi B* 215: 39–45 (1999).

7. Perkins, J.D., Mascarenhas, A., Zhang, Y., Geisz, J.F., Friedman, D.J., Olson, J.M., and Kurtz, S.R., "Nitrogen-Activated Transitions, Level Repulsion, and Band-gap Reduction in $GaAs_{1-x}N_x$ with $x < 0.03$," *Phys. Rev. Lett.* 82: 3312–3315 (1999).

8. Klar, P.J., Grüning, H., Heimbrodt, W., Koch, J., Höhnsdorf, F., Stolz, W., Vicente, P.M.A., and Camassel, J., "From N Isoelectronic Impurities to N-Induced Bands in the GaN_xAs_{1-x} Alloy," *Appl. Phys. Lett.* 76: 3439–3441 (2000).

9. Shan, W., Walukiewicz, W., Ager III, J.W., Haller, E.E., Geisz, J.F., Friedman, D.J., Olson, J.M., and Kurtz, S.R., "Band Anticrossing in GaInNAs Alloys," *Phys. Rev. Lett.* 82: 1221–1224 (1999).

10. Perkins, J.D., Mascarenhas, A., Geisz, J.F., and Friedman, D.J., "Conduction-Band-Resonant Nitrogen-Induced Levels in $GaAs_{1-x}N_x$ with $x < 0.03$," *Phys. Rev. B* 64: 121301(R)/01-04 (2001).

11. Kent, P.R.C., and Zunger, A., "Theory of Electronic Structure Evolution in GaAsN and GaPN," *Phys. Rev. B* 64: 115208/01-23 (2001).

12. Kent, P.R.C., and Zunger, A., "Evolution of III-V Nitride Alloy Electronic Structure: The Localized to Delocalized Transition," *Phys. Rev. Lett.* 86: 2613–2616 (2001).

13. Gonzalez Szwacki, N., and Boguslawski, P., "GaAs:N vs. GaAs:B Alloys: Symmetry-Induced Effects," *Phys. Rev. B* 64: 161201(R)/01-04 (2001).

14. Lindsay, A., and O'Reilly, E.P., "Influence of Nitrogen Resonant States on the Electronic Structure of GaN_xAs_{1-x}," *Solid State Commun.* 118: 313–315 (2001).

15. Jones, E.D., Modine, N.A., Allerman, A.A., Kurtz, S.R., Wright, A.F., Tozer, S.T., and Wei, X., "Bandstructure of $In_xGa_{1-x}As_{1-y}N_y$ Alloys and Effects of Pressure," *Phys. Rev. B* 60: 4430–4433 (1999).

16. Mattila, T., Wei, S.-H., and Zunger, A., "Localization and Anticrossing of Electron Levels in $GaAs_{1-x}N_x$," *Phys. Rev. B* 60: R11245–11248 (1999).

17. Lindsay, A., and O'Reilly, E.P., "Theory of Enhanced Band-gap Nonparabolicity in GaN_xAs_{1-x} and Related Alloys," *Solid State Commun.* 112: 443–447 (1999).

18. Bellaiche, L., Wei, S.-H., and Zunger, A., "Band Gaps of GaPN and GaAsN Alloys," *Appl. Phys. Lett.* 70: 3558–3560 (1997).

19. Bellaiche, L., Wei, S.-H., and Zunger, A., "Localization and Percolation in Semiconductor Alloys: GaAsN vs. GaAsP," *Phys. Rev. B* 54: 17568–17576 (1996).

20. Wei, S.-H., and Zunger, A., "Giant Composition-Dependent Optical Bowing Coefficient in GaAsN Alloys," *Phys. Rev. Lett.* 76: 664–667 (1996).

21. Rubio, A., and Cohen, M.L., "Quasiparticle Excitations in $GaAs_{1-x}N_x$ and $AlAs_{1-x}N_x$ Ordered Alloys," *Phys. Rev. B* 51: 4343–4346 (1995).

22. Neugebauer, J., and Van de Walle, C.G., "Electronic Structure and Phase Stability of $GaAs_{1-x}N_x$ Alloys," *Phys. Rev. B* 51: 10568–10571 (1995).

23. Kim, K., and Zunger, A., "Spatial Correlations in GaInAsN Alloys and Their Effects on Band-Gap Enhancement and Electron Localization," *Phys. Rev. Lett.* 86: 2609–2612 (2001).

24. Kent, P.R.C., Bellaiche, L., and Zunger, A., "Pseudopotential Theory of Dilute III-V Nitrides," *Semicond. Sci. Technol.* 17: 851–859 (2002).

25. O'Reilly, E.P., Lindsay, A., Tomic, S., and Kamal-Saadi, M., "Tight-Binding and k.P Models for the Electronic Structure of Ga(In)NAs and Related Alloys," *Semicond. Sci. Technol.* 17: 870–879 (2002).

26. Bellaiche, L., and Zunger, A., "Effects of Atomic Short-Range Order on the Electronic and Optical Properties of GaAsN, GaInN and GaInAs Alloys," *Phys. Rev. B* 57: 4425–4431 (1998).

27. Pan, Z., Li, L.H., Zhang, W., Lin, Y.W., Wu, R.H., and Ge, W., "Effect of Rapid Thermal Annealing on GaInNAs/GaAs Quantum Wells Grown by Plasma-Assisted Molecular-Beam Epitaxy," *Appl. Phys. Lett.* 77: 1280–1282 (2000).

28. Francoeur, S., Sivaraman, G., Qiu, Y., Nikishin, S., and Temkin, H., "Luminescence of As-Grown and Thermally Annealed GaAsN/GaAs," *Appl. Phys. Lett.* 72: 1857–1859 (1998).

29. Buyanova, I.A., Pozina, G., Hai, P.N., Thing, N.Q., Bergman, J.P., Chen, W.M., Xin, H.P., and Tu, C.W., "Mechanism for Rapid Thermal Annealing Improvements in Undoped GaN_xAs_{1-x}/GaAs Structures Grown by Molecular Beam Epitaxy," *Appl. Phys. Lett.* 77: 2325–2327 (2000).

30. Kitatani, T., Nakahara, K., Kondow, M., Uomi, K., and Tanaka, T., "Mechanism Analysis of Improved GaInNAs Optical Properties through Thermal Annealing," *J. Cryst. Growth* 209: 345–349 (2000).

31. Gilet, P., Chenevas-Paule, A., Duvaut, P., Grenouillet, L., Hollinger, P., Million, A., Rolland, G., and Vannuffel, C., "Growth and Characterization of Thick GaAsN Epilayers and GaInNAs/GaAs Multiquantum Wells," *Phys. Stat. Solidi A* 176: 279–283 (1999).

32. Xin, H.P., Tu, C.W., and Geva, M., "Observation of Quantum Dot-like Behavior of GaInNAs in GaInNAs/GaAs Quantum Wells," *Appl. Phys. Lett.* 74: 2337–2339 (1999).

33. Kageyama, T., Miyamoto, T., Makino, S., Koyama, F., and Iga, K., "Thermal Annealing of GaInNAs/GaAs Quantum Wells Grown by Chemical Beam Epitaxy and Its Effect on Photoluminescence," *Jpn J. Appl. Phys.* 38: L298–L300 (1999).

34. Klar, P.J., Grüning, H., Koch, J., Schäfer, S., Volz, K., Stolz, W., Heimbrodt, W., Kamal Saadi, A.M., Lindsay, A., and O'Reilly, E.P., "(Ga,In)(N,As): Fine Structure of the Band-gap due to Nearest-Neighbor Configurations of the Isovalent Nitrogen," *Phys. Rev. B* 64: 121203(R)/01-04 (2001).

35. Albrecht, M., Grillo, V., Remmele, T., Strunk, H.P., Egorov, A.Y., Dumitras, G., Riechert, H., Kaschner, A., Heitz, R., and Hoffmann, A., "Effect of Annealing on the In and N Distribution in InGaAsN Quantum Wells," *Appl. Phys. Lett.* 81: 2719–2721 (2002).

36. Wagner, J., Geppert, T., Köhler, K., Ganser, P., and Maier, M., "Bonding of Nitrogen in Dilute GaInAsN and AlGaAsN Studied by Raman Spectroscopy," *Solid-State Electronics* 47: 461–466 (2003).

37. Wagner, J., Geppert, T., Köhler, K., Ganser, P., and Herres, N., "N-Induced Vibrational Modes in GaAsN and GaInAsN Studied by Resonant Raman Scattering," *J. Appl. Phys.* 90: 5027–5031 (2001).

38. Kitatani, T., Kondow, M., and Kudo, M., "Transition of Infrared Absorption Peaks in Thermally Annealed GaInNAs," *Jpn J. Appl. Phys. (Part 2)* 40: L750–L752 (2001).

39. Kurtz, S., Webb, J., Gedvilas, L., Friedman, D., Geisz, J., Olson, J., King, R., Joslin, D., and Karam, N., "Structural Changes during Annealing of GaInAsN," *Appl. Phys. Lett.* 78: 748–750 (2001).

40. Lordi, V., Gambin, V., Friedrich, S., Funk, T., Takizawa, T., Uno, K., and Harris, J.S., "Nearest-Neighbor Configuration in (GaIn)(NAs) Probed by X-ray Absorption Spectroscopy," *Phys. Rev. Lett.* 90: 145505/01-04 (2003).

41. Alt, H.C., Egorov, A.Y., Riechert, H., Wiedemann, B., Meyer, J.D., Michelmann, R.W., and Bethge, K., "Local Vibrational Mode Absorption of Nitrogen in GaAsN and InGaAsN Layers Grown by Molecular Beam Epitaxy," *Physica B* 302–303: 282–290 (2001).

42. Samuelson, L., Nilsson, S., Wang, Z.-G., and Grimmeiss, H.M., "Direct Evidence for Random-Alloy Splitting of Cu Levels in GaAs$_{1-x}$P$_x$," *Phys. Rev. Lett.* 53: 1501–1504 (1984).

43. Mariette, H., Chevallier, J., and Leroux-Hugon, P., "Local-Environment Effect on the Nitrogen Bound State in GaP$_x$As$_{1-x}$ Alloys: Experiments and Coherent-Potential Approximation Theory," *Phys. Rev. B* 21: 5706–5716 (1980).

44. Matsuoka, T., Sasaki, T., and Katsui, A., "Growth and Properties of Wide-Gap Semiconductor InGaN," *Optoelectron., Devices and Technol.* 5: 53 (1990).

45. Amore Bonapasta, A., Filippone, F., Giannozzi, P., Capizzi, M., and Polimeni, A., "Structure and Passivation Effects of Mono- and Dihydrogen Complexes in GaAs$_{1-x}$N$_x$ Alloys," *Phys. Rev. Lett.* 89: 216401/01-04 (2002).

46. Polimeni, A., Baldassarri Höger von Högersthal, G., Bissiri, M., Capizzi, M., Fischer, M., Reinhardt, M., and Forchel, A., "Effect of Hydrogen on the Electronic Properties of $In_xGa_{1-x}As_{1-y}N_y$/GaAs Quantum Wells," *Phys. Rev. B* 63: 201304(R)/01-04 (2001).

47. Baldassarri Höger von Högersthal, G., Bissiri, M., Polimeni, A., Capizzi, M., Fischer, M., Reinhardt, M., and Forchel, A., "Hydrogen-Induced Band-gap Tuning of (InGa)(AsN)/GaAs Single Quantum Wells," *Appl. Phys. Lett.* 78: 3472–3474 (2001).

48. Capizzi, M., Polimeni, A., Baldassarri Höger von Högersthal, G., Bissiri, M., Amore Bonapasta, A., Jiang, F., Stavola, M., Fischer, M., Forchel, A., Sou, I.K., and Ge, W.K., "Photoluminescence and Infrared Absorption Study of Isoelectronic Impurity Passivation by Hydrogen," *Mat. Res. Soc. Symp. Proc.* 719: 251–256 (2003).

49. Klar, P.J., Grüning, H., Güngerich, M., Heimbrodt, W., Koch, J., Torunski, T., Stolz, W., Polimeni, A., and Capizzi, M., "Global Changes of the Band Structure and the Crystal Lattice of GaAsN due to Hydrogenation," *Phys. Rev. B* 67: 121206(R)/01-04 (2003).

50. O'Reilly, E.P., and Lindsay, A., "k·P Model of Ordered GaN_xAs_{1-x}," *Phys. Stat. Solidi B* 216: 131–134 (1999).

51. Skierbiszewski, C., Perlin, P., Wisniewski, P., Knap, W., Suski, T., Walukiewicz, W., Shan, W., Yu, K.M., Ager, J.W., Haller, E.E., Geisz, J.F., and Olson, J.M., "Large Nitrogen-Induced Increase of the Electron Effective Mass in $In_yGa_{1-y}N_xAs_{1-x}$," *Appl. Phys. Lett.* 76: 2409–2411 (2000).

52. Buyanova, I.A., Pozina, G., Hai, P.N., Chen, W.M., Xin, H.P., and Tu C.W., "Type I Band Alignment in the GaN_xAs_{1-x}/GaAs Quantum Wells," *Phys. Rev. B* 63: 033303/01-04 (2000).

53. Hai, P.N., Chen, W.M., Buyanova, I.A., Xin, H.P., and Tu, C.W., "Direct Determination of Electron Effective Mass in GaNAs/GaAs Quantum Wells," *Appl. Phys. Lett.* 77: 1843–1845 (2000).

54. Wu, J., Shan, W., Walukiewicz, W., Yu, K.M., Ager III, J.W., Haller, E.E., Xin, H.P., and Tu, C.W., "Effect of Band Anticrossing on the Optical Transitions in $GaAs_{1-x}N_x$/GaAs Multiple Quantum Wells," *Phys. Rev. B* 64: 085320/01-04 (2001).

55. Polimeni, A., Bissiri, M., Baldassarri Höger von Högersthal, G., Capizzi, M., Giubertoni, D., Barozzi, M., Bersani, M., Gollub, D., Fischer, M., and Forchel, A., "Hydrogen as a Probe of the Electronic Properties of (InGa)(AsN)/GaAs Heterostructures," *Solid State Electronics* 47: 447–454 (2003).

56. Masia, F., Polimeni, A., Baldassarri Höger von Högersthal, G., Bissiri, M., Capizzi, M., Klar, P.J., and Stolz, W., "Early Manifestation of Localization Effects in Ga(AsN)," *Appl. Phys. Lett.* 82: 4474–4476 (2003).

57. Young, D.L., Geisz, J.F., and Coutts, T.J., "Nitrogen-Induced Decrease of the Electron Effective Mass in $GaAs_{1-x}N_x$ Thin Films Measured by Thermomagnetic Transport Phenomena," *Appl. Phys. Lett.* 82: 1236–1238 (2003).

58. Wang, Y.J., Wei, X., Zhang, Y., Mascarenhas, A., Xin, H.P., Hong, Y.G., and Tu, C.W., "Evolution of the Electron Localization in a Nonconventional Alloy System $GaAs_{1-x}N_x$ Probed by High-Magnetic-Field Photoluminescence," *Appl. Phys. Lett.* 82: 4453–4455 (2003).

59. Choulis, S.A., Hosea, T.J.C., Tomic, S., Kamal-Saadi, M., Adams, A.R., O'Reilly, E.P., Weinstein, B.A., and Klar, P.J., "Detailed Electronic Structure of In$_y$Ga$_{1-y}$As$_{1-x}$N$_x$/GaAs Multiple Quantum Wells in the Dilute-N Regime from Pressure and k.P Studies," *Phys. Rev. B* 66: 165321/01-09 (2002).

59a. Tomic S., O'Reilly E.P., Fehse R., Sweeney S.J., Adams A.R., Andreev A.D., Choulis S.A., Hosea T.J.C., and Riechert H., "Theoretical and Experimental Analysis of 1.3 mm InGaAsN/GaAs Lasers", *IEEE Journal of Selected Topics in Quantum Electronics* 9: 1228–1238 (2003).

60. Tomic, S., O'Reilly, E.P., Klar, P.J., Grüning, H., Heimbrodt, W., Chen, W.M., and Buyanova, I.A., "The Influence of Conduction-Band Nonparabolicity on Electron Confinement and Effective Mass in Ga(N,As)/GaAs Quantum Wells," *Phys. Rev. B* 69:24 (2003).

61. Klar, P.J., Grüning, H., Heimbrodt, W., Koch, J., Stolz, W., Vicente, P.M.A., Kamal Saadi, A.M., Lindsay, A., and O'Reilly, E.P., "Pressure and Temperature Dependent Studies of GaN$_x$As$_{1-x}$/GaAs Quantum Well Structures," *Phys. Stat. Solidi B* 223: 163–169 (2001).

62. Grüning, H., Klar, P.J., Heimbrodt, W., Koch, J., Stolz, W., Lindsay, A., Tomi, S., and O'Reilly, E.P., "N-Composition and Pressure Dependence of the Inter Band Transitions of Ga(N,As)/GaAs Quantum Wells," *High Pressure Res.* 22: 293 (2002).

63. Klar, P.J., Grüning, H., Heimbrodt, W., Koch, J., Stolz, W., Tomi, S., and O'Reilly, E.P., "Monitoring the Strong Non-Parabolicity of the Conduction Band in (Ga,In)(N,As)/GaAs Quantum Wells," *Solid State Electronics* 47: 437–442 (2003).

64. Kitatani, T., Kondow, M., Kikawa, T., Yazawa, Y., Okai, M., and Uomi, K., "Analysis of Band Offset in GaNAs/GaAs by X-ray Photoelectron Spectroscopy," *Jpn. J. Appl. Phys. (Part 1)* 38: 5003–5006 (1999).

65. Sun, B.Q., Jiang, D.S., Luo, X.D., Zu, Z.Y., Pan, Z., Li, L.H., and Wu, R.H., "Intersubband Luminescence and Absorption of GaNAs/GaAs Single-Quantum-Well Structures," *Appl. Phys. Lett.* 76: 2862–2864 (2000).

66. Krispin, P., Spruytte, S.G., Harris, J.S., and Ploog, K.H., "Admittance Dispersion of n-Type GaAs/Ga(As,N)/GaAs Heterostructures Grown by Molecular Beam Epitaxy," *J. Appl. Phys.* 90: 2405–2410 (2001).

67. Krispin, P., Spruytte, S.G., Harris, J.S., and Ploog, K.H., "Electrical Depth Profile of p-Type GaAs/Ga(As,N)/GaAs Heterostructures Determined by Capacitance-Voltage Measurements," *J. Appl. Phys.* 88: 4153–4158 (2000).

68. Klar, P.J., Grüning, H., Heimbrodt, W., Weiser, G., Koch, J., Volz, K., Stolz, W., Koch, S.W., Tomic, S., Choulis, S.A., Hosea, T.J.C., O'Reilly, E.P., Hofmann, M., Hader, J., and Moloney, J.V., "Interband Transitions of Quantum Wells and Device Structures Containing Ga(N,As) and (Ga,In)(N,As)," *Semicond. Sci. Technol.* 17: 830–842 (2002).

69. Kane, E.O., "The k·P Method," *Semicond. Semimetal.* 1: 75 (1966).

70. Meney, A.T., Gonul, B., and O'Reilly, E.P., "Evaluation of Different Approximations Used in Envelope-Function Method," *Phys. Rev. B* 50: 10893–10904 (1994).

71. Sakai, S., Ueta, Y., and Terauchi, Y., "Band-gap Energy and Band Lineup of III-V Alloy Semiconductors Incorporating Nitrogen and Boron," *Jpn. J. Appl. Phys. (Part 1)* 32: 4413–4417 (1993).

72. Bellaiche, L., Wei, S.-H., and Zunger, A., "Composition Dependence of Interband Transition Intensities in GaPN, GaAsN, and GaPAs Alloys," *Phys. Rev. B* 56: 10233–10240 (1997).

73. Ekenberg, U., "Enhancement of Nonparabolicity Effects in a Quantum Well," *Phys. Rev. B*, 36: 6152–6155 (1987).

74. Kangarlu, A., Chandrasekhar, H.R., Chandrasekhar, M., Kapoor, Y.M., Chambers, F.A., Vojak, B.A., and Meese, J.M., "High Pressure Studies of GaAs-Al$_x$Ga$_{1-x}$As Quantum Wells at 300 and 80 K Using Photoreflectance Spectroscopy," *Phys. Rev. B* 38: 9790–9796 (1988).

75. Kim, K., Lambrecht, W.R.L., and Segall, B., "Elastic Constants and Related Properties of Tetrahedrally Bonded BN, AlN, GaN, and InN," *Phys. Rev. B* 53: 16310–16236 (1996).

76. Kim, K., Lambrecht, W.R.L., and Segall, B., "Erratum: Elastic Constants and Related Properties of Tetrahedrally Bonded BN, AlN, GaN, and InN," *Phys. Rev. B* 56: 7018–7021 (1997).

77. Vurgaftman, I., Meyer, J.R., and Ram-Mohan, L.R., "Band Parameters for III-V Compound Semiconductors and Their Alloys," *J. Appl. Phys.* 89: 5815–5875 (2001).

78. Pan, Z., Li, L.H., Lin, Y.W., Sun, B.Q., Jiang, D.S., and Ge, W.K., "Conduction Band Offset and Electron Effective Mass in GaInNAs/GaAs Quantum Well Structures with Low Nitrogen Concentration," *Appl. Phys. Lett.* 78: 2217–2219 (2001).

79. Dumitras, G., Riechert, H., Porteanu, H., and Koch, F., "Surface Photovoltage Studies of In$_x$Ga$_{1-x}$As and In$_{1-x}$Ga$_x$As$_{1-y}$N$_y$ Quantum Well Structures," *Phys. Rev. B* 66: 205324/01-08 (2002).

80. Sun, H.D., Hetterich, M., Dawson, M.D., Egorov, A.Y., Bernklau, D., and Riechert, H., "Optical Investigations of GaInNAs/GaAs Multi-Quantum Wells with Low Nitrogen Content," *J. Appl. Phys.* 92: 1380–1385 (2002).

81. Hetterich, M., Dawson, M.D., Egorov, A.Y., Bernklau, D., and Riechert, H., "Electronic States and Band Alignment in GaInNAs/GaAs Multi-Quantum Well Structures with Low Nitrogen Content," *Appl. Phys. Lett.* 76: 1030–1032 (2000).

82. Sun, H.D., Dawson, M.D., Othman, M., Yong, J.C.L., Rorison, J.M., Gilet, P., Grenouillet, L., and Million, A., "Optical Transitions in GaInNAs/GaAs Multi-Quantum Wells with Varying N Content Investigated by Photoluminescence Excitation Spectroscopy," *Appl. Phys. Lett.* 82: 376–378 (2003).

83. Kavokin, A.V., and Nesvizhskii, A.I., "Stark Effect Near the Type I–Type II Transition Point in Semiconductor Quantum Wells," *Phys. Rev. B* 49: 17055–17058 (1994).

84. Ohtani, N., Domoto, C., Egami, N., Mimura, H., Ando, M., Nakayama, M., and Hosoda, M., "Electric-Field-Induced Combination of Wannier-Stark Localization and Type I–Type II Crossover in a Marginal-Type I GaAs/AlAs Superlattice," *Phys. Rev. B* 61: 7505–7510, (2000).

85. Nakayama, M., Nakanishi, T., Piao, Z.S., Nishimura, H., Takahashi, M., and Egami, N., "Oscillator Strength of Type-II Light-Hole Exciton in Strained In$_{1-x}$Ga$_x$As/GaAs Single Quantum Wells," *Physica E* 7: 567–571 (2000).

86. Klar, P.J., Watling, J.R., Wolverson, D., Davies, J.J., Ashenford, D.E., and Lunn, B., "Magnetic-Field Induced Type I to Type II Transition in $Zn_{1-x}Mn_x$te/ZnTe Multiple Quantum Well Samples," *Semicond. Sci. Technol.* 12: 1240–1251 (1997).

87. Chen, P., Nicholls, J.E., Hogg, J.H.C., Stirner, T., Hagston, W., Lunn, B., and Ashenford, D.E., "Determination of the Band Offset of CdTe/$Cd_{1-x}Mn_x$Te Multiple Quantum Cells of Very Low x Values," *Phys. Rev. B* 50: 4732–4735 (1995).

CHAPTER 6

Role of Hydrogen in Dilute Nitrides

A. POLIMENI and M. CAPIZZI

INFM and Dipartimento di Fisica, Università di Roma, "La Sapienza," Italy

1-591-69019-6/04/$0.00+$1.50
© 2004 by CRC Press LLC

6.1 INTRODUCTION

Hydrogen is the smallest and one of the most chemically active atoms. It dramatically affects the electronic properties of crystalline semiconductors, as first discovered in ZnO [1]. Hydrogen easily diffuses in semiconductors [2] and neutralizes dangling bonds, thus cleaning up the band-gap from defect energy levels. Because hydrogen is present in the plasmas, etchants, precursors, and transport gases of most growth processes and device mass-production steps, a great interest has been focused on the effects of hydro-gen in the semiconductors that are used to produce modern devices [3]. Hydrogen-related centers were recognized in ultrapure germanium [4], and neutralization of boron acceptors has been discovered in silicon [5, 6], which suggests the existence of H-related donor levels in the band-gap [7, 8]. It was later shown that hydrogen can passivate donors as well, due to its amphoteric impurity behavior in both elemental and compound semi-conductors [9, 10]. This impurity passivation was explained in terms of the formation of complexes made of one H atom and one impurity [11, 12], and hydrogen-impurity bonds were detected, indeed, by infrared absorption spectroscopy [13–16].

Among the number of theoretical and experimental studies of defect passivation by H in semiconductors [11, 12, 17, 18], only a few treat the interaction between hydrogen and deep centers [19, 20]. Indeed, deep centers are characterized by a short-range potential, which is untreatable within the effective-mass approximation and leads to three major differ-ences with respect to the case of shallow impurities:

1. The electronic wavefunction corresponding to the defect is highly localized.
2. The properties of the center are sensitive to the effects of local-charge rearrangements and lattice relaxations.
3. The electronic level of the defect may be located anywhere in the band-gap [21].

The latter point is critical, since the passivation of a deep defect depends strictly on the relative positions of the H and defect levels. Finally, one needs a detailed description of the deep-defect short-range potential, which makes it difficult to identify them or describe them microscopically.

Among all deep, localized centers, isoelectronic impurities are a class by themselves [22]. It is well known that, in most cases, the substitution of an atom with an isovalent one can be treated in the virtual-crystal approxi-mation, where the electronic and transport properties of an $A_x B_{1-x} C$ ternary alloy are given by a linear interpolation of the properties of the AB and AC binary compounds. This is the case, e.g., of $Al_x Ga_{1-x} As$ and $GaAs_{1-x} P_x$. How-ever, when there is a great difference among the electronegativity and size of the interchanged atoms, quite dramatic changes in the properties of the host material are observed, as for N in $GaAs_{1-y}N_y$ and $GaP_{1-y}N_y$. In these cases, for small percentages of substituting N, the energy gap of the host strongly decreases. This is contrary to what is expected for the substitution of a heavy anion with a lighter one. At the same time, the electron effective mass increases and the lattice parameter decreases. Moreover, states due to single N atoms appear in the energy gap of the host material, as in GaP, or are resonant with the conduction band (CB) states, as in GaAs, while states due to clusters of N atoms appear in the energy gap [23].

While there is a general consensus that the aforementioned effects are related to the quite high carrier localization around the N atom produced by its high electronegativity, different theoretical models account for the effects N has on the host material. In a first theoretical approach, the strong perturbation of the translational symmetry of the host-material potential gives rise to perturbed host states, with an ensuing downward shift of the CB minimum (CBM) for increasing N concentration and a progressive disappearance of the energy levels of N clusters in the band-gap [24]. In a band anticrossing model, instead, a phenomenological repulsive interaction between the CBM and a single N level resonant with the CB continuum of states accounts for most N-insertion effects [25]. This model is supported by recent tight-binding calculations [26]. Finally, the effects of the interaction among nitrogen atoms and/or clusters and the ensuing impurity-band formation at high N concentration are highlighted in a third model [27].

In this chapter, we will show that hydrogen passivates most of the effects of N introduction in $GaAs_{1-y}N_y$ and $GaP_{1-y}N_y$. In particular, H irradiation of these materials leads to a full recovery of the band-gap energy, electron effective mass, temperature coefficient of the band-gap thermal shift, and lattice parameter of the N-free material.

The chapter is organized as follows. In Section 6.2, we describe the samples investigated and the experimental techniques employed for sample characterization. In Section 6.3.1, a description of the effects of N incorporation in GaAs for different ranges of N concentrations is given. In Section 6.3.2.1, the effects of hydrogen irradiation on the optical properties of N-containing samples with N concentration varying from $y < 0.01\%$ to $y \approx 1\%$ are presented. The process of H removal from the samples is analyzed in great detail in Section 6.3.2.2. A measurement of the electron effective mass and exciton wavefunction extent in $GaAs_{1-y}N_y$ is presented in Section 6.3.2.3 along with a description of the effects H has on these quantities. Section 6.3.2.4 regards the study of the effect H has on the response of the band-gap to temperature variation in $GaAs_{1-y}N_y$. The effects of H on the lattice properties of $GaAs_{1-x}N_x$ are described in 6.3.2.5. In Section 6.4, we describe the effects of H in $GaP_{1-y}N_y$. Finally, we address the microscopic mechanism leading to the passivation of N by H in $GaAs_{1-y}N_y$ and $GaP_{1-y}N_y$ and draw the conclusions of our work.

6.2   EXPERIMENTAL: HYDROGENATION AND
CHARACTERIZATION TECHNIQUES

The samples considered in this review were grown by different techniques. One set of samples was grown by solid-source molecular-beam epitaxy (MBE) and consists of $GaAs_{1-y}N_y/GaAs$ and $In_xGa_{1-x}As_{1-y}N_y/GaAs$ single quantum wells (QWs) having In concentration $x = 25$ to 42%, N concentration $y = 0.7$ to 5.2%, and QW thickness $L = 6.0$ to 8.2 nm. MBE-grown $GaAs_{1-y}N_y/GaAs$ epilayers were also considered (layer thickness $t = 310$ nm, with $y < 0.01$ and $y = 0.81$ and 1.3%). Another set of samples consists of four 0.5-μm-thick $GaAs_{1-y}N_y/GaAs$ epilayers ($y = 0.043$, 0.1, 0.21, and 0.5%) grown by metalorganic vapor-phase epitaxy [28, 29]. Finally, $GaP_{1-y}N_y$ epilayers were grown by gas-source MBE on GaP. In this case, nitrogen concentrations are $y = 0.05$, 0.12, 0.6, 0.81, and 1.3%. The $GaP_{1-y}N_y$ epilayer thickness is 250 nm for all samples, except for the $y = 1.3\%$ epilayer, which is 750-nm thick. In all cases, the composition and layer thickness of the samples have been derived by X-ray diffraction measurements.

Hydrogenation was obtained by ion-beam irradiation from a Kaufman source with the samples held at 300°C [30]. The ion energy was about 100 eV, and the current density was a few tens of μA/cm^2. Several hydrogen doses ($d_H = 10^{14}$ to 10^{20} ions/cm^2) were used. Posthydrogenation thermal annealing was performed at 1.0×10^{-6} torr at temperatures, T_a, ranging between 220°C and 550°C and for various durations, t_a, ranging between 1 and 50 h.

The electronic properties of the samples were investigated mainly by means of photoluminescence (PL) spectroscopy. For $In_xGa_{1-x}As_{1-y}N_y$ and $GaAs_{1-y}N_y$ samples, PL was excited by the 515-nm line of an Ar^+ laser or by an Nd-vanadate laser (excitation wavelength equal to 532 nm). For $GaP_{1-y}N_y$, the 458-nm line of an Ar^+ laser was used, instead. PL was dispersed then by a 1-m single-grating monochromator or a 0.75-m double monochromator and detected by a liquid-nitrogen-cooled Ge detector or (InGa)As linear array, or by a photomultiplier with a GaAs/Cs cathode.

Details on other experimental techniques employed will be given in the following sections.]

6.3   EFFECTS OF NITROGEN AND HYDROGEN ON
THE ELECTRONIC PROPERTIES OF
$In_xGa_{1-x}As_{1-y}N_y$

6.3.1   The N Isoelectronic Impurity in $GaAs_{1-y}N_y$

Figure 6.1 shows the low-temperature ($T = 10$ K) photoluminescence spectra of a representative series of as-grown samples whose N concentration varies over about two orders of magnitude. At the very early stage of N incorporation in GaAs (N concentration less than 0.01%, bottommost spec-

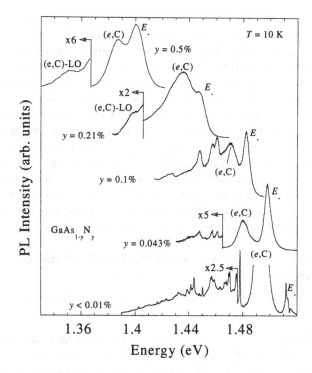

Figure 6.1 Low-temperature (10 K) PL spectra of GaAs$_{1-y}$N$_y$ epilayers with different levels of y. E_- and (e,C) indicate the free-exciton and free-electron to neutral-carbon acceptor recombinations, respectively. LO indicates the longitudinal optical phonon replica of the (e,C) transition.

trum in Figure 6.1), the low-temperature PL spectra are characterized by a number of sharp lines (line width $\approx$ 0.5 meV) between 1.40 and 1.48 eV, which are due to the recombination of excitons localized on N complexes [31–33]. These lines are attributed to carrier recombination from electronic levels due to N pairs and/or clusters [28, 34–38] and are superimposed on a broad band also related to N doping. The luminescence intensity associated with these transitions varies from line to line and increases with y (not shown here). An exact assignment of each line to a given N complex is made rather difficult by the strong dependence of the material optical properties on the growth conditions, as extensively reported in the literature [34–38].

Free-electron to neutral-carbon acceptor (e,C) and free-exciton (E_-) recombinations of GaAs are observed at 1.493 eV and 1.515 eV, respectively. As the nitrogen concentration is increased further ($y = 0.043$ and 0.1%), the energy of the excitonic recombination from the material's band gap, E_-, as well as the (e,C) recombination band start redshifting very rapidly, coexisting with and taking in the levels associated with the N complexes, whose energies do not change with N concentration [28, 36, 37]. This highlights the strongly localized character of the N isoelectronic traps, contrary to that of shallow impurities, whose wavefunctions overlap at smaller concentrations (10^{16}–10^{18} cm^{-3}). Eventually, at higher N concentrations (alloy limit, $y > 0.1\%$), the GaAs$_{1-y}$N$_y$ band-gap keeps redshifting along with the C-related states [39]. The dramatic variation of the GaAs host band-gap with y is accompanied by other major effects on the electronic and optical properties of GaAs$_{1-y}$N$_y$. Indeed, with increasing N concentration, the electron effective mass increases [27, 40–48], a sizable Stokes shift between emission and absorption is observed [29, 36, 49], the band-gap dependence on temperature [50, 51] and hydrostatic pressure [28, 52, 53] decreases, and N resonant states move to higher energy [25, 28, 54]. In the following discussion, we show that most of the above-mentioned changes can be fully and reversibly counteracted by irradiation with atomic hydrogen.

6.3.2   Effects of Hydrogen on the Electronic Properties of In$_x$Ga$_{1-x}$As$_{1-y}$N$_y$

6.3.2.1   Restoration of the host band-gap by hydrogenation: dilute, amalgamation, and alloy limits

We describe first the effects of hydrogen irradiation on the optical properties of GaAs$_{1-y}$N$_y$/GaAs epilayers in the very dilute nitrogen limit ($y < 0.01\%$).

Figure 6.2 shows the effect of hydrogen irradiation on the sample. Hydrogenation at various H doses, d_H, leads to a progressive and finally complete quenching of the N-related lines as well as of the broad underlying band [31–33]. The $d_H = 5 \times 10^{15}$ ions/cm^2 spectrum closely reproduces that of pure GaAs, where only two bands are observed, namely, the longitudinal optical (LO) phonon replicas of the C-related free-to-bound transition at 1.4934 eV. This H-induced passivation has never been reported before for any isoelectronic impurity, except for a weak reduction in the luminescence intensity of a few N-related lines in GaP:N [55]. Note that a 100% passivation of impurity luminescence bands is hardly attainable even in the common case of H passivation of shallow impurities in GaAs or Si [3].

We now move to the so-called amalgamation limit, corresponding to the existence of both localized and extended (or Bloch-like) states in the material electronic structure [56]. Figure 6.3 illustrates such a case for GaAs$_{1-y}$N$_y$ with $y = 0.1\%$. The bottom curve shows the PL spectrum of the H-free sample. H irradiation leads first to a passivation of the N cluster states and then to an apparent reopening of the GaAs$_{1-y}$N$_y$ band-gap toward that of the GaAs reference (top curve). As a matter of fact, both the (e,C) and the E_- recombination bands converge to those of the GaAs reference. The energy separation between these two transitions decreases with increasing N concentration, most likely due to the increase of the tensile strain with increasing x. Indeed, for increasing N concentration, the top of the

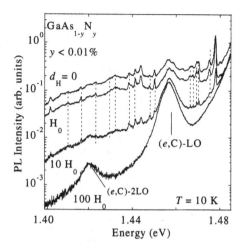

Figure 6.2 PL spectra ($T = 10$ K) of a GaAs$_{1-y}$N$_y$ epilayer in the very dilute N limit after irradiation at different H doses, d_H. LO indicates phonon replica transitions. (H$_0$ = 5 × 10^{13} ions/cm^2; laser power density $P = 10$ W/cm^2.) (Copyright (2002) by the American Physical Society.)

valence band acquires a more pronounced light-hole character and, in turn, the binding energy of the acceptor impurity decreases.

Similar results have been observed in the full alloy limit as shown in the following section.

6.3.2.2 Activation energies for the removal of hydrogen: evidences for multiple N-containing complexes

The effects of H described above are fully reversible by thermal annealing of the samples. This is nicely exemplified in Figure 6.4 for an In$_x$Ga$_{1-x}$As$_{1-y}$N$_y$ QW ($L = 7.0$ nm, $x = 34\%$, and $y = 0.7\%$). By increasing the H dose (see spectra from [a] to [c] in Figure 6.4), the energy of the QW free exciton blueshifts toward that of the corresponding N-free reference sample (see lower dashed line in the figure), whose peak energy does not vary much upon hydrogenation (see upper dashed line in the figure). The PL band line width broadens sizably in all In- and N-containing hydrogenated samples due to an increased microscopic disorder, which is much lower in the case of GaAs$_{1-y}$N$_y$ [32, 57, 58].

By annealing the sample hydrogenated with $d_H = 5 × 10^{17}$ ions/cm^2 for 1 h at increasing temperature (see spectra from [b′] to [d] in Figure 6.4), the QW exciton energy redshifts continuously until the PL line shape nearly overlaps the spectrum that the QW showed before hydrogenation. This demonstrates the full reversibility of the hydrogenation process and rules out any role played by ion bombardment or heating during hydrogenation in the band-gap variation of In$_x$Ga$_{1-x}$As$_{1-y}$N$_y$.

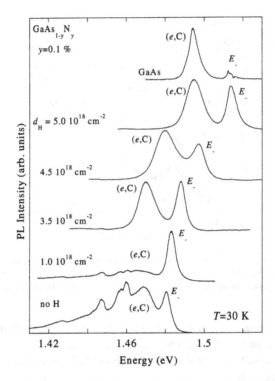

Figure 6.3 PL spectra (T = 30 K) of a GaAs$_{0.999}$N$_{0.001}$ alloy treated with different hydrogen doses, d_H. The bottom and top spectra refer to an untreated GaAs$_{1-y}$N$_y$ and a reference GaAs sample, respectively. Different laser power densities were used for the different samples to highlight the presence of both (e,C) and E_- bands.

We now focus on a quantitative analysis of the annealing experiments performed on several hydrogenated N-containing samples. This analysis allows us to estimate the dissociation energy E_D of the N–H bonds, which are most likely responsible for the dramatic changes induced by H in N-containing III-V compounds and alloys.

Figure 6.5 (a) shows the PL spectra at T = 10 K of a GaAs$_{1-y}$N$_y$ epilayer in the case of very dilute N limit (y < 0.01%) for different postgrowth treatments. The sharp recombination lines due to N complexes — and visible in the untreated sample (see top curve in the figure) — are quenched after hydrogenation (see bottom trace) [31–33]. By annealing the hydrogenated sample at T_a = 330°C, the H-N-complex bonds responsible for the N passivation begin to dissociate, and the sharp luminescence lines are progressively recovered for increasing annealing times with sizably different rates; compare, for example, lines i and h in the figure.

The probability per unit time for the dissociation process of N–H bonds can be written as $p = \nu \cdot \exp(-E_D/k_B T_a)$ [59], where k_B is the Boltzmann constant and the attempt frequency ν is set equal to 93 THz, namely, the

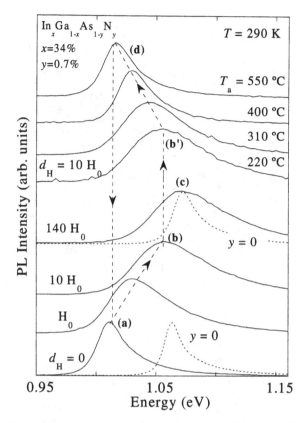

Figure 6.4 Graphs (a) to (c): Room-temperature peak-normalized PL spectra of hydrogenated $In_{0.34}Ga_{0.66}As_{1-y}N_y$/GaAs QWs with $y = 0.007$ (continuous lines) and $y = 0$ (dashed lines) for different hydrogen doses, d_H ($H_0 = 5 \times 10^{16}$ ions/cm^2). Graphs (b') to (d): Room-temperature peak-normalized PL spectra of the $In_{0.34}Ga_{0.66}As_{0.993}N_{0.007}$ QW hydrogenated at $d_H = 10\ H_0$ (5×10^{17} ions/cm^2) and annealed for 1 h at different temperatures. (Laser power density $P = 70$ W/cm^2 for all spectra.) (Copyright by Elsevier.)

value of the N–H local vibrational mode in $In_xGa_{1-x}As_{1-y}N_y$ [60]. In turn, the percent of surviving bonds is given by $n = \exp(-pt_a)$ if one assumes that N–H bonds are irreversibly broken [59]. Since the intensity of the PL sharp lines is proportional to the number of the corresponding N complexes optically active, $n = [1 - I(T_a,t_a)/I_{sat}]$ for each PL line, where $I(T_a,t_a)$ is the line intensity after an annealing time t_a at T_a and I_{sat} is its saturation value for t_a (or T_a) $\to \infty$. The intensity of most of the lines emitting in the energy range 1.46–1.48 eV reaches a well-defined saturation value, with E_D ranging between 2.00 and 2.25 eV. See the full and open circles in Figure 6.5b, where the dependence of I/I_{sat} ($=1 - n$) on t_a is shown for a few representative cases. The PL intensity of lines emitting below 1.46 eV does not saturate, even for $t_a > 23$ h at 330°C. It saturates, instead, for $T_a \geq 400°$C, leading to

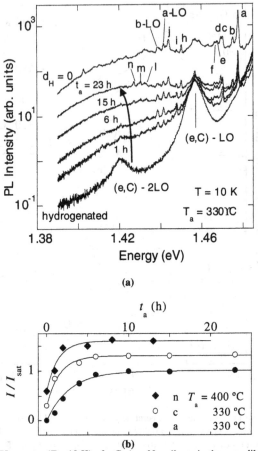

(a)

(b)

Figure 6.5 (a) PL spectra ($T = 10$ K) of a $GaAs_{1-y}N_y$ epilayer in the very dilute N limit ($y <$ 0.01%) for different postgrowth treatments. The top trace ($d_H = 0$) corresponds to an untreated sample, and the bottom trace corresponds to a hydrogenated sample ($d_H = 5 \times 10^{15}$ ions/cm²). Intermediate spectra have been recorded on the same hydrogenated sample annealed at $T_a =$ 330°C for different annealing times, t_a. Transitions corresponding to N complexes whose activation energies have been determined are labeled with letters. X-nLO are the phonon replicas of the X line. (Laser power density $P = 10$ W/cm².) (b) Dependence of the percent of broken N–H bonds, I/I_{sat}, on the annealing time t_a for three PL lines shown in Figure 6.5(a). The continuous lines are fits to the data meant to determine the activation energy, E_D, for the N–H bond dissociation ($E_D = 2.12$, 2.15, and 2.36 eV for lines c, a, and n, respectively). Data have been offset along the vertical axis for clarity purposes. (Copyright (2002) by the American Physical Society.)

E_D values varying between 2.28 and 2.40 eV, as shown in Figure 6.5b for line n. Such an increase in E_D for low-energy-emitting complexes indicates that these centers are most likely N clusters or chains with a large number of N atoms [56], where H can be caged more tightly than in small N

complexes. Hydrogen recapture by long N-atom chains may result in a
higher effective value of E_D, as well.

We now turn to the alloy limit ($y > 0.1\%$) and look for hints of N
complexes with features similar to those observed in the dilute limit. The
dependence of the PL peak energy, E_p, on annealing time is shown in Figure
6.6 (upper panel) for a 9.0-nm-thick $GaAs_{0.97}N_{0.03}/GaAs$ QW annealed at
330°C (full dots). We note that E_p saturates to a value about 70 meV higher
than that of the band-gap of the nonhydrogenated sample (horizontal dashed
line). This suggests the persistence of some N–H bonds that are broken by
further annealing for 1 h at $T_a \geq 350$°C (open symbols in Figure 6.6) and
points to the existence of *different types* of H-N complexes in the N alloy
limit with activation-energy values similar to those found in the dilute limit
for N complexes emitting below 1.46 eV.

The activation energy for the N–H bond dissociation is now obtained
quantitatively from isochronal thermal annealing measurements. We
assume that the *effective* N concentration, $y_{eff}(T_a, t_a)$, in a sample irradiated
with H and thermally annealed at temperature T_a for a time t_a can be deduced
from $E_p(y_{eff} \neq 0) = E_p(y = 0) - ay^b_{eff}$. The two parameters a and b are found
by a fit of this formula to the $GaAs_{1-y}N_y$ ($In_xGa_{1-x}As_{1-y}N_y$, $x \approx 30\%$) band-
gap values obtained at $T=150$ K, which gives $a = 3.5$ (2.7) eV and $b = 0.63$
(0.77). (The empirical formulas reported here have been derived in the
$GaAs_{1-y}N_y$ case by using a large set of data from the literature [61]. For
$In_xGa_{1-x}As_{1-y}N_y$, we used a combined set of optical and X-ray diffraction
data taken in our samples.) If y is the as-grown N concentration and $y_{eff}[0,0]$
is that of the N atoms electronically active after an H irradiation and no
annealing, the percent of residual N–H bonds after an annealing is given
by $n = [y - y_{eff}(T_a, t_a)]/[y - y_{eff}[0,0]]$. Figure 6.6 (lower panel) shows the
dependence of n on T_a ($t_a = 1$ h) for the same sample shown in the upper
panel (full dots). The dashed line is a fit of the relation $n = \exp[-t_a \cdot v \cdot \exp
(-E_D/k_BT_a)]$ to the data ($E_D = 2.20$ eV), as done in the case of very dilute
N concentration. In this case as well as for all the samples in the alloy limit,
data are poorly reproduced by this simple model. A much better fit is
obtained when we assume that a *distribution* of N complexes contribute to
the sample band-gap reduction. This gives rise to a Gaussian distribution
for the N–H bond-dissociation energies, $G(E - <E_D>)$, with a mean value
$<E_D>$ and a standard deviation σ. In turn, this leads to the formula $n(t_a,T_a)$

$$= \int_0^\infty e^{-vt_a \cdot \exp(-E/k_BT_a)} G(E - <E_D>) \, dE,$$ whose fit to the data is shown by the

continuous line ($<E_D> = 2.20$ eV and $\sigma = 0.35$ eV). PL measurements at
$T = 150$ K are used in order to avoid the contribution of localized states to
the determination of the sample effective band-gap [51, 62]. Also, we point

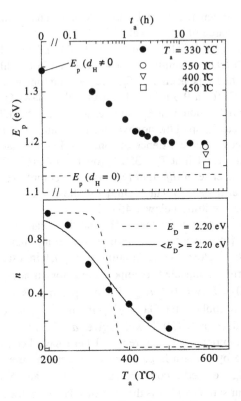

Figure 6.6 Upper panel: Dependence of the PL peak energy, E_p, on annealing time, t_a, at T = 10 K for a hydrogenated (d_H = 5 ×10^{17} ions/cm^2) GaAs$_{0.97}$N$_{0.03}$/GaAs QW annealed at T_a = 330°C (full dots). Open symbols are the PL peak energies for the same sample after further annealing for 1 h at higher T_a. The PL peak energy of the untreated sample (d_H = 0) is indicated by the horizontal dashed line. Lower panel: Residual N–H bonds n after annealing for 1 h at temperature T_a for the hydrogenated GaAs$_{0.97}$N$_{0.03}$/GaAs QW shown in the upper panel. The dashed line is a fit of a single-valued activation energy (E_D = 2.20 eV) model to the data. The continuous line is a fit to the data by a model that allows for different N complexes with a Gaussian distribution of activation energies around $\langle E_D \rangle$ = 2.20 eV and standard deviation σ = 0.35 eV. (Copyright (2002) by the American Physical Society.)

out that the $\langle E_D \rangle$ values found are independent of the PL measurement temperature as long as $T \geq 100$ K.

The values of $\langle E_D \rangle$ derived from PL measurements at T = 150 K are shown in Figure 6.7 as a function of nitrogen concentration (upper x-axis) for In$_x$Ga$_{1-x}$As$_{1-y}$N$_y$ and GaAs$_{1-y}$N$_y$ alloys (full black dots and full gray triangle, respectively). The increase in $\langle E_D \rangle$ with y can be accounted for by the relative statistical increase in the number of clusters with an increasing number of N atoms, which bind H more stiffly. The E_D values already derived for the different single N-H complexes in the dilute N limit are shown as open triangles as a function of ΔE (lower x-axis), the energy

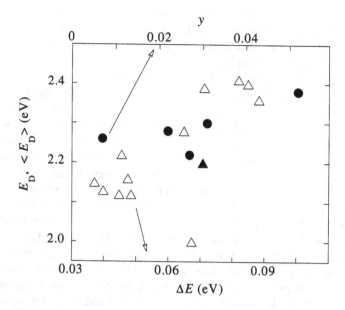

Figure 6.7 Dependence of $<E_D>$ on the nitrogen concentration y (upper x-axis) in In_xGa_{1-x} $As_{1-y}N_y$ and $GaAs_{1-y}N_y$ QWs (full black dots and full gray triangle, respectively). The E_D single values (open triangles) obtained for various N-related sharp PL lines in the $GaAs_{1-y}N_y$ epilayer with $y < 0.01\%$ shown in Figure 6.5 are shown vs. the energy difference ΔE (lower x-axis) between the bulk GaAs free exciton and N-related PL lines. (Copyright (2002) by the American Physical Society.)

difference between the PL peak energies of the N complex and the bulk GaAs free exciton. The activation energies derived in both dilute N and alloy limits cover the same range of values. Moreover, E_D increases with the number of atoms inside the N clusters (or ΔE according to Kent and Zunger [56]), and $<E_D>$ increases with the N concentration. Therefore, a quasi-continuum distribution of N clusters affects the PL spectrum in the dilute limit as well as the band-gap energy in the "alloy" limit.

6.3.2.3 Restoration of the host conduction band structure by hydrogenation: electron effective mass and exciton Bohr radius

Of all the effects nitrogen exerts on the conduction band of GaAs, the variation in the electron effective mass, m_e, is one of the most controversial. The data reported in the literature [27, 40–48, 63, 64] show a great deal of scatter. We describe now a method for deriving the electron effective mass and exciton wavefunction extent, r_{exc}, in $GaAs_{1-y}N_y$. The method is based on measuring the effect of a magnetic field on the electronic states associated with the material's conduction band. In particular, we show how a controlled introduction of H in the lattice allows monitoring the evolution of the $GaAs_{1-y}N_y$ electron effective mass and exciton degree of localization upon variation in the effective concentration of N present in the samples.

Figure 6.8a shows the PL spectra of a $GaAs_{1-y}N_y$ epilayer with $y = 0.1\%$ recorded under different magnetic field values, B. The magnetic field was applied along the growth direction of the sample. The sample was excited by an Nd-vanadate laser (excitation wavelength = 532 nm) and the luminescence was dispersed by a 3/4 m monochromator and detected by a N-cooled (InGa)As linear array. The spectral resolution of our experimental apparatus was 0.1 nm. Figure 6.8b shows the magnetic-field dependence of the energy of different recombination lines, whose attribution has been discussed previously (see Sections 6.3.1 and 6.3.2.1). PL spectra were taken at about 30 K to reduce the contribution from possible N-related localized states and donor–acceptor-pair recombination [51, 62], the latter being ionized with temperature before the (e,C) recombination. Upon application of a magnetic field, all PL lines below 1.462 eV (open circles) remain pinned in energy, consistent with the highly localized nature expected for N cluster states. In contrast, the E_- (full circles) and (e,C) (full squares) bands blueshift with increasing B. The behavior of these bands with B is consistent with the extended nature of the states involved.

Similar findings were observed for all samples shown in Figure 6.1. As an example, the PL spectra of a sample with $y = 0.043\%$ is shown in Figure 6.9 for different B values (lower panel). The B-induced shift, ΔE_d, of the E_- and (e,C) recombination lines is shown as a function of B in the upper panel. The E_- band shifts with B at a lower rate than the (e,C) band, owing to the larger coulomb attraction between the electron and hole in the former case.

If one assumes that the C-related level stays fixed in energy, as reported by Rossi et al. [65], while the conduction-band bottom shifts upon application of B, then the shift of the (e,C) transition can be ascribed entirely to the shift of the conduction-band bottom. Then, the electron effective mass, m_e, can be estimated by fitting (see dashed line in upper panel in Figure 6.9) the formula for the magnetic-field dependence of the bottommost Landau level of the conduction band, $\Delta E_d = \beta B = (-he/2m_e)B$, to the shift of the (e,C) transition in the B-linear region of ΔE_d [65]. At zero magnetic field, ΔE_d extrapolates to a negative value, on the order of $k_B T/2$, as found in other magneto-PL measurements of the B-induced shift of free-electron to neutral-acceptor recombination [66–68]. This behavior is usually attributed to the change in the density of states of the system from three- to one-dimensional, which is due to the applied magnetic field. By using this method for all samples, we find that the electron effective mass increases from 0.065 m_0 ($y = 0\%$, where m_0 is the electron mass in vacuum) to 0.074 m_0 ($y = 0.043\%$) before tending to saturation at ≈ 0.13 m_0 for $y \geq$

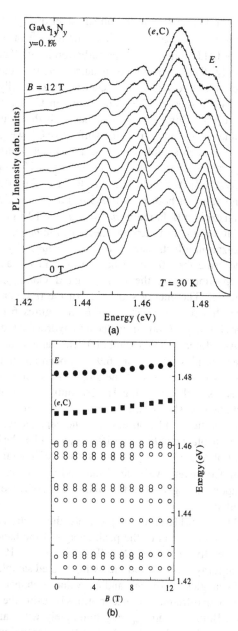

Figure 6.8 (a) PL spectra (T = 30 K) of a GaAs$_{0.999}$N$_{0.001}$ epilayer recorded under different magnetic fields, B. The term (e,C) indicates the free-electron to neutral-carbon recombination, and E_- indicates the free-exciton recombination. Other recombination lines below 1.462 eV are related to N-complex states. (b) B dependence of the peak energy of the different recombination bands observed in Figure 6.8a. Note that only the (e,C) and E_- transitions display a sizable shift with B.

0.1% [64]. The continuous line in the upper panel of Figure 6.9 is a fit of $\Delta E_d = \alpha B^2 = e^2 < r_{eh}^2 > /(8\mu)B^2$ to the diamagnetic shift of the free-exciton energy in the low-field regime (small-perturbation limit). The terms r_{eh} and μ are, respectively, the electron-hole distance and the reduced effective mass of the exciton. At very low N concentration, α rapidly decreases by a factor of ≈ 2 with respect to the value it has in GaAs, and it tends to saturate for $y > 0.1\%$. This behavior matches well that found for m_e. By using the m_e values determined from the shift of (e,C), we get an estimate of $r_{exc} = \sqrt{<r_{eh}^2>}$ for each sample from the diamagnetic-shift formula $\Delta E_d = \alpha B^2$.[*] The r_{exc} values decrease monotonically from 15.3 nm ($y = 0\%$) to 9.5 nm ($y = 0.5\%$) [64]. We point out that the procedure described above provides a sound value of both m_e ($=0.065\ m_0$) and r_{exc} ($=15.3$ nm, which corresponds to an exciton Bohr radius $a_0^{exc} = r_{exc}/\sqrt{2} = 10.8$ nm) for the GaAs reference. In addition, the fast increase (decrease) in m_e (r_{exc}) indicates that N-induced localization effects start at very low values of y.

We now describe the effects H has on m_e and r_{exc} as derived from magneto-PL experiments on the hydrogenated GaAs$_{0.999}$N$_{0.001}$ sample shown in Figure 6.3. The energy shift, ΔE_d, of the (e,C) and E_- recombination lines are shown as a function of B in Figures 6.10a and 6.10b, respectively, for GaAs$_{1-y}$N$_y$ (both untreated and hydrogenated, full symbols) and for the GaAs reference (open symbols). The same analysis performed for the untreated samples (see Figure 6.9) has been applied to the hydrogenated samples. We point out that the slope of the line fitting the (e,C) transitions increases with increasing H dose until the value of the GaAs reference is obtained, as shown in Figure 6.10a. A similar recovery of the pristine GaAs properties can be observed in the dependence of the exciton diamagnetic shift displayed in Figure 6.10b. In particular, the inset of Figure 6.10b shows a fit of $\Delta E_d = [e^2 < r_{eh}^2 > /(8\mu)]B^2$ to the E_- data in the quadratic, low-field region. Consistent with the H-induced m_e^* decrease, the exciton size increases upon hydrogenation, thus indicating a sizable change in the conduction-band shape.

Figures 6.11a and 6.11b show, respectively, the electron effective mass and exciton size as a function of the peak energy of the band-gap exciton, E_-. Full dots refer to the GaAs$_{1-y}$N$_y$ alloy with $y = 0.1\%$ for both the untreated sample (gray symbol) and the hydrogenated samples (black symbols). Gray full triangles are the m_e^* and r_{exc} values obtained for nonhydrogenated samples with different N concentrations by using the same methods described above. Both m_e^* and r_{exc} vary biuniquely with sample emission energy, namely, they depend on the *effective* N concentration in the crystal

---

[*] As for the hole effective mass, the increasing tensile strain with increasing y modifies the top of the valence-band character from prevalently heavy to light. We get an estimate of the hole-mass variation, m_h, from the measured change in the acceptor binding energy roughly estimated by the energy difference between the free exciton and the (e,C) recombination and then setting $m_h = 0.45\ m_0$ as the starting point for $y = 0\%$.

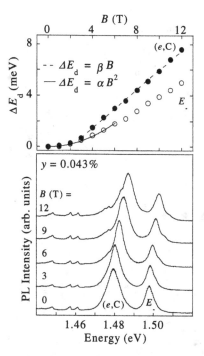

Figure 6.9 Lower panel: PL spectra of a $GaAs_{1-y}N_y$ epilayer with $y = 0.043\%$ taken at different magnetic fields ($T \approx 30$ K). Upper panel: B-induced shift value, ΔE_d, of the (e,C) (filled symbols) and of the E_- (open symbols) peak energy, as a function of the magnetic field. The continuous (dashed) line is a fit of a quadratic (linear) B dependence on the ΔE_d values of the E_- (e,C) peak. (From Ref. [64], copyright (2003) by the American Institute of Physics.)

regardless of how this concentration has been achieved (either by N incorporation in GaAs or by H irradiation in $GaAs_{1-x}N_x$). Most importantly, these findings allow monitoring the evolution of the electronic properties of $GaAs_{1-x}N_x$ in a virtually continuous manner. In particular, the electron effective mass increases very rapidly from the GaAs value ($= 0.065\ m_0$ as derived here) to $\approx 0.09\ m_0$ for $E_- \approx 1.49$ eV and reaches a value of $\approx 0.13\ m_0$ for $E_- \approx 1.45$ eV. Accordingly, the exciton size undergoes a shrinking with an increase in the N effective concentration and varies from $r_{exc} = 15.3$ nm in GaAs to 10.3 nm in $GaAs_{0.998}N_{0.002}$.

6.3.2.4 Restoration of the temperature coefficient

Nitrogen changes not only the electronic properties of $In_xGa_{1-x}As$, but it affects also the response of the host lattice band-gap to external perturbations such as hydrostatic pressure [28, 52, 53] and lattice temperature, T [50, 51, 69–76]. In both cases, a strong deviation from the behavior found in $In_xGa_{1-x}As$ is observed. The temperature dependence of the $In_xGa_{1-x}As_{1-y}$

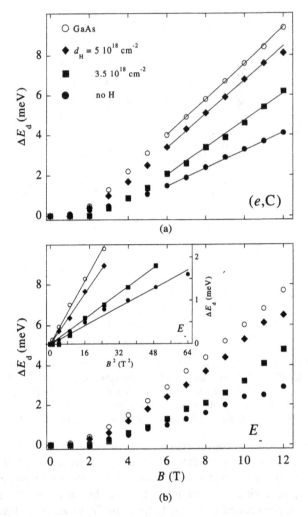

Figure 6.10 (a) Energy shift with magnetic field of the free-electron to neutral-carbon recombination (e,C) for some of the samples shown in Figure 6.1. The continuous lines are a fit to the data by $\Delta E_d = (\bar{h}e/2m_e^*)B$. The electron effective mass m_e^* is derived directly from the line slope. (b) Energy shift with magnetic field of the free-exciton recombination E_-. The samples considered are the same as in part (a). The inset highlights the low-field part of the graph and displays the E_- energy dependence on B^2. The straight lines are a fit to the data through $\Delta E_d = [e^2 < r_{eh}^2 >/(8\mu)]B^2$ using $< r_{eh}^2 >$ as a fit parameter.

N_y optical gap shows a pronounced slowdown with respect to that of the N-free material, this effect being larger for $x = 0$ and increasing y. Figure 6.12 shows the shift, R_S, of the PL peak position with respect to its value at $T = 10$ K as a function of temperature in a GaAs$_{1-y}$N$_y$/GaAs single QW ($y = 2.9\%$, $L = 9.0$ nm, full circles).[*]

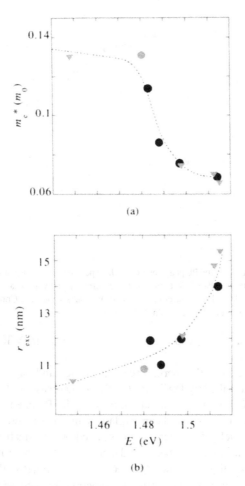

Figure 6.11 (a) Electron effective-mass dependence on the free-exciton peak energy ($T = 10$ K) and (b) exciton size dependence on the free-exciton peak energy ($T = 10$ K). Filled dots refer to an untreated (gray) and hydrogenated (black) GaAs$_{0.999}$N$_{0.001}$ samples; gray triangles refer to untreated GaAs$_{1-x}$N$_x$ epilayers with different x values. Dashed lines are guides to the eye.

The $y = 0$ data refer to the GaAs substrate emission as measured in the same sample (open circles). First, we note that at fixed temperature R_S sizably decreases with increasing effective N concentration. We quantify this slowdown in the T dependence of the GaAs$_{1-y}$N$_y$ gap by the thermal shrinkage rate S defined as the slope of R_S. If we disregard localization effects by measuring the slope of R_S for $T \geq 300$ K, the thermal shrinkage

---

*In the analysis of the temperature dependence of the gap energy, we neglect the variation of the QW potential offset due to the thermal variation of the well and barrier material band gaps. In fact, the change of carrier confinement energy between 10 K and 540 K produced by this effect is quite small [≈5 meV].

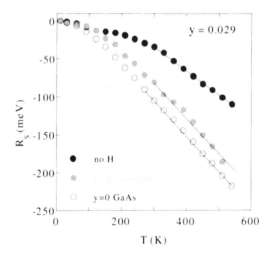

Figure 6.12 Shift, R_S, of the PL peak position with respect to its $T = 10$ K value vs. temperature for a GaAs$_{1-y}$N$_y$ (y = 2.9%) QW untreated (black circles) or hydrogenated with d_H = 10^{17} ions/cm^2 (gray circles). The open circles refer to the GaAs band gap. Continuous lines are best fits to the data in the T range from 300 to 540 K.

rate S of the GaAs$_{1-y}$N$_y$ gap is equal to 0.48 and 0.32 meV/K for y = 0 and 2.9%, respectively.

The reduction in the thermal redshift of the band-gap in In$_x$Ga$_{1-x}$As$_{1-y}$N$_y$ has been ascribed by Perlin et al. [69] to a decrease of the deformation potentials in N-containing material or, in works by Polimeni et al. [62] and Suemune et al. [72], it has been explained in the framework of a band anticrossing model [25]. Other authors pointed out the sizable contribution of localized states to the temperature dependence of the photoluminescence peak energy position for low T [51, 62, 70–76]. Finally, the increasing localized character of the electron wavefunction for increasing N concentration has been invoked to explain the smaller S values in N-containing samples compared with N-free samples [77]. Indeed, following Allen and Heine [78], a cancellation of the Debye-Waller and self-energy contributions may take place for narrow energy bands such as those of impurities with wavefunctions strongly localized in real space. This is the case of GaAs$_{1-y}$N$_y$. Therefore, a full recovery of the thermal band-gap shrinkage measured in GaAs could be expected upon hydrogenation, as indeed is shown in Figure 6.12 by the gray circles.

6.3.2.5   Effect of hydrogen on the lattice structure of GaAs$_{1-y}$N$_y$

In this section, we describe the consequences on the lattice properties of nitrogen passivation by hydrogen in GaAs$_{1-y}$N$_y$. For this purpose we performed X-ray diffraction measurements on samples treated under different conditions. These measurements on $GaAs_{1-y}N_y$/GaAs epilayers have been performed in the $\theta - 2\theta$ geometry by exploiting a molybdenum-rotating anode as X-ray generator and setting the incident wavelength at the $K_{\alpha 1}$ Mo fluorescence line (0.7092 Å) by means of a Si [1,1,1] channel-cut monochromator. The scattered intensity was detected by a NaI(Tl) scintillation detector. The X-ray diffraction data have been recorded in the vicinity of the [0,0,4] crystal plane reflection.

Figure 6.13a shows the photoluminescence spectra of a $GaAs_{1-y}$ N_y epilayer with $y = 0.81\%$ exposed to different postgrowth treatments and showing the band-gap increase (decrease) upon H introduction in (removal from) the lattice, extensively described in the previous sections. The X-ray diffraction data of the same $GaAs_{1-y}N_y$ epilayers whose PL spectra are shown in Figure 6.13a are displayed in Figure 6.13b. In the as-grown $GaAs_{1-y}N_y$ sample (bottom solid curve), two diffraction peaks in the rocking curve are observed. The higher and lower intensity peaks originate from the GaAs substrate and $GaAs_{1-y}N_y$ epilayer, respectively. The value of the angular separation between the two peaks allows us to measure the N concentration by using the empirical formula reported by Wei et al. [79]. The positive angular shift of the N-containing epilayer peak indicates that it has a smaller lattice constant along the growth direction ($a_{GaAsN}^{\perp} = 5.636$ Å) than it has in the plane where it is lattice-matched to the GaAs substrate ($a_{GaAsN}^{//} = a_{GaAs} = 5.653$ Å). Remarkably, the X-ray diffraction data recorded on the hydrogenated $GaAs_{1-y}N_y$ epilayer show a disappearance of the diffraction peak associated with the $GaAs_{1-y}N_y$ epilayer (see middle curve of Figure 6.13b). In addition, a shoulder can be detected at slightly smaller angles with respect to the GaAs peak, which indicates the presence of compressive strain in the hydrogenated $GaAs_{1-y}N_y$ epilayer. As a result, the value of the lattice constant of the hydrogenated sample can exceed even that of the GaAs [80, 81]. A heat treatment similar to the one the sample is subjected to during the hydrogenation process does not vary the material lattice properties (see gray dotted line, bottom of Figure 6.13b), reproducing the lack of variation in the optical properties shown at the bottom of Figure 6.13a by the gray dotted curve. One might wonder if the recovery of the GaAs lattice constant arises from a randomization of the lattice due to H bombardment or from nitrogen diffusion out of the lattice. These possibilities are ruled out by the diffraction data recorded on an identical piece of sample that was previously hydrogenated and then annealed until all H was removed. A full restoration of the $GaAs_{1-y}N_y$ lattice properties is observed, together with a full recovery of the $GaAs_{1-y}N_y$ band-gap (see top of Figures 6.13b and 6.13a). All of these results show that the crystal unit cell of $GaAs_{1-y}N_y$ undergoes a large variation of its size upon H insertion and that this processes is, once more, reversible.

It should be mentioned that similar effects have been observed in hydrogenated Si:B, where a partial relaxation of the Si:B lattice toward that

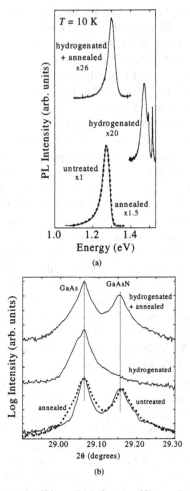

Figure 6.13 (a) PL spectra of a 300-nm-thick GaAs$_{0.9919}$N$_{0.0081}$ epilayer after different post-growth treatments. Bottom (continuous line): untreated sample; middle: same sample after exposure to $d_H = 3.0 \times 10^{18}$ ions/cm^2 (narrow bands on the high-energy side of the main PL band are due to carrier recombination from GaAs states); top: hydrogenated sample with the same H dose of 3.0×10^{18} ions/cm^2 but annealed at 500°C for 30 min. The spectrum of a sample subjected to the same heat treatment (temperature 300°C and duration 6 h) as the hydrogenated sample — but in the absence of hydrogen — is shown by the thick gray dotted line superimposed over the spectrum of the untreated sample (bottom continuous line). Note that the hydrogenation/annealing process introduces some nonradiative defects responsible for the PL intensity decrease. Normalization factors are given for each PL spectrum. (b) X-ray diffraction curves for the same samples whose PL spectra are displayed in Figure 6.13a. The X-ray diffraction data have been recorded in the vicinity of the [0,0,4] crystal plane reflection. The curves have been vertically offset for clarity. (Copyright (2003) by the American Physical Society.)

of the undoped Si lattice was observed [82]. This effect was attributed to the formation of B-H complexes, i.e., the same complexes responsible for the electrical passivation of B.

6.4   EFFECT OF HYDROGEN ON THE ELECTRONIC PROPERTIES OF $GaP_{1-y}N_y$

We now discuss the effect of hydrogenation on $GaP_{1-y}N_y$, a system that has attracted much interest since the 1960s, when it was shown that nitrogen incorporation gives rise to efficient light emission by carrier recombination from levels associated with single and multiple nitrogen complexes [83].

The PL spectra of a $GaP_{1-y}N_y$ epilayer with $y = 0.6\%$ are shown in Figure 6.14 for exposures to hydrogen doses increasing from bottom to top (continuous lines). The line shapes of the samples irradiated with $d_H = 10^{17}$ and 1.2×10^{18} ions/cm^2 reproduce those of as-grown epilayers with $y = 0.12$ and 0.05%, respectively (see dashed lines in Figure 6.14). In particular, the center of mass of the PL spectra blueshifts with increasing hydrogen dose, and a new group of lines appears where the untreated samples with smaller N concentration show carrier recombination from third-neighbor N pairs and related phonon replicas (≈ 2.26 eV) [84]. It should be noticed that the N cluster states are pinned at fixed energy regardless of the mechanism (either N incorporation or H irradiation) that exposes them in the forbidden gap of the host lattice. Finally, a reduction in the PL emission intensity and a line-width broadening at the highest d_H employed is observed, as in the case of hydrogenated $In_xGa_{1-x}As_{1-y}N_y$/GaAs. Similar behaviors are observed in the hydrogenated epilayers with higher N concentration ($y = 0.8$ and 1.3%), as well as in those with lower N content ($y = 0.05$ and 0.12%). Moreover, in the $y = 0.05$ sample — where the emission lines due to carrier recombination in different N complexes are spectrally well distinguished, as shown in Figure 6.15 — one can observe:

1. A clear passivation of the background signal underlying the narrow lines, most likely localized states due to distorted bonds in the sample arising from lattice relaxation around N atoms
2. A shift in the spectral weight from below to above 2.3 eV
3. An increase by more than two orders of magnitude of the integrated intensity of the emission line due to a single nitrogen atom at 2.318 eV at the highest H dose ($d_H = 2.8 \times 10^{18}$ ions/cm^2)

The last feature, which is different from what is found in $GaAs_{1-y}N_y$ where all N-related lines were fully passivated by H (see Figure 6.2), could be attributed to a reduction in the carrier transfer toward lower-energy states. Alternatively, it may be due to a passivation of a single N atom in a pair, which leaves the other N atom acting as a single N, or to a more effective

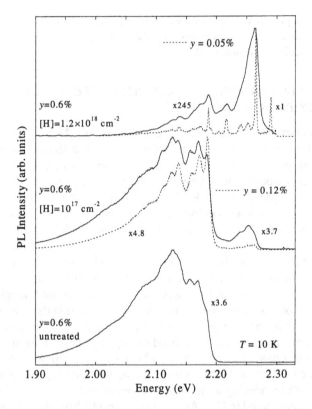

Figure 6.14 Peak-normalized PL spectra (T = 10 K) of a GaP$_{1-y}$N$_y$ epilayer with y = 0.6% after exposure to different H doses (continuous lines). PL spectra of as-grown GaP$_{1-y}$N$_y$ epilayers having lower N concentration (dashed lines) are shown for comparison. PL normalization factors are given. (Laser power density P = 10 W/cm^2.) (Copyright (2003) by the American Physical Society.)

H binding to deeper N complexes and closer N pairs, as suggested also by Singh and Weber [55].

The effects induced by H on the optical properties of GaP$_{1-y}$N$_y$ are fully reversible. This is shown, e.g., in Figure 6.16, where the PL spectra of the GaP$_{0.994}$N$_{0.006}$ samples hydrogenated with d_H = 1.2 × 10^{18} ions/cm^2 and annealed at different temperatures are displayed. Both the PL line shape and efficiency change sizably between T_a = 300°C and 400°C, and they are almost completely recovered at the highest T_a (550°C). An analysis of the dependence of the PL center of mass (see filled dots in Figure 6.16) and integrated intensity on T_a provides the same activation energy for both processes, E_a = 2.36 eV (a value quite similar to those found in In$_x$Ga$_{1-x}$As$_{1-y}$N$_y$). This analysis suggests also that H may form complexes with N, thus passivating the electronic activity of the latter, and that these N-H

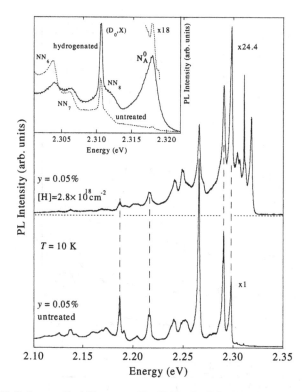

Figure 6.15 Peak-normalized PL spectra ($T = 10$ K) of a GaP$_{1-y}$N$_y$ epilayer with $y = 0.05\%$ before (lower curve) and after (upper curve) H irradiation with $d_H = 2.8 \times 10^{18}$ ions/cm^2. In the latter case, the horizontal dashed line sets the spectrum baseline. The vertical dashed lines highlight some lines whose relative intensity changes after hydrogenation. PL normalization factors are given. (Laser power density $P = 10$ W/cm^2.) Inset: Enlarged part of the spectra shown in the main part of the figure. The intensity of the PL signals can be compared directly. N$_A^0$ is the recombination line associated with a single N level; (D$_0$,X) indicates the line due to a sulfur-bound exciton recombination; NN$_i$ indicates the levels of the ith nearest-neighbor N pairs. (Copyright (2003) by the American Physical Society.)

complexes may act as nonradiative recombination centers, leading to a decrease in the emission efficiency [85].

Finally, we would like to point out that light emission continuously varying from red to green can be obtained on the same chip with $y = 0.81\%$ (not shown here) by suitable H irradiation doses.

6.5   ROLE OF THE N-H$_2^*$ COMPLEX

The experimental results presented above show that hydrogen is a local probe of the charge distribution around N atoms. The results also indicate that H and N atoms interact strongly in the lattice, leading to an apparent passivation of the electronic activity of N. The energetics of H in semicon-

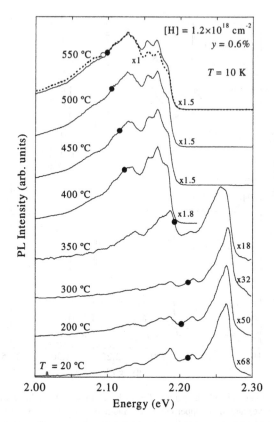

Figure 6.16 Peak-normalized PL spectra (continuous lines) ($T = 10$ K) of a GaP$_{0.994}$N$_{0.006}$ epilayer hydrogenated with 1.2×10^{18} ions/cm^2 after thermal annealing for 1 h at different temperatures, T_a. PL spectrum of the same sample without hydrogen (thick dotted line). Filled and open dots indicate, respectively, the energy position of the spectrum center of mass of the annealed and untreated samples. PL normalization factors are given. (Laser power density $P = 10$ W/cm^2.) (Copyright (2003) by the American Physical Society.)

ductors has been the object of intense theoretical studies due to the strong capability of H to affect the electronic and structural properties of many materials [86]. The case of H in GaAs$_{1-y}$N$_y$ has recently been addressed by several theoretical groups in an effort to understand the microscopic mechanism responsible for the band-gap reopening of GaAs$_{1-y}$N$_y$ upon H incorporation [87–90]. It has been shown that the N-H$_{BC}$ and N-H$_2^*$ complexes are, respectively, the most stable mono- and di-hydrogen complexes in GaAs$_{1-y}$N$_y$. In the N-H$_{BC}$ complex, there is only a H$_{BC}$ atom bonded to N at a bond center (BC) position between the Ga and N atoms. In the N-H$_2^*$ complex, two strong Ga–H$_{BC}$ and N–H$_{AB}$ bonds are formed, which involve, respectively, a H$_{BC}$ atom in bond center position between the Ga and N atoms and a H$_{AB}$ atom in opposite position with respect to the same nitrogen

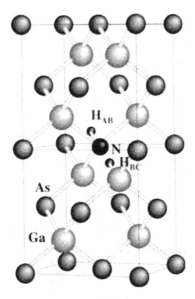

Figure 6.17 Sketch of the N-H$_2^*$ complex. One hydrogen atom, H$_{BC}$, is bound to Ga and located at the bond center position between the Ga and N atoms; the other hydrogen atom, H$_{AB}$, is placed in an antibonding site with respect to the N atom.

atom (see Figure 6.17). Most importantly, theoretical calculations have demonstrated that the N-H$_2^*$ accounts for the electronic passivation of N, while a monohydride complex does not lead to any appreciable effect on the band-gap of GaAs$_{1-y}$N$_y$ [87–90]. Moreover, the recovery of the N-free-material lattice parameter and electron effective mass can also be accounted for by the formation of N-H$_2^*$ complexes.

Although no direct experimental evidence of N-H$_2^*$ complexes has been provided yet, far-infrared absorption measurements indicate clearly that the N–H bonds are broken in GaAs$_{1-y}$N$_y$ by H introduction. Figure 6.18 shows the infrared spectra of untreated (bottom curve), hydrogenated (middle curve), and hydrogenated plus annealed (top curve) GaAs$_{0.9919}$N$_{0.0081}$ epilayers in the energy region of the local vibrational mode of the Ga–N bond (472 cm^{-1}) [80, 90, 91]. The decrease in the absorption intensity of this mode found in the hydrogenated sample indicates a decrease in the number of Ga–N bonds present in the sample. Subsequent thermal annealing restores the prehydrogenation bond number (see top curve in Figure 6.18). These data provide further evidence for strong changes in the lattice environment around the N atoms in GaAs$_{1-y}$N$_y$ upon hydrogenation. They show that the microscopic complex responsible for the band-gap reopening has to involve the breaking of Ga–N bonds. In addition, new modes at higher energy appear concomitantly with the disappearance of the Ga–N bonds, as reported previously [91, 92]. These infrared absorption measurements

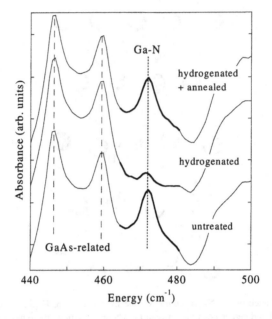

Figure 6.18 Far-infrared absorption spectra ($T = 4.2$ K) of the same samples whose PL spectra and X-ray diffraction curves are shown in Figure 6.13. Bottom line: untreated $GaAs_{0.9919}N_{0.0081}$ epilayer; middle line: same sample after hydrogen irradiation with $d_H = 3.0 \times 10^{18}$ ions/cm²; top line: sample hydrogenated with $d_H = 3.0 \times 10^{18}$ ions/cm² and annealed afterward at 500°C for 30 min. The Ga–N local vibrational mode at 472 cm⁻¹ is highlighted by thick lines. The other modes in the spectrum are due to the GaAs substrate. (From Ref. [83], copyright (2003) by the American Physical Society.)

indicate that the new complexes formed upon hydrogenation involve two H atoms [91].

6.6   CONCLUSIONS

Hydrogen irradiation of $In_xGa_{1-x}As_{1-y}N_y$ and $GaP_{1-y}N_y$ results in dramatic changes in the electronic properties of these materials for a wide range of N concentrations. In particular, we have shown that H counteracts in a controlled way the effects of N in $In_xGa_{1-x}As_{1-y}N_y$, leading to a fully tunable recovery of the band gap, electron effective mass, exciton wavefunction extent, lattice constant, and temperature coefficient that the host lattice had before N incorporation. All the H-induced changes in $In_xGa_{1-x}As_{1-y}N_y$ can be reversed by thermal annealing, which provides the activation energy (≈ 2.3 eV) for the breaking of the bonds of the N-H complexes responsible for the observed effects. From the above results, it emerges that the charge distribution around N atoms maintains a strong localized character upon going from the impurity to the alloy limit. Hydrogenation of $GaP_{1-y}N_y$ shows a band-gap reopening similar to that observed in $GaAs_{1-y}N_y$. Hydrogen removal in $GaP_{1-y}N_y$ has been achieved by thermal annealing. First-principle calculations indicate that a N-dihydride complex can account for all the effects of N in $In_xGa_{1-x}As_{1-y}N_y$ and $GaP_{1-y}N_y$. Finally, far-infrared absorption measurements show strong changes in the lattice environment of N atoms.

Acknowledgments

The authors thank A. Frova for his support throughout this work. We also would like to acknowledge G. Baldassarri Höger von Högersthal, M. Bissiri, F. Masia, and M. Felici for their valuable contribution at different stages of this work. We thank M. Stavola and F. Jiang (Leigh University) for far-infrared measurements and fruitful discussions. We are grateful to A. Amore Bonapasta and F. Filippone (CNR-ISM, Roma, Italy) and V. Fiorentini and S. Sanna (Università di Cagliari, Italy) for sharing their theoretical calculations. We acknowledge the most fruitful collaboration of P.J. Klar and W. Stolz (Philipps University, Marburg, Germany) and I.A. Buyanova and W.M. Chen (Linköping University, Sweden). We thank G. Ciatto (ESRF, Grenoble) and F. Boscherini (Università di Bologna, Italy) for X-ray diffraction measurements. We thank M. Geddo (Università di Parma, Italy) and A. Vinattieri (Università di Firenze) for valuable collaboration. We are grateful to A. Forchel (Wuerzburg University, Germany) and C.W. Tu (University of California at La Jolla, CA) for providing some of the samples studied. We thank A. Zunger and Y. Zhang (National Renewable Energy Laboratory, Golden, CO) for exchanging ideas. We are grateful to A. Miriametro and L. Ruggieri (Dipartimento di Fisica, Università di Roma, "La Sapienza") for very valuable technical assistance.

This work has been funded by Progetto Giovani Ricercatori, COFIN 2001 (MIUR), and FIRB.

References

1. Mellwo, E., "Die Wirkung von Wasserstoff auf die Leitfähigkeit und Lumineszenz von Zinkoxydkristallen," *Zeitschrift für Physik* 138: 478–488 (1954).

2. Van Weiringen, A., and Warmholtz, N., "On the Permeation of Hydrogen and Helium in Single Crystal Silicon and Germanium at Elevated Temperatures, *Physica* 22: 849–865 (1956).

3. Pankove, J., and Johnson, N.M., eds., *Hydrogen in Semiconductors*, Semiconductors and Semimetals, vol. 34 (Boston: Academic Press, 1991).

4. Haller, E.E., "Isotope Shifts in the Ground State of Shallow Hydrogenic Centers in Pure Germanium," *Phys. Rev. Lett.* 40: 584–586 (1978); Kahn, J.M., McMurray, R.E., Jr., Haller, E.E., and Falicov, L., "Trigonal Hydrogen-Related Acceptor Complexes in Germanium," *Phys. Rev. B* 36: 8001–8014 (1987).

5. Pankove, J.I., Carlson, D.E., Berkeyheiser, J.E., and Wance, R.O., "Neutralization of Shallow Acceptor Levels in Silicon by Atomic Hydrogen," *Phys. Rev. Lett.* 51: 2224–2225 (1983).

6. Sah, C.-T., Sun, J.Y.-C., and Tzou, J.J.-T., "Deactivation of the Boron Acceptor in Silicon by Hydrogen," *Appl. Phys. Lett.* 43: 204–206 (1983).

7. Capizzi, M., and Mittiga, A., "Hydrogen in Crystalline Silicon: A Deep Donor?" *Appl. Phys. Lett.* 50: 918–920 (1987).

8. Pantelides, S.T., "Effect of Hydrogen on Shallow Dopants in Crystalline Silicon," *Appl. Phys. Lett.* 50: 995–997 (1987).

9. Chevallier, J., Dautremont-Smith, W.C., Tu, C.W., and Pearton, S.J., "Donor Neutralization in GaAs(Si) by Atomic Hydrogen," *Appl. Phys. Lett.* 47: 108–110 (1985).

10. Johnson, N.M., Herring, C., and Chadi, D.J., "Interstitial Hydrogen and Neutralization of Shallow-Donor Impurities in Single-Crystal Silicon," *Phys. Rev. Lett.* 56: 769–772 (1986).

11. DeLeo, G.G., and Fowler, W.B., "Computational Studies of Hydrogen-Containing Complexes in Semiconductors," in *Hydrogen in Semiconductors*, ed. J. Pankove and N.M. Johnson, Semiconductors and Semimetals, vol. 34 (Boston: Academic Press, 1991), 511–546.

12. Van de Walle, C.G., "Theory of Isolated Interstitial Hydrogen and Muonium in Crystalline Semiconductors," in *Hydrogen in Semiconductors*, ed. J. Pankove and N.M. Johnson, Semiconductors and Semimetals, vol. 34 (Boston: Academic Press, 1991), 585–622.

13. Johnson, N.M., "Mechanism for Hydrogen Compensation of Shallow-Acceptor Impurities in Single-Crystal Silicon," *Phys. Rev. B* 31: 5525–5528 (1985).

14. Pankove, J.I., Zanzucchi, P.J., Magee, C.W., and Lucovsky, G., "Hydrogen Localization near Boron in Silicon," *Appl. Phys. Lett.* 46: 421–423 (1985).

15. Stavola, M., and Pearton, S.J., "Vibrational Spectroscopy of Hydrogen-Related Defects in Silicon," in *Hydrogen in Semiconductors*, ed. J. Pankove and N.M. Johnson, Semiconductors and Semimetals, vol. 34 (Boston: Academic Press, 1991), 139–184.

16. Chevallier, J., Clerjaud, B., and Pajot, B., "Neutralization of Defects and Dopants in III-V Semiconductors," in *Hydrogen in Semiconductors*, ed. J. Pankove and N.M. Johnson, Semiconductors and Semimetals, vol. 34 (Boston: Academic Press, 1991), 447–510.

17. Haller, E.E., "Hydrogen in Crystalline Semiconductors," *Semicond. Sci Technol.* 6: 73–84 (1991).

18. Estreicher, S.K., "Hydrogen-Related Defects in Crystalline Semiconductors: A Theorist's Perspective," *Mater. Sci. Eng. R-Reports* 14: 319–412 (1995).

19. Pearton, S.J., "Neutralization of Deep Levels in Silicon," in *Hydrogen in Semiconductors*, ed. J. Pankove and N.M. Johnson, Semiconductors and Semimetals, vol. 34 (Boston: Academic Press, 1991), 65–90.

20. Amore Bonapasta, A., and Capizzi, M., "Hydrogen as a Deep Impurity in Semiconductors and Its Interaction with Deep Centers in III-V Compounds," *Defect Diffusion Forum* 157–159: 133–174 (1998).

21. Bachelet, G.B., "Electronic States and Structural Properties of Deep Centers in Semiconductors," in *Crystalline Semiconducting Materials and Devices*, ed. P.N. Butcher, N.M. March, and M.P. Tosi (New York: Plenum Press, 1986), 243-304.

22. Czaja, W., "Isoelectronic Impurities in Semiconductors," *Festkörperprobleme* 11: 65–85 (1971).

23. Ager III, J.W., and Walukiewicz, W., eds., "III-V Semiconductor Alloys," *Semicond. Sci. Technol.* 17: 741–906 (2002).
24. Kent, P.R.C., and Zunger, A., "Evolution of III-V Nitride Alloy Electronic Structure: The Localized to Delocalized Transition," *Phys. Rev. Lett.* 86: 2613–2616 (2001).
25. Walukiewicz, W., Shan, W., Wu, J., and Yu, K.M., "Band Anticrossing in III-N-V Alloys: Theory and Experiments," chap. 2 in *Physics and Applications of Dilute Nitrides*, ed. I.A. Buyanova and W.M. Chen (New York: Taylor & Francis, 2004).
26. O'Reilly, E.P., "k·P Theory of Dilute Nitrides," chap. 3 in *Physics and Applications of Dilute Nitrides*, ed. I.A. Buyanova and W.M. Chen (New York: Taylor & Francis, 2004).
27. Zhang, Y., Mascharenas, A., Xin, H.P., and Tu, C.W., "Formation of an Impurity Band and Its Quantum Confinement in Heavily Doped GaAs:N," *Phys. Rev. B* 61: 7479–7482 (2000).
28. Klar, P.J., Grüning, H., Heimbrodt, W., Koch, J., Höhnsdorf, F., Stolz, W., Vicente, P.M.A., and Camassel, J., "From N Isoelectronic Impurities to N-Induced Bands in the GaN$_x$As$_{1-x}$ Alloy," *Appl. Phys. Lett.* 76: 3439–3441 (2000).
29. Grüning, H., Chen, L., Hartmann, T., Klar, P.J., Heimbrodt, W., Höhnsdorf, F., Koch, J., and Stolz, W., "Optical Spectroscopic Studies of Nitrogen Related Bands in Ga(N,As)," *Phys. Stat. Solidi B* 215: 39–45 (1999).
30. Seager, C.H., "Hydrogenation Methods," in *Hydrogen in Semiconductors*, ed. J. Pankove and N.M. Johnson, Semiconductors and Semimetals, vol. 34 (Boston: Academic Press, 1991), 17–33.
31. Bissiri, M., Baldassarri Höger von Högersthal, G., Polimeni, A., Gaspari, V., Ranalli, F., Capizzi, M., Amore Bonapasta, A., Jiang, F., Stavola, M., Gollub, D., Fischer, M., Reinhardt, M., and Forchel, A., "Hydrogen Induced Passivation of Nitrogen in GaAs$_{1-x}$N$_x$," *Phys. Rev. B* 65: 235210/1-5 (2002).
32. Polimeni, A., Baldassarri Höger von Högersthal, G., Bissiri, M., Capizzi, M., Fischer, M., Reinhardt, M., and Forchel, A., "Role of Hydrogen in III-N-V Compound Semiconductors," *Semicond. Sci. Technol.* 17: 797–802 (2002).
33. Bissiri, M., Baldassarri Höger von Högersthal, G., Polimeni, A., Capizzi, M., Gollub, D., Fischer, M., Reinhardt, M., and Forchel, A., "Role of N Clusters in the In$_x$Ga$_{1-x}$As$_{1-y}$N$_y$ Band-gap Reduction, *Phys. Rev. B* 66: 033311/1-4 (2002).
34. Makimoto, T., and Kobayashi, N., "Sharp Photoluminescence Lines from Nitrogen Atomic-Layer-Doped GaAs," *Appl. Phys. Lett.* 67: 688–690 (1995).
35. Shima, T., Makita, Y., Kimura, S., Sanpei, H., Fukuzawa, Y., Sandhu, A., and Nakamura, Y., "Effect of Low-Energy Nitrogen Molecular-Ion Impingement during the Epitaxial Growth of GaAs on the Photoluminescence Spectra," *Appl. Phys. Lett.* 74: 2675–2677 (1999).
36. Makimoto, T., Saito, H., Nishida, T., and Kobayashi, N., "Excitonic Luminescence and Absorption in Dilute GaAs$_{1-x}$N$_x$ Alloy (x < 0.3%)," *Appl. Phys. Lett.* 70: 2984–2986 (1997).
37. Zhang, Y., Mascarenhas, A., Geisz, J.F., Xin, H.P., and Tu, C.W., "Discrete and Continuous Spectrum of Nitrogen-Induced Bound States in Heavily Doped GaAs$_{1-x}$N$_x$," *Phys. Rev. B* 63: 085205/1-8 (2001).
38. Francoeur, S., Nikishin, S.A., Jin, C., Qiu, Y., and Temkin, H., "Excitons Bound to Nitrogen Clusters in GaAsN," *Appl. Phys. Lett.* 75: 1538–1540 (1999).
39. Tisch, U., Finkman, E., and Salzman, J., "The Anomalous Bandgap Bowing in GaAsN," *Appl. Phys. Lett.* 81: 463–465 (2002).

40. Chen, W.M., and Buyanova, I.A., "Electron Effective Masses of Dilute Nitrides: Experiment," chap. 4 in *Physics and Applications of Dilute Nitrides*, ed. I.A. Buyanova and W.M. Chen (New York: Taylor & Francis, 2004).

41. Heroux, J.B., Yang, X., and Wang, W.I., "Photoreflectance Spectroscopy of Strained (InGa)(AsN)/GaAs Multiple Quantum Wells," *J. Appl. Phys.* 92: 4361–4366 (2002).

42. Wu, J., Shan, W., Walukiewicz, W., Yu, K.M., Ager III, J.W., Haller, E.E., Xin, H.P., and Tu, C.W., "Effect of Band Anticrossing on the Optical Transitions in $GaAs_{1-x}N_x$/GaAs Multiple Quantum Wells," *Phys. Rev. B* 64: 85320/1-4 (2001).

43. Jones, E.D., Allerman, A.A., Kurtz, S.R., Modine, N.A., Bajaj, K.K., Tozer, S.W., and Wie, X., "Photoluminescence-Linewidth-Derived Reduced Exciton Mass for $In_yGa_{1-y}As_{1-x}N_x$ Alloys," *Phys. Rev. B* 62: 7144–7149 (2000).

44. Gorczyca, I., Skierbiszewski, C., Suski, T., Christensen, N.E., and Svane, A., "Pressure and Composition Dependence of the Electronic Structure of $GaAs_{1-x}N_x$," *Phys. Rev. B* 66: 081106(R)/1-4 (2002).

45. Duboz, J.-Y., Gupta, J.A., Byloss, M., Aers, G.C., Liu, H.C., and Wasilewski, Z.R., "Intersubband Transitions in InGaNAs/GaAs Quantum Wells," *Appl. Phys. Lett.* 81: 1836–1838 (2002).

46. Wang, Y.J., Wei, X., Zhang, Y., Mascarenhas, A., Xin, H.P., Hong, Y.G., and Tu, C.W., "Evolution of the Electron Localization in a Nonconventional Alloy System $GaAs_{1-x}N_x$ Probed by High-Magnetic-Field Photoluminescence," *Appl. Phys. Lett.* 82: 4453–4455 (2003).

47. Pan, Z., Li, L.H., Lin, Y.W., Sun, B.Q., Jiang, D.S., and Ge, W.K., "Conduction Band Offset and Electron Effective Mass in GaInNAs/GaAs Quantum-Well Structures with Low Nitrogen Concentration," *Appl. Phys. Lett.* 78: 2217–2219 (2001).

48. Hetterich, M., Dawson, M.D., Egorov, A.Y., Bernklau, D., and Riechert, H., "Electronic States and Band Alignment in GaInNAs/GaAs Quantum-Well Structures with Low Nitrogen Content," *Appl. Phys. Lett.* 76: 1030–1032 (2000).

49. Buyanova, I.A., Chen, W.M., and Monemar, B., "Electronic Properties of Ga(In)NAs Alloys," *MRS Internet J. Nitride Semicond. Res.* 6 (2): 1–19 (2001).

50. Uesugi, K., Suemune, I., Hasegawa, T., Akutagawa, T., and Nakamura, T., "Temperature Dependence of Band-gap Energies of GaAsN Alloys," *Appl. Phys. Lett.* 76: 1285–1287 (2000).

51. Polimeni, A., Capizzi, M., Geddo, M., Fischer, M., Reinhardt, M., and Forchel, A., "Effect of Temperature on the Optical Properties of (InGa)(AsN)/GaAs Single Quantum Wells," *Appl. Phys. Lett.* 77: 2870–2872 (2000).

52. Jones, E.D., Modine, N.A., Allerman, A.A., Kurtz, S.R., Wright, A.F., Tozer, S.T., and Wei, X., "Band Structure of $In_xGa_{1-x}As_{1-y}N_y$ Alloys and Effects of Pressure," *Phys. Rev. B* 60: 4430–4433 (1999).

53. Tsang, M.S., Wang, J.N., Ge, W.K., Li, G.H., Fang, Z.L., Chen, Y., Han, H.X., Li, L.H., and Pan, Z., "Hydrostatic Pressure Effect on Photoluminescence from a $GaN_{0.015}As_{0.985}$/GaAs Quantum Well," *Appl. Phys. Lett.* 78: 3595–3597 (2001).

54. Perkins, J.D., Mascarenhas, A., Zhang, Y., Geisz, J.F., Friedman, D.J., Olson, J.M., and Kurtz, S.R., "Nitrogen-Activated Transitions, Level Repulsion, and Band-gap Reduction in $GaAs_{1-x}N_x$ with $x < 0.03$," *Phys. Rev. Lett.* 82: 3312–3315 (1999).

55. Singh, M., and Weber, J., "Shallow Impurity Neutralization in GaP by Atomic Hydrogen," *Appl. Phys. Lett.* 54: 424–426 (1989).

56. Kent, P.R.C., and Zunger, A., "Theory of Electronic Structure Evolution in GaAsN and GaPN Alloys," *Phys. Rev. B* 64: 115208/1-23 (2001).

57. Polimeni, A., Baldassarri Höger von Högersthal, G., Bissiri, M., Capizzi, M., Fischer, M., Reinhardt, M., and Forchel, A., "Effect of Hydrogen on the Electronic Properties of $In_xGa_{1-x}As_{1-y}N_y$/GaAs Quantum Wells," *Phys. Rev. B* 63: 201304/1-4 (2001).

58. Baldassarri Höger von Högersthal, G., Bissiri, M., Polimeni, A., Capizzi, M., Fischer, M., Reinhardt, M., and Forchel, A., "Hydrogen Induced Band-Gap Tuning of (InGa)(AsN/GaAs) Single Quantum Wells," *Appl. Phys. Lett.* 78, 3472–3474 (2001).

59. Bergman, K., Stavola, M., Pearton, S.J., and Lopata, J., "Donor-Hydrogen Complexes in Passivated Silicon," *Phys. Rev. B* 37: 2770–2773 (1988).

60. Kurtz, S.R., Webb, J., Gedvillas, L., Friedman, D., Geisz, J., Olson, J., King, R., Joslin, D., and Karam, K., "Structural Changes during Annealing of GaInAsN," *Appl. Phys. Lett.* 78: 748–750 (2001).

61. Zhang, Y., Mascharenas, A., Xin, H.P., and Tu, C.W., "Scaling of Band-Gap Reduction in Heavily Nitrogen Doped GaAs," *Phys. Rev. B* 63: 161303(R)/1-4 (2001).

62. Polimeni, A., Capizzi, M., Geddo, M., Fischer, M., Reinhardt, M., and Forchel, A., "Effect of Nitrogen on the Temperature Dependence of the Energy Gap in $In_xGa_{1-x}As_{1-y}N_y$/GaAs Single Quantum Wells, *Phys. Rev. B* 63: 195320/1-5 (2001).

63. Baldassarri Höger von Högersthal, G., Polimeni, A., Masia, F., Bissiri, M., Capizzi, M., Gollub, D., Fischer, M., and Forchel, A., "Magneto-Photoluminescence Studies in (InGa)(AsN)/GaAs Heterostructures," *Phys. Rev. B* 67: 233304/1-4 (2003).

64. Masia, F., Polimeni, A., Baldassarri Höger von Högersthal, G., Bissiri, M., Capizzi, M., Klar, P.J., and Stoltz, W., "Early Manifestation of Localization Effects in Highly Diluted Ga(AsN)," *Appl. Phys. Lett.* 82: 4474–4476 (2003).

65. Rossi, J.A., Wolfe, C.M., and Dimmock, J.O., "Acceptor Luminescence in High-Purity n-Type GaAs," *Phys. Rev. Lett.* 25: 1614–1617 (1970).

66. Rühle, W., and Göbel, E., "New Aspects of the Magnetophotoluminescence of a Band to Acceptor Transition in GaAs," *Phys. Stat. Solidi B* 78: 311–317 (1976).

67. Dean, P.J., Venghaus, H., and Simmonds, P.E., "Conduction-Band-to-Acceptor Magnetoluminescence in Zinc Telluride," *Phys. Rev. B* 18: 6813–6823 (1978).

68. Zemon, S., Norris, P., Koteles, E.S., and Lambert, G., "Studies of Free-to-Bound Acceptor Photoluminescence in an Applied Magnetic Field for Undoped GaAs Grown by Metalorganic Vapor-Phase Epitaxy and Molecular-Beam Epitaxy," *J. Appl. Phys.* 59: 2828–2832 (1986).

69. Perlin, P., Subramanya, S., Mars, D.E., Kruger, J., Shapiro, N.A., Siegle, H., and Weber, E.R., "Pressure and Temperature Dependence of the Absorption Edge of a Thick $Ga_{0.92}In_{0.08}As_{0.985}N_{0.015}$ Layer," *Appl. Phys. Lett.* 73: 3703–3705 (1998).

70. Fan, J.C., Hung, W.K., Chen, Y.F., Wang, J.S., and Lin, H.H., Mechanism for Photoluminescence in an $In_yAs_{1-y}N/In_xGa_{1-x}As$ Single Quantum Well," *Phys. Rev. B* 62: 10990–10994 (2000).

71. Grenouillet, L., Bru-Chevallier, C., Guillot, G., Gilet, P., Duvaut, P., Vannuffel, C., Million, A., and Chevanes-Paule, A., "Evidence of Strong Carrier Localization below 100 K in a GaInNAs/GaAs Single Quantum Well," *Appl. Phys. Lett.* 76: 2241–2243 (2000).

72. Suemune, I., Uesugi, K., and Walukiewicz, W., "Role of Nitrogen in the Reduced Temperature Dependence of Band-Gap Energy in GaNAs," *Appl. Phys. Lett.* 77: 3021–3023 (2000).

73. Pinault, M.-A., and Tournié, E., "On the Origin of Carrier Localization in $Ga_{1-x}In_xN_yAs_{1-y}$/GaAs Quantum Wells," *Appl. Phys. Lett.* 78: 1562–1564 (2001).

74. Kaschener, A., Lüttgert, T., Born, H., Hoffmann, A., Egorov, A.Y., and Riechert, H., "Recombination Mechanisms in GaInNAs/GaAs Multiple Quantum Wells," *Appl. Phys. Lett.* 78: 1391–1393 (2001).

75. Shirakata, S., Kondow, M., and Kitatani, T., "Photoluminescence and Photoreflectance of GaInNAs Single Quantum Wells," *Appl. Phys. Lett.* 79: 54–56 (2001).

76. Luo, X.D., Xu, Z.Y., Ge, W.K., Pan, Z., Li, L.H., and Lin, Y.W., "Photoluminescence Properties of a $GaN_{0.015}As_{0.985}$/GaAs Single Quantum Well under Short Pulse Excitation," *Appl. Phys. Lett.* 79: 958–960 (2001).

77. Polimeni, A., Bissiri, M., Augieri, A., Baldassarri Höger von Högersthal, G., Capizzi, M., Gollub, D., Fischer, M., Reinhardt, M., and Forchel, A., "Reduced Temperature Dependence of the Band-gap in Ga(AsN) Investigated with Photoluminescence," *Phys. Rev. B* 65: 235325/1-5 (2002).

78. Allen, P., and Heine, V., "Theory of the Temperature Dependence of Electronic Band Structures," *J. Phys. C* 9: 2305–2312 (1976).

79. Wei, L., Pessa, M., and Likonen, J., "Lattice Parameter in GaNAs Epilayers on GaAs: Deviation from Vegard's Law," *Appl. Phys. Lett.* 78: 2864–2866 (2001).

80. Klar, P.J., Grüning, H., Güngerich, M., Heimbrodt, W., Koch, J., Torunski, T., Stolz, W., Polimeni, A., and Capizzi, M., "Global Changes of the Band Structure and the Crystal Lattice of Ga(N,As) due to Hydrogenation," *Phys. Rev. B* 67: 121206(R)/1-4 (2003).

81. Polimeni, A., Ciatto, G., Ortega, L., Jiang, F., Boscherini, F., Filippone, F., Amore Bonapasta, A., Stavola, M., and Capizzi, M., "Lattice Relaxation by Atomic Hydrogen Irradiation of III-N-V Semiconductor Alloys," *Phys. Rev. B* 68: 085204/1-5 (2003).

82. Stutzmann, M., Harsanyi, J., Breitschwerdt, A., and Herrero, C.P., "Lattice Relaxation due to Hydrogen Passivation in Boron-Doped Silicon," *Appl. Phys. Lett.* 52: 1667–1669 (1988).

83. Thomas, D.G., and Hopfield, J.J., "Isoelectronic Traps due to Nitrogen in Gallium Phosphide," *Phys. Rev.* 150: 680–689 (1966).

84. Zhang, Y., Fluegel, B., Mascarenhas, A., Xin, H.P., and Tu, C.W., "Optical Transitions in the Isoelectronically Doped Semiconductor GaP:N: An Evolution from Isolated Centers, Pairs, and Clusters to an Impurity Band," *Phys. Rev. B* 62: 4493–4500 (2000).

85. Polimeni, A., Bissiri, M., Felici, M., Capizzi, M., Buyanova, I.A., Chen, W.M., Xin, H.P., and Tu, C.W., "Nitrogen Passivation Induced by Atomic Hydrogen: The GaPN Case," *Phys. Rev. B* 67: 201303(R)/1-4 (2003).

86. Van de Walle, G.C., and Neugebauer, J., "Universal Alignment of Hydrogen Levels in Semiconductors, Insulators and Solutions," *Nature* 423: 626–628 (2003).

87. Kim, Y.-S., and Chang, K.J., "Nitrogen-Monohydride versus Nitrogen-Dihydride Complexes in GaAs and $GaAs_{1-x}N_x$ Alloys," *Phys. Rev. B* 66: 073313/1-4 (2002).

88. Zhang, S.B., and Wei, S.-H., "Theory of Defects in Dilute Nitrides," chap. 7 in *Physics and Applications of Dilute Nitrides*, ed. I.A. Buyanova and W.M. Chen (New York: Taylor & Francis, 2004).

89. Amore Bonapasta, A., Filippone, F., Giannozzi, P., Capizzi, M., and Polimeni, A., "Structure and Passivation Effects of Mono- and Di-Hydrogen Complexes in $GaAs_y N_{1-y}$ Alloys," *Phys. Rev. Lett.* 89: 216401/1-4 (2002).

90. Orellana, W., and Ferraz, A.C., "Stability and Electronic Structure of Hydrogen–Nitrogen Complexes in GaAs," *Appl. Phys. Lett.* 81: 3816–3818 (2002).

91. Alt, H.C., Egorov, A.Y., Riechert, H., Wiedemann, B., Meyer, J.D., Michelmann, R.W., and Bethge, K., *Appl. Phys. Lett.* 77: 3331–3333 (2000).

92. Capizzi, M., Polimeni, A., Baldassarri Höger von Högersthal, G., Bissiri, M., Amore Bonapasta, A., Jiang, F., Stavola, M., Fischer, M., Forchel, A., Sou, I.K., and Ge, W.K., "Photoluminescence and Infrared Absorption Study of Isoelectronic Impurity Passivation by Hydrogen," *Mater. Res. Soc. Symp. Proc.* 719: 251–256 (2002).

CHAPTER 7

Raman Scattering in Dilute GaAsN and GaPN Alloys

A. MASCARENHAS and S. YOON

National Renewable Energy Laboratory, Golden, Colorado

7.1   INTRODUCTION

There have been several attempts made during the past few years to provide an explanation for the giant band-gap bowing observed in dilute GaAsN alloys [1]. The recent discovery [2] of a resonant level, E_+, in the conduction band of the alloy and an anticrossing behavior between this level and the conduction-band minimum (CBM), E_- exhibited in pressure-dependent photoreflectivity studies [3] led to speculation about the origin [2, 3] of the resonant level and the role it plays in the giant band-gap bowing. Shan et al. [3] used the anticrossing behavior as compelling evidence for level repulsion between E_+ and the CBM (a two-level interaction model) as the reason for the giant band-gap lowering. But pseudopotential calculations indicated that the band-gap lowering was a result of (a) electronic coupling between the CBM at Γ, (b) the conduction-band extrema at the L and X points of the Brillouin zone induced by the isolated nitrogen impurity level, and (c) the ensuing repulsion between all these levels [4–7]. The isolated nitrogen impurity level resonant at 170 meV above the CBM of GaAs had been predicted using tight-binding calculations over two decades ago [8]. Over a decade ago, in studies of the low-temperature photoluminescence of GaAsN ($<6 \times 10^{17}$ cm^{-3} N), this level, denoted as N_X, was directly observed as it emerged into the band-gap at high pressures [9, 10]. The two-level repulsion model had assumed that it was from the interaction of this level with the CBM at Γ that the resonant impurity level E_+ had evolved. However, pseudopotential calculations indicated that E_+ originated from nitrogen-induced splitting of the conduction band at the L point [7]. The different opinions on the origin of E_+ have led to a great deal of controversy regarding the mechanisms underlying the giant band-gap bowing.

The isoelectronic traplike behavior of substitutional nitrogen impurities in GaP drew a great deal of attention several decades ago [11] because it enabled the early design of visible-light-emitting diodes (LEDs). Due to the steep central-cell potential of the impurity and its consequent localization and trapping of electrons, the states N_X (generated by the isolated nitrogen impurity) and NN_m (generated by pairs of impurities) are delocalized in **k**-space, allowing for efficient photo- and electroluminescence from the otherwise indirect gap material GaP. More recently, success in increasing the nitrogen doping level from 10^{17} to 10^{21} cm^{-3} has led to excitement about the possibility of fabricating ultrabright LEDs from this material, which is closely lattice-matched to silicon. At these high N doping levels, the photoluminescence transitions from the N_X level (referred to as the A-line) and the various NN_m levels broaden and merge due to formation of impurity bands [12]. The apparent lowering of this band-edge transition energy [13] is reminiscent of the abnormally large band-gap bowing recently observed for GaAsN [1].

However, there are distinct differences between these two materials. In GaP, the conduction-band minimum (CBM) occurs near the X-point, and all the N impurity states are located in the forbidden gap, whereas in GaAs, the CBM occurs at the Γ point, and the N_X as well as several of the NN_m states are located above the CBM. A simple two-band model based on level repulsion with N_X and the CBM at Γ has been proposed to explain the giant band-gap lowering in GaAsN [3] as well as in GaPN [14]. In this model, N_X and the CBM at Γ interact and evolve into two new states, E_+ and E_-, where the lower energy state forms the new conduction-band minimum. This would imply that the A-line evolves into the new conduction-band minimum, which exhibits a rapid lowering in GaPN. However, absorption measurements have indicated that the A-line barely moves with increased nitrogen doping, raising doubts about the validity of the two-band repulsion model [12, 15]. Since the A-line and the NN_m pair lines rapidly broaden with increased N doping and become indiscernible from the absorption background, it is difficult to track their evolution at higher N concentrations, thus making conclusions drawn from such measurements less reliable. On the other hand, based on their pseudopotential calculations, Kent et al. [6, 16] proposed that the large band-gap reduction in these materials is caused by the downward-moving conduction edge overtaking the localized nitrogen states.

In this article, we present Raman and resonant Raman scattering results on GaAsN and GaPN, hoping to elucidate some of the critical issues that remain unanswered, such as the mechanism of the large band-gap bowing and the nature of the nitrogen-induced states and their couplings to the host band extrema.

7.2   EXPERIMENTAL STUDIES

Raman scattering spectra in these studies were measured in a pseudoback-scattering geometry using a SPEX 1877 Triplemate spectrometer equipped with a nitrogen-cooled charge-coupled device (CCD) array detector. The samples were excited with either Ti:Sapphire laser (1.54–1.80 eV), dye lasers with the DCM (1.82–1.98 eV) dye, the R6G (1.96–2.18 eV) dye, or the R110 (2.15–2.26 eV) dye; or with a frequency-doubled Nd:YAG (yttrium aluminum garnet) laser (2.33 eV), or the single lines of an Ar^+ ion laser (2.41–2.71 eV), focused to a line of dimensions 5 mm × 100 μm. Unless stated otherwise, the spectra were obtained in the $z(YY)\bar{z}$ scattering geometry, where z and Y are the [001] and [110] crystal directions, respectively. In this scattering configuration, the zone-center longitudinal optical (LO_Γ) phonon is Raman-active, but the zone-center transverse optical (TO_Γ) phonon is not Raman-active for the T_d space group of zinc-blende structure crystals. In the next section, we briefly review the Raman scattering spectroscopy technique.

7.2.1   General Discussion of Raman Scattering

Raman scattering refers to the *inelastic* scattering of light by various elementary excitations of the material being probed [17]. Incident photons can couple to lattice vibrations, electronic excitations, and magnetic excitations simultaneously, and hence the resulting Raman scattering spectra can provide information regarding the energy, lifetime, and symmetry properties of these excitations. Raman scattering, in the first order, mainly probes the Brillouin zone center due to momentum-conservation restrictions. The wave vector k_I of general excitation source (visible light) is $\approx 10^5$ cm^{-1}, which is much smaller than the Brillouin zone-boundary wave vector ($\approx 10^8$ cm^{-1}). Hence, the wave vector q ($= k_I - k_S$) of an excitation of interest is small (~ 0) compared with the Brillouin zone-boundary wave vector, where k_S is the wave vector of the scattered light. This restriction can, however, be removed for higher-order processes, such as two-phonon or two-magnon scattering. In these cases, only the sum of the two wave vectors of the phonons (magnons) needs to be zero. There is another circumstance when the rule of momentum conservation is relaxed so that phonons (*not* from the higher-order process) with nonzero wave vector can participate in the scattering process. Indeed, this exception is crucial to the discussion of concepts presented in this chapter and will be elaborated further in the sections below.

7.2.2   Resonant Raman Scattering

As will be seen below (Equation 7.2), the Raman scattering intensity is enhanced near a resonance where the incoming excitation matches the transition energy between an initial (ground) state and an electronic level, such as a conduction-band-edge state. Resonant Raman scattering can, in principle, provide information regarding the electron-phonon interaction, the electron-photon interaction, and the nature of the host electronic states involved in these interactions from the resonance enhancement of the scattering *intensity* response [18]. In this study, however, we focus on another aspect of resonant Raman scattering, namely, the phonon line-width resonance, and show that the phonon line-width resonance is a very powerful tool to study strongly localized impurity states in semiconductors, such as is the case of nitrogen impurities in GaAs(GaP).

7.2.3   Materials

7.2.3.1   GaAsN

GaAs$_{1-x}$N$_x$ epilayers with thickness of 0.15–1.5 μm were grown by low-pressure metalorganic chemical vapor deposition. High-resolution X-ray diffraction (XRD) was used to determine the lattice parameter of the strained GaAs$_{1-x}$N$_x$ layers from the symmetric (0,0,4) rocking curve. The thickness of the GaAs$_{1-x}$N$_x$ layers studied in this work was kept below the critical thickness for strain relaxation, as was evidenced by the lack of crosshatching or misfit lines on the sample surfaces and by the observation of thickness fringes in the XRD rocking curves. The nitrogen concentration was then determined from the measured $GaAs_{1-x}N_x$ lattice parameter using the lattice parameters and elastic constants of GaAs and cubic GaN [19] assuming Vegard's law is valid.

7.2.3.2 GaPN

The samples used in our study were 1-μm-thick $GaP_{1-x}N_x$ epilayers ($x = 0.25$, 0.5, 1.0, 1.5, and 2.0), which were lifted off from the (100) GaP substrates. The samples were grown by metalorganic chemical vapor deposition with a thin AlP release layer between the epilayer and the substrate. This AlP layer was subsequently etched away in 10% HF, lifting off the $GaP_{1-x}N_x$ epilayers in order to remove the unwanted photo response from the GaP substrate. The nitrogen concentration x was estimated by either XRD or secondary-ion mass spectroscopy (SIMS) measurements on these samples.

7.3 RESULTS AND DISCUSSIONS

7.3.1 GaAsN

The results of off-resonant and resonant Raman scattering studies on GaAsN are presented below. By monitoring the nitrogen local vibration mode in off-resonant Raman scattering, one can directly estimate nitrogen concentration, even in quaternary compounds such as $Ga_{1-y}In_yAs_{1-x}N_x$, [20], where only an indirect estimation of the nitrogen composition is otherwise possible. Resonant Raman scattering studies reveal that a resonant level, E_+, which is attributed to the split singlet state (a_1) from the fourfold-degenerate L-point conduction band, contains significant L and X characters, in addition to Γ character [21]. Analysis of the LO_Γ phonon line-width resonance reveals that the remaining threefold-degenerate L conduction band (t_2) further splits into a doublet (e_2) state and an additional singlet (a_1) state [22].

7.3.1.1 Nitrogen local vibration mode studies

The isoelectronic substitution of only 1% nitrogen in GaAs reduces the fundamental band-gap (E_0) by ≈ 200 meV, leading to the so-called giant band-gap bowing. In the context of band-gap engineering for specific applications, e.g., the 1.3-μm semiconductor laser, it is very important to accurately determine the nitrogen concentration in $GaAs_{1-x}N_x$, and X-ray diffraction (XRD) measurements have been widely used for this purpose. However, for quaternaries such as $Ga_{1-y}In_yAs_{1-x}N_x$, XRD measurements alone cannot directly determine the N concentration due to the lattice constant compensation by the presence of In. Instead, one has to measure the In and N compositions separately in the respective ternaries $Ga_{1-y}In_yAs$ and $GaAs_{1-x}N_x$ under specific growth conditions and indirectly estimate x and y in $Ga_{1-y}In_yAs_{1-x}N_x$ grown under the same conditions.

It has been known that local vibration mode (LVM) absorption is directly proportional to the substitutional impurity concentration. Alt et al. [23] have recently reported nitrogen concentrations in $GaAs_{1-x}N_x$ determined by the infrared absorption technique (x_{LVM}) and compared them with those determined by XRD (x_{XRD}). They found that x_{LVM} x_{XRD} for $x_{LVM} <$ 0.01, but a large deviation from the linear behavior was observed for x_{LVM} > 0.02. In this section, we report results of our recent investigation on the nitrogen LVM intensity and its frequency (ω_{LVM}) as a function of x_{LVM} in $GaAs_{1-x}N_x$ (x 0.04) at 300 K. We show that the intensity of the nitrogen LVM with respect to the GaAs-LO phonon intensity exhibits a remarkable linear dependence on the nitrogen concentration for x 0.03.

Figure 7.1 illustrates the room-temperature Raman spectra of $GaAs_{1-x}N_x$ for x = 0, 0.006, 0.012, 0.025, and 0.032. Fast deterioration of the sharp second-order Raman features of GaAs near 513 and 540 cm^{-1} and gradual blueshift of ω_{LVM} are observed [24]. The normalized LVM intensity is observed to be increasing with nitrogen concentration. Since the integrated intensity of a phonon is proportional to the corresponding oscillator strength, the integrated intensity of the GaAs-LO phonon (I_{LO}) in $GaAs_{1-x}N_x$ is proportional to the As-composition $1-x$, and that of the nitrogen LVM (I_{LVM}) is proportional to the N-composition x. Thus, one can calculate the N composition from Raman measurements (x_{Raman}) as follows,

$$x_{Raman} = \frac{I_{LVM}}{fI_{LO} + I_{LVM}}, \tag{7.1}$$

where f represents the relative nitrogen LVM oscillator strength with respect to the GaAs-LO phonon oscillator strength. The values of x_{Raman} calculated using Equation 7.1 as a function of x_{XRD} is shown in Figure 7.2. In order to obtain the decomposed integrated intensity of the LO phonon, we first subtracted the linear baseline in the spectral range from 260 to 310 cm^{-1} and then fit the spectrum with two Lorentzian oscillators (LO and TO). Similar peak-fitting analysis was done from 420 to 500 cm^{-1} with two Lorentzian oscillators, corresponding to the LVM and the second-order Raman scattering of GaAs at 455 cm^{-1}, which appears as a shoulderlike feature for x > 0.006 but was distinctly observed in GaAs, as shown in Figure 7.1.

The triangular data points in Figure 7.2 were calculated using Equation 7.1 with f = 1.0, i.e., assuming equal oscillator strengths for the Ga-As and Ga-N modes. In this case, $x_{Raman} - x_{XRD}$ increases with nitrogen concentration, which indicates that the Ga-N mode has much stronger oscillator strength than the Ga-As mode. We also calculated x_{Raman} for f = 1.3 (circles), which gives a better linear fit to x_{XRD} for $x_{XRD} <$ 0.03. However, a deviation

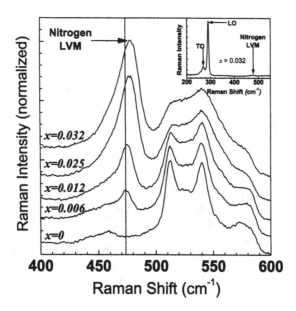

Figure 7.1 Room-temperature Raman spectra of $GaAs_{1-x}N_x$ for $x = 0$, 0.006, 0.012, 0.025, and 0.032, excited with Ar^+ 514-nm line. Raman intensity is normalized with respect to the GaAs-LO phonon intensity for each sample, and the baselines for each spectrum are shifted vertically for clarity. Inset illustrates the TO and LO phonons as well as the nitrogen LVM for the $x = 0.032$ sample.

from a linear dependence is observed for $x_{XRD} > 0.03$. An empirical quadratic fit to the data for $x_{XRD} < 0.04$ is also shown in Figure 7.2 for comparison.

Figure 7.3 shows the nitrogen LVM Raman frequency ω_{LVM} in $GaAs_{1-x}N_x$ as a function of nitrogen concentration. Just as for x_{Raman}, ω_{LVM} exhibits an excellent linear dependence on nitrogen concentration for $x_{XRD} < 0.03$, and some deviation from the linear dependence is seen for $x_{XRD} > 0.03$. The linear as well as the empirical quadratic fits of x_{Raman} and ω_{LVM} to x_{XRD} are summarized in Table 7.1. The deviation from a linear dependence observed in Figures 7.2 and 7.3 for the high-nitrogen-concentration samples can be attributed to the fact that only the substitutional nitrogen atoms at the anion sublattice (N_{As}) contribute to its LVM intensity, whereas not only N_{As} but also other nitrogen-related defects, e.g., interstitial nitrogen, can produce the nitrogen-induced lattice-parameter change in XRD measurements. In this context, it is noteworthy that Spruytte et al. [25] have recently reported that a significant amount of interstitial nitrogen exists in their as-grown $GaAs_{1-x}N_x$ samples with $x > 0.029$.

In summary, from our Raman scattering study of the nitrogen LVM intensity in $GaAs_{1-x}N_x$ for $x < 0.04$, we show that both the relative intensity of the nitrogen LVM with respect to that of the GaAs-LO phonon as well

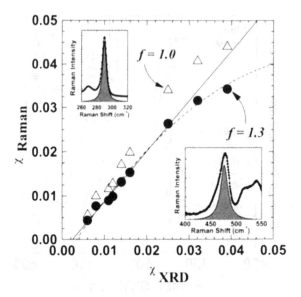

Figure 7.2 Nitrogen compositions, x_{Raman}, calculated using Equation 7.1 as a function of nitrogen concentration determined by X-ray measurement, x_{XRD}, for $f = 1.0$ (triangles) and $f = 1.3$ (circles). The solid line is a linear fit to the data, with $f = 1.3$ for $x_{XRD} < 0.03$, and the dashed line is an empirical quadratic fit. Error bars are approximately the same size as the diameter of the circles. Insets show the integrated intensities (Lorentzian oscillators) of LO phonon (≈ 292 cm^{-1}) and LVM (≈ 473 cm^{-1}) after baseline corrections.

as ω_{LVM} exhibit excellent linear dependence on x_{XRD} for $x_{XRD} < 0.03$. We have also obtained an empirical fit to x_{Raman} as a function of x_{XRD} for $x_{XRD} < 0.04$, providing a reliable calibration for determining nitrogen composition not only in this ternary alloy, but also in the quaternaries such as $Ga_{1-y}In_yAs_{1-x}N_x$, where only indirect estimation of the nitrogen concentration was possible.

7.3.1.2   Resonant Raman scattering studies

7.3.1.2.1   Symmetry of the Intermediate State   Resonant Raman scattering (RRS) enables one to microscopically probe electronic states by monitoring the resonance enhancements for various lattice vibrations that occur at critical points in the joint interband electronic density of states. Thus it is a very powerful spectroscopic technique that involves the electron-phonon interaction in a crystal. Although there have been a few Raman scattering studies on GaAs$_{1-x}$N$_x$ [24, 26–28], most of them have not discussed the symmetry of E_+ state. Cheong et al. [29] recently studied the nature of the E_+ state by performing RRS measurements on GaAs$_{1-x}$N$_x$ at room temperature. They observed that L-point zone-boundary longitudinal acoustic (LA$_L$) and longitudinal optical (LO$_L$) phonons are activated only

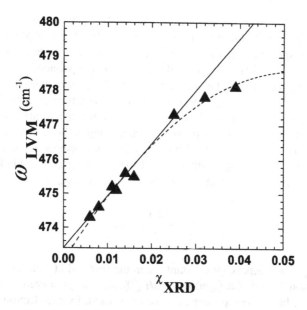

Figure 7.3 Nitrogen LVM Raman frequency as a function of nitrogen concentration x_{XRD}. The solid line represents a linear fit for $x < 0.03$, and the dashed line is a quadratic fit: $\omega_{LVM} = -990x_{XRD}^2 + 210.9x_{XRD} + 473.0$.

Table 7.1 Linear and Empirical Fits of x_{Raman} and ω_{LVM} to x_{XRD}

	Linear ($x_{XRD} < 0.03$)	Empirical ($x_{XRD} < 0.04$)
x_{Raman} (for $f = 1.3$)	$1.145x_{XRD} - 0.003$	$-12.2x_{XRD}^2 + 1.51x_{XRD} - 0.005$
ω_{LVM} (cm^{-1})	$153.4x_{XRD} + 473.4$	$-1990x_{XRD}^2 + 210.9x_{XRD} + 473.0$

near resonance with the E_+ level, and they concluded that the E_+ state contains a significant L component. In this section, we report the results of a recent RRS study near the E_+ transition for two GaAs$_{1-x}$N$_x$ samples, with nominal values of $x = 0.008$ and 0.01, at 80 K and with special emphasis on all the zone-boundary phonons to further investigate the symmetry of the E_+ state.

The Raman scattering response near resonance can be approximated using the initial electronic state $|0\rangle$ for the valence band maximum at Γ and the intermediate state $|m\rangle$ for the level at resonance with energy E_m and lifetime Γ_m^{-1} as [30]

$$I(\omega) \left| \frac{\langle 0|H_{eR}|m\rangle\langle m|H_{ep}|m\rangle\langle m|H_{eR}|0\rangle}{(\hbar\omega - E_m - i\Gamma_m)(\hbar\omega - E_m - \hbar\omega_p - i\Gamma_m)} \right|^2, \qquad (7.2)$$

where $h\omega$ is an incident photon energy, $h\omega_p$ is a phonon energy, and H_{eR} and H_{ep} are the electron-radiation and electron-phonon interaction Hamiltonians, respectively. If the intermediate state $|m\rangle$ is a Bloch state with a well-defined momentum $\mathbf{k}$, then only zone-center (Γ) phonons with momentum $\mathbf{q} = 0$ can be involved in the Raman scattering process within the dipole approximation. However, if $|m\rangle$ is not a Bloch state, then electron momentum $\mathbf{k}$ is no longer a good quantum number for the Raman scattering process, and the momenta of phonons participating in the scattering process near resonance are determined by the various $\mathbf{k}$ components of $|m\rangle$ [31]. For example, if $|m\rangle$ is spatially localized so that it is delocalized in $\mathbf{k}$ space, it can be written as the sum of various $\mathbf{k}$ states $|\mathbf{k}\rangle$ e.g., $|\Gamma\rangle$, $|L\rangle$, and $|X\rangle$:

$$|m\rangle = \sum_{\mathbf{k}} a_{\mathbf{k}} |\mathbf{k}\rangle, \qquad (7.3)$$

where $a_{\mathbf{k}}$ is a $\mathbf{k}$-dependent constant. Then the first and last matrix elements in Equation 7.2, i.e., $\langle 0|H_{eR}|m\rangle$ and $\langle m|H_{eR}|0\rangle$, are always nonzero, since both $|m\rangle$ and $|0\rangle$ have Γ components. On the other hand, for the electron-phonon interaction matrix element $\langle m|H_{ep}|m\rangle$, both the initial and the final states have Γ, L, and X components, and there exist possible combinations that allow activation of the zone-boundary phonons, as summarized in Table 7.2. Therefore, the symmetry of an electronic state can be probed by monitoring the activation of various zone-boundary phonons near resonance.

In Figure 7.4, two representative Raman spectra are shown [32]. According to the usual Raman scattering selection rules in a zinc-blende crystal, the LO_Γ phonon is allowed, whereas the TO_Γ phonon is forbidden in this scattering geometry, $z(YY)\bar{z}$. The lower spectrum in Figure 7.4 is obtained for off-resonance excitation at 1.797 eV. The TO_Γ and TO_X phonons are observed at 272 and 257 cm^{-1}, respectively, and broad disorder-activated transverse acoustic (DATA) and longitudinal acoustic (DALA) phonons are weakly observed around 80 and 200 cm^{-1}, respectively. The activation of the forbidden TO phonons as well as the broad acoustic phonon features originate from the relaxation of the Raman scattering selection rule and crystal momentum conservation, both produced by nitrogen-induced alloy disorder. The upper spectrum in Figure 7.4, which is resonantly excited near E_+, is strikingly different from the lower one. On one hand, all the phonon modes observed in the lower spectrum with off-resonance excitation get stronger in the upper spectrum due to resonance enhancement, as expected. On the other hand, many sharp zone-boundary phonons such as LO_L, LA_X, LA_L, TA_X, and TA_L, whose traces cannot be found in the lower spectrum, appear quite strong in the upper spectrum. This striking contrast in their emergence between off- and near-resonance excitations strongly suggests that the alloy-disorder-induced effects are not responsible

Table 7.2 Possible Combinations of Initial and Final States Resulting in Nonzero Matrix Element $\langle m|H_{ep}|m\rangle$

$\langle a|$	Phonon Involved	$|a\rangle$
Γ	Γ	Γ
L	Γ	L
X	Γ	X
L	L	Γ
X	X	Γ
Γ	L	L
Γ	X	X

for the activation of the sharp zone-boundary phonons distinctly observed in the upper spectrum in Figure 7.4. Rather, it is the symmetry of the E_+ state ($|m\rangle$) that plays a crucial role through the matrix element $\langle m|H_{ep}|m\rangle$ in Equation 7.2 in activating these sharp zone-boundary phonons.

Figure 7.5 illustrates the resonant Raman scattering intensity profiles for various longitudinal phonons. They show distinct maxima near 1.88 eV (±0.01 eV), which is 50 meV higher than the energy position of E_+ (1.83 eV) measured using the electromodulation technique at $T = 90$ K [33]. This indicates that the outgoing resonance, rather than the incoming resonance, dominates the light-scattering process. It should be emphasized here that the intensity resonance profiles for all the acoustic phonons in Figure 7.5 do not include any contribution from the broad DATA and DALA background, but instead they represent sharp zone-boundary phonon contributions only. The resonance enhancement of zone-boundary phonons at L and X is also distinctly observed for transverse vibrational modes, as shown in Figure 7.6. While the intensity resonance enhancement of the LO_Γ and the TO_Γ phonons at the E_+ state can be explained without invoking non-Γ components of E_+ (see Equation 7.2), those for zone-boundary (L and X) phonons cannot be accounted for, unless E_+ does include L and X components in its wavefunction.

In summary, our resonant Raman scattering results provide strong evidence of significant L and X components in the wavefunction of the nitrogen-induced E_+ state in GaAs$_{1-x}$N$_x$. We find that various zone-boundary phonons at L and X not only emerge as strong and sharp Raman features near the E_+ resonance, but they also exhibit the same intensity resonance enhancement as that for zone-center phonons, LO_Γ and TO_Γ.

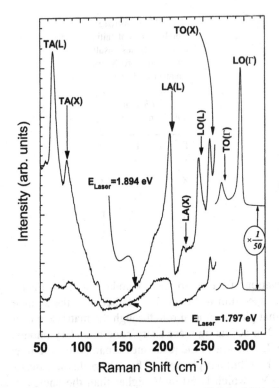

Figure 7.4 Raman spectra of GaAs$_{1-x}$N$_x$ with $x = 0.8\%$ at $T = 80$ K for 1.894-eV (upper curve) and 1.797-eV (lower curve) excitations. The upper spectrum at 1.894 eV corresponds to resonant excitation near E_+, whereas the spectrum at 1.797 eV represents off-resonance excitation. Curves are vertically shifted for clarity, and the intensity of the zone-center (Γ) phonons (260 cm^{-1}) are scaled down by 50. The observed phonon energies are TA$_L$ = 65 cm^{-1}, TA$_X$ = 83 cm^{-1}, LA$_L$ = 209 cm^{-1}, LA$_X$ = 226 cm^{-1}, LO$_L$ = 244 cm^{-1}, TO$_X$ = 257 cm^{-1}, TO$_\Gamma$= 272 cm^{-1}, and LO$_\Gamma$ = 295 cm^{-1}.

7.3.1.2.2 *Electronic States*

In this section we present new resonant Raman scattering results on GaAsN for the perturbation of the conduction-band electronic states near the L point that is induced by nitrogen impurities, which helps shed light on the origin of E_+.

A typical resonance profile of the LO$_\Gamma$ phonon intensity in GaAs$_{1-x}$N$_x$ is shown in Figure 7.7a, where a strong resonance enhancement occurs at the excitation energy E_I that is approximately near the E_+ transition energy measured by electroreflectance. A very weak LO phonon intensity maximum is also observed near the $E_0 + \Delta_0$ transition energy of GaAs$_{1-x}$N$_x$. The LO phonon Raman signatures in ternary semiconductor alloys exhibit line-width broadening due to alloy-induced disorder, but the phonon line width shows no excitation-energy dependence in resonant Raman scattering. Recently, Cheong et al. [29] reported that the LO Raman signature in GaAs$_{1-x}$N$_x$ showed a very unusual line-width broadening for resonant exci-

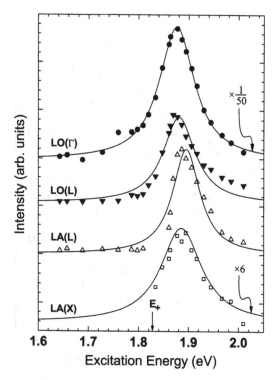

Figure 7.5 Resonance profiles of various longitudinal phonon modes in GaAs$_{1-x}$N$_x$ ($x = 0.8\%$) measured at $T = 80$ K. Curves are vertically shifted for clarity, and the intensity of the LO$_\Gamma$ phonon is scaled down by a factor of 50, while the LA$_X$ phonon intensity is scaled up by a factor of 6. E_+ measured in modulated reflectivity at $T = 90$ K is shown for reference. Lines are guides for the eye.

tation energies below the E_+ transition. As illustrated in the inset of Figure 7.7, the width broadening is highly asymmetric toward the lower-energy side of the LO Raman line. Since the LO phonon energy is the highest at the zone center and continuously decreases as the phonon wave vector $\mathbf{q}$ moves away from the Γ point, the observed asymmetric broadening strongly suggests that it is due to activation of LO phonons with nonzero wave vectors that span a significant portion of the Brillouin zone. The selective activation of non-Γ phonons only for certain excitation energies indicates that the asymmetric line-width broadening is not due to nitrogen-induced disorder, but rather is closely related to the nature of the intermediate electronic state participating in the resonant Raman scattering process. Since an electronic state that is delocalized in the Brillouin zone must be strongly localized in real space, the observed asymmetric LO phonon line-width broadening indicates that the intermediate electronic state involved in this resonant Raman scattering process is strongly spatially localized.

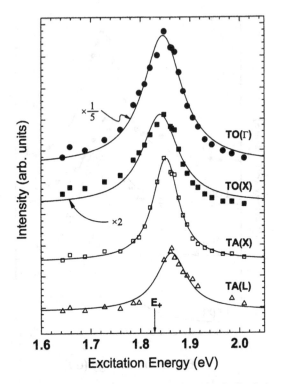

Figure 7.6 Resonance profiles of various transverse phonon modes in GaAs$_{1-x}$N$_x$ with $x =$ 0.8%, measured at $T = 80$ K. Curves are vertically shifted for clarity, and the intensity of the TO$_\Gamma$ phonon is scaled down by a factor of five, while the TO$_X$ phonon intensity is scaled up by a factor of two. E_+ measured in modulated reflectivity at $T = 90$ K is shown for reference. Lines are guides for the eye.

Similar arguments can be applied to account for the concomitant res-onance enhancement of the TO/LO phonon intensity ratio: the non-Γ com-ponents of the localized electronic state involved in the Raman scattering process relax the momentum-conservation rule, resulting in further strengthening of the forbidden TO phonon intensity aside from the excita-tion-energy-independent alloy-disorder effect. Resonance enhancement of the LO$_\Gamma$ phonon line width and the concomitant enhancement of TO/LO intensity ratio for 0.78% GaAsN are displayed in Figure 7.7b, where two distinct maxima (labeled as E_W and E_W') for both the LO FWHM (full width at half maximum) and TO/LO intensity ratio are observed. Similar double maxima E_W and E_W' are distinctly observed in GaAsN samples with several different nitrogen compositions exceeding $\approx 0.35\%$, and their peak positions are plotted as a function of nitrogen concentration in Figure 7.8. E_W and E_W' are not clearly resolved for $x < 0.3\%$, presumably due to their proximity.

It should be pointed out here that Cheong *et al.* [29] used data obtained, as a first approximation, in an unpolarized configuration in their paper.

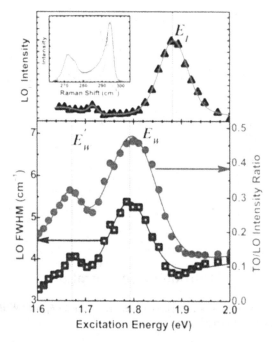

Figure 7.7 Upper panel: LO$_\Gamma$ phonon intensity resonance profile for $x = 0.78\%$; Raman spectra measured at $T = 80$ K. Inset: Normalized (with respect to the GaAs LO phonon) Raman spectra excited with photon energies of 1.789 eV (dotted curve) and 1.907 eV (solid curve). Lower panel: Resonance profile ($T = 80$ K) for the FWHM (empty square) of the LO$_\Gamma$ phonon and TO$_\Gamma$/LO$_\Gamma$ intensity ratio (filled circle) for $x = 0.78\%$. Solid lines are guides for the eye.

However, since the alloy-disorder-induced (symmetric) line-width broadening effect is stronger in a forbidden geometry than in an allowed one, we have used only allowed scattering geometry in the present work in order to minimize the alloy-disorder-induced line-width broadening effect. This could explain why the E_W observed in the unpolarized configuration (Cheong et al. [29]) and the E_W measured in the allowed scattering geometry in the present work exhibit a larger difference for high-concentration samples (>1%), where the alloy-disorder effect is increasingly important. It is also worth mentioning that the quality of recently grown samples studied in the current work have improved compared with the samples studied previously (Cheong et al. [29]).

The excitation-energy positions for the LO phonon intensity maximum (E_I) are shown in Figure 7.8. There is clearly a strong correlation between E_I and E_+ transition observed in modulated electroreflectance [2, 34]. In contrast, none of the E_W and E'_W exhibits any correlation with E_+, implying that they have a completely different origin than that for E_+. It is important to note that the line width of the LO$_\Gamma$ phonon in pure GaAs remains almost constant despite changing the excitation energy from 1.55 eV to 2.0 eV, a

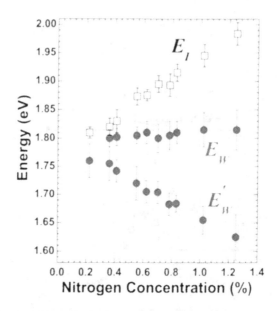

Figure 7.8 Energy positions of LO_Γ phonon intensity maximum E_I and FWHM maxima E_W and E'_W as a function of nitrogen concentration.

range that encompasses the $E_0 + \Delta_0$ transition of pure GaAs and the E_+ transition of $GaAs_{1-x}N_x$ for $x < 0.03$ at $T = 300$ K [29]. Thus, the observed resonance maxima E_W and E'_W for the LO phonon line width are solely due to nitrogen-induced changes to the electronic states.

We will first discuss the origin of the E_+ state. Numerous theoretical [4, 5, 7, 16] and experimental [21, 29] studies have indicated that the E_+ state originates from the nitrogen-induced coupling of the conduction-band minima at the Γ, L, and X points that break the L-valley degeneracy, resulting in a triplet state L_{3c} and a singlet state L_{1c} [4, 5, 7, 16] that has been attributed to the origin of E_+ [7]. As evidenced by the activation and strong resonance enhancement of the sharp L and X zone-boundary phonons near the E_+ transition [21, 29], E_+ comprises wavefunctions with definite Γ, L, and X components induced by the interconduction-band mixing. In contrast, strongly localized states such as E_W and E'_W comprise wavefunctions with k-components from throughout the entire Brillouin zone (unlike L_{3c}, they have relatively very little of Γ, L, and X components [16]), leading to a significant asymmetric line-width broadening of the LO-phonon Raman signature for resonance at these localized states. If the two-level interaction of N_X with the CBM at Γ and their resulting hybridization into the E_+ and E_- levels, whose consequent level repulsion is the principal mechanism for the giant band-gap bowing, then one would have expected to observe the asymmetric line-width broadening phenomenon of the LO phonon under resonance at the E_+ and E_- levels. But this has turned out precisely not to be true for the resonance Raman scattering studies at E_+ as well as at E_- [35].

Instead, conventional Raman intensity resonances are observed at E_+ and E_-, whereas asymmetric line-width broadening resonances are observed at E_W and E_W', indicating that these states are much more delocalized in **k**-space. At zero nitrogen doping, both E_W and E_W' extrapolate to a value located below the E_1 transition in GaAs by 1.2 eV, which is precisely the energy difference between the valence-band maxima at the Γ and L points. The E_1 transition in GaAs refers to the transition between the conduction- and valence-band extrema that occur at the **k**-point located along the [111] direction of the Brillouin zone, which is close to the L point. This suggests that both those transitions emerge as a result of the perturbation of the L-point conduction band edge by nitrogen impurities. In electromodulated reflectance studies of dilute GaAsN alloys, transitions denoted as E_1 (rises slowly with increased nitrogen), E_+ (rapidly rises with increasing nitrogen), and E_* (falls rapidly with increased nitrogen) have been observed [33]. The first transition has also been observed in ellipsometry studies, and the last two have been recently corroborated in more-precise electromodulated reflectance studies as a function of nitrogen concentration [33–35]. At zero nitrogen doping, the data in Figure 7.8 indicate that all the transitions originate from near the conduction band edge at the L point of GaAs. Evidently, the quadruply degenerate E_1 transition in GaAs splits into three levels — E_+, E_W, and E_W' — as a result of perturbation by nitrogen impurities.

Since this model explaining the data in Figure 7.8 implies a splitting of the GaAsN conduction band near the L point that differs from that assumed in the theoretical models discussed earlier [7, 16], we will further elaborate on this. The theoretical models assume that with the origin chosen to be located at the site of a nitrogen impurity, the point group symmetry remains unchanged from T_d. One then looks for the compatibility of the quadruply degenerate conduction-band states near the L point of GaAs with the irreducible representations of the T_d point group whose point operations are centered on the nitrogen impurity site, and one obtains $4a_1 \rightarrow a_1 + t_2$. Note, however, that in the presence of an additional nitrogen impurity lying in the vicinity of the first impurity, that the point symmetry is no longer T_d but is reduced and depends on the choice of the site for the second impurity. Corresponding to each site m for the second impurity is an NN_m pair state, and several of these pairs that generate bound states have been observed in dilute GaAsN together with triplet states and larger clusters. Although less abundant than the isolated nitrogen impurity, these states are far more spatially localized. The additional splitting of the triplet state t_2 should therefore not come as a surprise.

In summary, we have observed two distinct maxima, E_W and E_W', in the resonant Raman scattering profile for the LO-phonon asymmetric line-width broadening in dilute GaAsN. The data lead to a new interpretation for the splitting of the quadruply degenerate conduction band near the L point. The study also reveals asymmetric line-width-broadening resonances to be a powerful signature for studying strongly localized impurity states in semiconductors.

7.3.2 GaPN

Raman scattering has been reported to be a useful tool to study structural properties of GaPN via monitoring Raman phonon behavior. (See, for example, Buyanova et al. [37] and references therein.) On the other hand, using resonant Raman scattering studies on GaPN, it is possible to observe the evolution of the X-point conduction band edge as a function of N-doping quite distinctly, even at higher N concentrations. This is not so clearly observed in absorption measurements. Our results reveal that with increased N-doping, this edge remains practically stationary, but it is the increasingly deeper NN_m pair and cluster levels that appear to track the position of the absorption edge [30]. The study reveals that the behavior of GaPN with increased N-doping is quite distinct from that of GaAsN, and that in the former, the band edge clearly evolves from nitrogen-impurity states.

7.3.2.1   *Resonant Raman scattering*

Figure 7.9 illustrates Raman spectra of $GaP_{1-x}N_x$ with $x = 0.01$ and 0.02 measured at two different excitation energies. Spectral features from samples with different x are essentially the same. It is clearly seen that there are notable differences between the Raman spectra of GaP and that of $GaP_{1-x}N_x$:

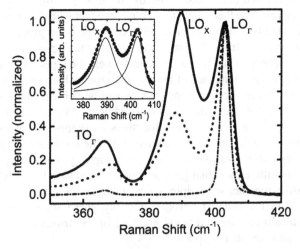

(a)

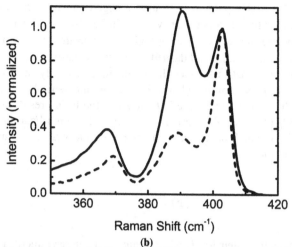

(b)

Figure 7.9 (a) Raman spectra of $GaP_{0.99}N_{0.01}$ in $z(YY)\bar{z}$ scattering geometry ($T = 300$ K) with excitation energies of 1.984 eV (dashed line) and 2.331 eV (solid line). Three phonon modes are seen in the region shown in the figure: TO_Γ phonon near 367 cm^{-1}, LO_X phonon near 388 cm^{-1}, and LO_Γ phonon near 403 cm^{-1}, in the order of increasing energy. Each spectrum is normalized to the LO_Γ phonon intensity. The $z(YY)\bar{z}$ Raman spectrum of GaP substrate ($T = 300$ K) with excitation energy of 2.331 eV (dashed-dotted line) is shown for comparison. Inset: Fitting result of phonons using Lorentzian (symmetric, LO_X phonon) and Fano (asymmetric, LO_Γ phonon) functions for $GaP_{0.99}N_{0.01}$ with an excitation energy of 2.331 eV (circles represent experimental data). (b) Raman spectra of $GaP_{0.98}N_{0.02}$ in $z(YY)\bar{z}$ scattering geometry ($T = 300$ K) with excitation energies of 1.986 eV (dashed line) and 2.331 eV (solid line).

1. There is an additional phonon mode (indicated as LO_X) in GaPN near 388 cm^{-1}. (The energy of this mode is very close to the X-point LO phonon energy, and it is generally referred to as LO_X mode. See, for example, Volicek et al. [38].)

2. The symmetry-forbidden TO_Γ phonon is relatively much stronger in GaPN.

3. The width of the LO_Γ phonon broadens significantly with N doping.

The first two differences can be explained by relaxation of the momentum-conservation rules and the symmetry-selection rules due to the symmetry lowering of the lattice introduced by N doping, and the third is due to the N-induced alloy-disorder effect in the lattice dynamics of GaP, which symmetrically broadens the LO_Γ phonon. Note that a small (≤ 2 cm^{-1}) shift is observed in the energies of LO_X and TO_Γ phonons as a function of the laser excitation energy, which is not completely understood at this point. Such a small energy shift, however, does not affect the main result of the current work. Note also that at the excitation energy of 2.331 eV, the intensity of the LO_X phonon is relatively large compared with that of the LO_Γ phonon, which is a peculiarity that will be discussed later.

Comparing the two spectra for $GaP_{1-x}N_x$ measured with two different excitation energies (solid and dashed lines), it is clear that the LO_Γ phonon exhibits an asymmetric line-width broadening, i.e., only the low-energy side of the LO_Γ phonon is enhanced when the incident photon energy is 2.331 eV. In the inset of Figure 7.9, we show the fitting result of the LO_X and the LO_Γ phonons. The best fit was obtained when a Fano profile (see Equation 7.4), which describes an asymmetric line shape, was used to fit the LO_Γ phonon

$$I(\omega) = I_0 \frac{(q+\varepsilon)^2}{1+\varepsilon^2}, \qquad (7.4)$$

where $I(\omega)$ is the intensity, I_0 is a constant, q is an asymmetry parameter that defines the "asymmetry" of the line shape, and $\varepsilon = (\omega - \omega_0)/\Gamma$ with phonon frequency ω_0 and phonon line width Γ. (Note that Equation 7.4 was used to parameterize the LO_Γ phonon line shape only. The physical process underlying the asymmetric line width is *not* Fano interference.)

This asymmetric line-width broadening is elaborated more clearly in the spectra of Figure 7.9a. As will be discussed later, this reveals that the line width of the LO_Γ phonon shows a "resonant" behavior with respect to the excitation energy, and this behavior cannot be simply explained by the N-induced alloy-disorder effect. In fact, a similar asymmetric line-width broadening resonance was also recently reported for $GaAs_{1-x}N_x$ [22, 29], as seen in Section 7.3.1.2, and the same explanation for that phenomenon can be applied to the present case. When the intermediate state has some degree of spatial localization, it can be expressed using Equation 7.3, and so phonons with nonzero wave vector q are allowed to participate in the electron-phonon interaction matrix element in Equation 7.2, thus resulting in zone-center phonon line-width broadening.

Figure 7.10a illustrates the excitation-energy dependence of the LO_Γ phonon FWHM (full width at half maximum) for $GaP_{0.985}N_{0.015}$. It reveals two distinct maxima in the line-width resonance profile, which are denoted as E_W and E'_W. Similar double maxima are also observed for all the $GaP_{1-x}N_x$ samples we have studied. In contrast, the FWHM of the GaP LO_Γ phonon (triangles) shows only simple monotonic energy dependence attributed to the spectral resolution of the measurement. We thus conclude that the double maxima in the resonance profile for the asymmetric LO_Γ phonon line-width broadening are solely a N-induced effect and, as discussed above, are due to the spatially localized nature of the electronic states, which serve as the intermediate states for the resonant Raman scattering process. Note that the intensity ratio between the LO_X and the LO_Γ phonons almost exactly follows the line-width resonance

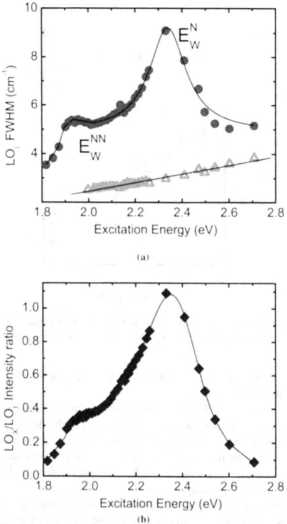

Figure 7.10 (a) Excitation energy dependence (resonance profile) of LO_Γ phonon FWHM for $GaP_{0.985}N_{0.015}$ (open circles) and-gap (filled triangles). The maxima are labeled as E_w and E'_w. (b) Intensity ratio of LO_X and LO_Γ phonons. Note that the ratio maxima coincide with the FWHM maxima (dotted lines). Lines are guides for the eye.

profile, as shown in Figure 7.10b. The reason for this is that the non-Γ components of these localized electronic states involved in the Raman scattering process relax the momentum-conservation rule (see Equations 7.2 and 7.3), resulting in a further strengthening of the forbidden LO_X phonon intensity in addition to the excitation-energy-independent alloy-disorder effect. Since N_X is located barely 10 meV below the conduction-band minimum, which is near the X-point, it generates a strong perturbation that introduces a delocalized

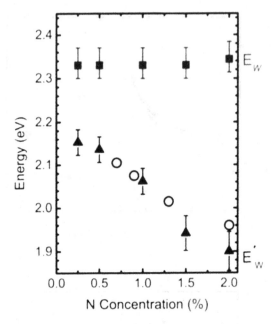

Figure 7.11 Energy positions of FWHM maxima E_W (circles) and E'_W (triangles) as a function of nitrogen concentration. Open squares are photomodulated transmission data from Shan et al. [14].

character for these band-edge states, thus leading to a selective resonance enhancement of LO_X.

Figure 7.11 shows the "resonant" energies for the line-width maxima as a function of nitrogen concentration. As seen in the figure, E_W remains stationary throughout the concentration range studied. We propose that E_W is closely related to the X-conduction band-edge level because E_W (2.33 eV) is very close to this level (2.27 eV) after subtracting the LO_Γ phonon energy (0.05 eV), assuming an outgoing resonance.

In contrast, E'_W decreases monotonically as the nitrogen concentration is increased. Note that E'_W closely follows the position of the conduction band edge (CBE) as measured at room temperature by photomodulated transmission measurements [14] shown in Figure 7.11. The behavior of E'_W can be explained as follows. The interaction between pairs of isolated N_X impurity states generates a variety of NN_m pair states situated below N_X, where m decreases with decreasing distance between the two interacting nitrogen atoms [39], as well as cluster states such as nitrogen triplets. [40] Since these NN pair states and cluster states are more strongly localized in real space than N_X, they too are expected to produce significant asymmetric line-width broadening of the LO phonon in resonant Raman scattering. It is well known that the random distribution of substitutional impurities follows a Poisson distribution, with the most probable nearest-neighbor distance $\langle r_{NN} \rangle$ approximately given by $0.54\ n^{-1/3}$, where n is the nitrogen concentration [41]. The term $\langle r_{NN} \rangle$ decreases with increasing nitrogen concentration, and the NN pair interaction between closely separated pairs becomes more dominant. Thus, the closer the NN pair separation, the more localized the NN pair state is, and the greater is its contribution to the asymmetric line-width broadening as the nitrogen concentration is increased. Hence, we attribute the observed line-width broadening maximum E'_W to originate from the proliferation of nitrogen cluster states and NN pair states with closer pair separation, which makes E'_W shift downward with increasing nitrogen concentration.

In summary, the ability to track the evolution of E_W quite distinctly from that of E'_W, even at high N concentrations, reveals that it is the closer-lying pairs and clusters that are contributing to the rapidly lowering absorption edge in GaPN, whereas the X-point conduction band edge is observed to remain almost stationary, even at higher N concentration. This study also reveals resonant Raman scattering to be a powerful tool for investigating strongly localized impurity states in semiconductors.

7.4   CONCLUSIONS

We have studied GaAsN and GaPN using Raman scattering spectroscopy. In GaAsN, we have been able to use the nitrogen local vibration mode as a measure of nitrogen concentration in the system, even where direct measurements of concentration were not previously possible. The activation of various forbidden zone-boundary phonons in resonant Raman scattering at the E_+ transition reveals that the nitrogen-induced resonant state E_+ in GaAsN contains significant components of L and X character, in addition to Γ character. It is also observed that hostlike LO_Γ phonons in both systems (≈ 292 cm^{-1} in GaAsN and ≈ 403 cm^{-1} in GaPN) show distinctive double line-width resonances due to the spatially localized intermediate states, though their origins are quite different. In GaAsN, the intermediate states originate from the splitting of L-point conduction band, whereas in GaPN, the intermediate states are X-point conduction band and nitrogen pair/cluster states. The results presented in this chapter demonstrate that asymmetric line-width-broadening resonances can provide a powerful means to study semiconductor systems with strongly localized impurity states.

Acknowledgments

We thank M. C. Hanna and J. F. Geisz for the samples used in our experiments and H. M. Cheong and M. J. Seong for contributions to this research. The support of the Office of Science/Basic Energy Sciences/Division of Material Sciences is gratefully acknowledged.

References

1. Weyers, M., Sato, M., and Ando, H., "Red Shift of Photoluminescence and Absorption in Dilute GaAsN Alloy Layers," *Jpn. J. Appl. Phys. (Part 2)* 31: L853–855 (1992).

2. Perkins, J.D., Mascarenhas, A., Zhang, Y., Geisz, J.F., Friedman, D.J., Olson, J.M., and Kurtz, S.R., "Nitrogen-Activated Transitions, Level Repulsion, and Band-gap Reduction in $GaAs_{1-x}N_x$ with $x < 0.03$," *Phys. Rev. Lett.* 82: 3312–3315 (1999).

3. Shan, W., Walukiewicz, W., Ager III, J.W., Haller, E.E., Geisz, J.F., Friedman, D.J., Olson, J.M., and Kurtz, S.R., "Band Anticrossing in GaInNAs Alloys," *Phys. Rev. Lett.* 82: 1221–1224 (1999); Walukiewicz, W., Shan, W., Wu, J., and Yu, K.M., "Band Anticrossing in III-N-V Alloys: Theory and Experiments," Ed. I.A. Buyanova and W.M. Chen (New York: Taylor & Francis, 2003), PAGES.

4. Jones, E.D., Modine, N.A., Allerman, A.A., Kurtz, S.R., Wright, A.F., Tozer, S.T., and Wei, X., "Bandstructure of $In_xGa_{1-x}As_{1-y}N_y$ Alloys and Effects of Pressure" *Phys. Rev. B* 60: 4430–4433 (1999).

5. Wang, L.W., "Large-Scale Local-Density-Approximation Band Gap-Corrected GaAsN Calculation," *Appl. Phys. Lett.* 78: 1565–1567 (2001).

6. Kent, P.R.C., and Zunger, A., "Evolution of III-V Nitride Alloy Electronic Structure: The Localized to Delocalized Transition," *Phys. Rev. Lett.* 86: 2613–2616, (2001).

7. Szwacki, N.G., and Boguslawski, P., "GaAs:N and GaAs:Bi Alloys: Symmetry-Induced Effects," *Phys. Rev. B* 64: 161201/1–4 (2001).

8. Hjalmarson, H.P., Vogl, P., Wolford, D.J., and Dow, J.D., "Theory of Substitutional Deep Traps in Covalent Semiconductors," *Phys. Rev. Lett.* 44: 810–813 (1980).

9. Wolford, D.J., Bradley, J.A., Fry, K., and Thompson, J., "The Nitrogen Isoelectronic Trap in GaAs," in *Proceedings of the 17th International Conference on the Physics of Semiconductors*, ed. J.D. Chadi and W.A. Harrison (New York: Springer, 1985), 627–634.

10. Liu, X., Pistol, M.E., Samuelson, L., Schwetlick, S., and Seifert, W., "Nitrogen Pair Luminescence in GaAs," *Appl. Phys. Lett.* 56: 1451–1453 (1990).

11. Czaja, W., "Isoelectronic Impurities in Semiconductors," *Festkoerperprobleme* 11: 65–85 (1971); Dean, P.J., "Isoelectronic Traps in Semiconductors (Experimental)," *J. Lumin.* 7: 51–78 (1973); Zhang, Y., and Ge, W., "Behavior of Nitrogen Impurities in III-V Semiconductors," *J. Lumin.* 85: 247–260 (2000).

12. Zhang, Y., Fluegel, B., Mascarenhas, A., Xin, H.P., and Tu, C.W., "Optical Transitions in the Isoelectronically Doped Semiconductor GaP:N: An Evolution from Isolated Centers, Pairs, and Clusters to an Impurity Band," *Phys. Rev. B* 62: 4493–4500 (2000).

13. Baillargeon, J.N., Cheng, K.Y., Hoefler, G.E., Pearah, P.J., and Hsieh, K.C., "Luminescence Quenching and the Formation of the $GaP_{1-x}N_x$ Alloy in GaP with Increasing Nitrogen Content," *Appl. Phys. Lett.* 60: 2540–2542 (1992); Miyoshi, S., Yaguchi, H., Onabe, K., Ito, R., and Shirako, Y., "Metalorganic Vapor Phase Epitaxy of $GaP_{1-x}N_x$ Alloys on GaP," *Appl. Phys. Lett.* 63: 3506–3508 (1993).

14. Shan, W., Walukiewicz, W., Yu, K.M., Wu, J., Ager III, J.W., Haller, E.E.H., Xin, H.P., and Tu, C.W., "Nature of Fundamental Band-gap in GaN_xP_{1-x} Alloys," *Appl. Phys. Lett.* 76: 3251–3253 (2000); Wu, J., Walukiewicz, W., Yu, K.M., Ager III, J.W., Haller, E.E., Hong, Y.G., Xin, H.P., and Tu, C.W., "Band Anticrossing in $GaP_{1-x}N_x$ Alloys," *Phys. Rev. B* 65: 241303/1–4 (2002).

15. Leibiger, G., Gottschalch, V., Schubert, M., Benndorf, G., and Schwabe, R., "Evolution of the Optical Properties of III-V Nitride Alloys: Direct Band-to-Band Transitions in GaN_yP_{1-y} ($0 \leq y \leq 0.029$)," *Phys. Rev. B* 65: 245207/1–6 (2002).

16. Kent, P.R.C., and Zunger, A., "Theory of Electronic Structure Evolution in GaAsN and GaPN Alloys," *Phys. Rev. B* 64: 115208/1–23 (2001).

17. Hayes, W., and Loudon, R., *Scattering of Light by Crystals* (New York: John Wiley & Sons, 1978).

18. Yu, P.Y., and Cardona, M., *Fundamentals of Semiconductors* (Heidelberg: Springer-Verlag, 2001).
19. Sherwin, M.E., and Drummond, T.J., "Predicted Elastic Constants and Critical Layer Thicknesses for Cubic Phase AlN, GaN, and InN on β-SiC," *J. Appl. Phys.* 69: 8423–8425 (1991).
20. Seong, M.J., Hanna, M.C., and Mascarenhas, A., "Compositional Dependence of Raman Intensity of the Nitrogen Localized Vibrational Mode in GaAs$_{1-x}$N$_x$," *Appl. Phys. Lett.* 79: 3974–3976 (2001).
21. Seong, M.J., Mascarenhas, A., and Geisz, J.F., "Γ-L-X-mixed Symmetry of Nitrogen-Induced States in GaAs$_{1-x}$N$_x$ Probed by Resonant Raman Scattering," *Appl. Phys. Lett.* 79: 1297–1299 (2001).
22. Mascarenhas, A., Seong, M.J., Yoon, S., Verley, J.C., Geisz, J.F., and Hanna, M.C., "Evolution of Electronic States in GaAs$_{1-x}$N$_x$ Probed by Resonant Raman Spectroscopy," *Phys. Rev.* B68:233201/1–3 (2004).
23. Alt, H.C., Egorov, A.Y., Riechert, H., Wiedemann, B., Meyer, J.D., Michelman, R.W., and Bethge, K., "Infrared Absorption Study of Nitrogen in N-implanted GaAs and Epitaxially Grown GaAs$_{1-x}$N$_x$ Layers," *Appl. Phys. Lett.* 77: 3331–3333 (2000).
24. Prokofyeva, T., Sauncy, T., Seon, M., Holtz, M., Qiu, Y., Nikishin, S.A., and Temkin, H., "Raman Studies of Nitrogen Incorporation in GaAs$_{1-x}$N$_x$," *Appl. Phys. Lett.* 73: 1409–1411 (1998).
25. Spruytte, S.G., Coldren, C.W., Harris, J.S., Wampler, W., Krispin, P., Ploog, K., and Larson, M.C., "Incorporation of Nitrogen in Nitride-Arsenides: Origin of Improved Luminescence Efficiency after Anneal," *J. Appl. Phys.* 89: 4401–4406 (2001).
26. Zhang, Y., Mascarenhas, A., Geisz, J.F., Xin, H.P., and Tu, C.W., "Discrete and Continuous Spectrum of Nitrogen-Induced Bound States in Heavily Doped GaAs$_{1-x}$N$_x$," *Phys. Rev. B* 63: 085205/1–8 (2001).
27. Mintairov, A.M., Blagnov, P.A., Melehin, V.G., Faleev, N.N., Merz, J.L., Qiu, Y., Nikishin, S.A., and Temkin, H., "Ordering Effects in Raman Spectra of Coherently Strained GaAs$_{1-x}$N$_x$," *Phys. Rev. B* 56: 15836–15841 (1997).
28. Wagner, J., Kohler, K., Ganser, P., and Herres, N., "GaAsN Interband Transitions Involving Localized and Extended States Probed by Resonant Raman Scattering and Spectroscopic Ellipsometry," *Appl. Phys. Lett.* 77: 3592–3594 (2000).
29. Cheong, H.M., Zhang, Y., Mascarenhas, A., and Geisz, J.F., "Nitrogen-Induced Levels in GaAs$_{1-x}$N$_x$ Studied with Resonant Raman Scattering," *Phys. Rev. B* 61: 13687–13690 (2000).
30. Yoon, S., Seong, M.J., Geisz, J.F., Duda, A., and Mascarenhas, A., "Evolution of Electronic States in GaP$_{1-x}$N$_x$ Studied by Resonant Raman Scattering Spectroscopy," *Phys. Rev. B* 67: 235209/1–4 (2003).
31. Trallero-Giner, C., Cantarero, A., Cardona, M., and Mora, M., "Impurity-Induced Resonant Raman Scattering," *Phys. Rev. B* 45: 6601–6613 (1992).
32. Hellwege, K.H., and Madelung, O., eds., *Semiconductors, Physics of Group IV Elements and III-V Compounds*, Landolt-Bornstein New Series, Group III, vol. 17, part A (Berlin: Springer-Verlag, 1982).
33. Perkins, J.D., Mascarenhas, A., Geisz, J.F., and Friedman, D.J., "Conduction-Band-Resonant Nitrogen-Induced Levels in GaAs$_{1-x}$N$_x$ with $x < 0.03$," *Phys. Rev. B* 64: 121301/1–4 (2001).
34. Francoeur, S., Seong, M.J., Hanna, M.C., Geisz, J.F., Mascarenhas, A., Xin, H.P., and Tu, C.W., "Origin of the Nitrogen-Induced Optical Transitions in GaAs$_{1-x}$N$_x$," *Phys. Rev. B* 68: 075207/1–5 (2003).
35. Seong, M.J., Cheong, H.M., Yoon, S., Geisz, J.F., and Mascarenhas, A., "Symmetry of GaAs$_{1-x}$N$_x$ Conduction Band Minimum Probed by Resonant Raman Scattering," *Phys. Rev. B* 67: 153301/1–4 (2001).

36. Leibiger, G., Gottschalch, V., Rheinlander, B., Sik, J., and Schubert, M., "Nitrogen Dependence of the GaAsN Interband Critical Points E_1 and $E_1+\Delta_1$ Determined by Spectroscopic Ellipsometry," *Appl. Phys. Lett.* 77: 1650–1652 (2000).

37. Buyanova, I.A., Chen, W.M., Goldys, E.M., Xin, H.P., and Tu, C.W., "Structural Properties of a GaN_xP_{1-x} Alloy: Raman Studies," *Appl. Phys. Lett.* 78: 3959–3961 (2001).

38. Volicek, V., Gregora, I., Riede, V., and Neumann, H., "Raman Scattering Study of GaP:N Epitaxial Layers," *J. Phys. Chem. Solids* 49: 797–805 (1988).

39. Thomas, D.G., and Hopfield, J.J., "Isoelectronic Traps due to Nitrogen in Gallium Phosphide," *Phys. Rev.* 150: 680–689 (1966).

40. Mariette, H., "Picosecond Spectroscopy in III-V Compounds and Alloy Semiconductors," *Physica B* 146: 286–303 (1987); Gil, B., and Mariette, H., "NN_2 Trap in GaP: A Reexamination," *Phys. Rev. B* 35: 7999–8004 (1987).

41. Thomas, G.A., Capizzi, M., DeRosa, F., Bhatt, R.N., and Rice, T.M., "Optical Study of Interacting Donors in Semiconductors," *Phys. Rev. B* 23: 5472–5494 (1981).

CHAPTER 8

Theory of Defects in Dilute Nitrides

S.B. ZHANG and SU-HUAI WEI

National Renewable Energy Laboratory, Golden, Colorado

Defects play very important role in semiconductors and, in many cases, they determine the electronic and optoelectronic properties of the host materials. This is especially true for the dilute nitrides, which are caused by the intrinsically large size and chemical mismatch between the first-row nitrogen and other group V elements such as P, As, and Sb. As a result, the equilibrium solubility of N in the dilute nitrides is extremely low, and the minority-carrier lifetime is notoriously small in comparison with conventional III-V semiconductor alloys, hindering many potential applications. Recently, first-principles theory has been developed to unveil the physical properties of the various defects such as vacancies, antisites, split-interstitials, as well as the complexes involving hydrogen. These studies show that the defect physics of the dilute nitrides is *qualitatively* different from that of conventional semiconductors, due to interaction with or between nitrogen atoms. This has led to a whole range of new physical phenomena, ranging from the existence of a new class of intrinsic deep traps involving N pairs (and possibly clusters) to a surprising modification of the band gap by hydrogen. Epitaxial growth conditions that enhance the N incorporation while maintaining a low density of intrinsic defects are also elucidated.

8.1   INTRODUCTION

Dilute nitrides are emerging materials for optoelectronics. These materials provide the opportunity to study semiconductor alloys where atoms of very large size mismatches are mixed purposely by modern epitaxial growth techniques. To date, extensive efforts have been made toward the understanding of the electronic properties of such alloys, in which size-mismatch-induced spatial localization of the electronic states appears to have played a very important role, in contrast to conventional alloys. In comparison, however, relatively little progress has been made toward the understanding of the defect and impurity properties in this class of materials, despite the fact that such studies are essential for realizing the vast technological potentials of the dilute nitrides. The electronic-state localization of the nitrogen atoms is also a key to understanding the defect/impurity properties because, in many aspects, the characteristics of these localized states are no different from any other impurity states.

Regarding defects and impurities, recent experiments have demonstrated the following unique properties:

1. Significantly low minority-carrier lifetimes due to some unspecified defect(s) that is most likely nitrogen related. Overly short minority-carrier lifetimes lead to diffusion lengths that are too small, hindering solar cell performance.
2. Exceptional modification of the fundamental band gap by donorlike impurities. For example, it was shown that Si could increase the band gap of GaAsN significantly. More strikingly, hydrogen could increase the gap by several hundred meV, fully recovering the optical gap of GaAs.
3. While hydrogen is typically an amphoteric impurity in conventional semiconductors, experimental evidence shows that H behaves exclusively as a donor in dilute GaAsN and GaInAsN.

4. Unexpected interactions between defects/impurities mediated by N in the host. For example, it was recently shown that the presence of hydrogen could increase the concentration of Ga vacancies in GaAs. This effect is significantly enhanced in GaAsN. Recent experiments also indicate that the concentration of anion antisite in dilute nitrides can be significantly higher than that in the corresponding binary semi-conductors.

In this chapter, we will address some of these basic issues concerning defects and impurities by reviewing some of the recent first-principles total-energy calculations. In Section 8.2, we discuss the theory of defects in epitaxial materials. It appears that epitaxy is the key for the growth of the dilute nitrides, and the physics of the defects/impurities in such materials could be qualitatively different from those semiconductors grown by more traditional methods. In Section 8.3, we will discuss the formation of nitro-gen and native defect complexes. Sections 8.4 and 8.5 will be devoted to the nitrogen and hydrogen complexes, and the nitrogen-hydrogen-defect complexes, respectively, followed by a brief summary in Section 8.6.

8.2   THEORY OF DEFECTS IN EPITAXIAL MATERIALS

Defect physics in epitaxial materials can be qualitatively different from that in melt-grown materials. This is because epitaxial films are often not in a true equilibrium with other bulk phases. The calculated thermodynamic solubility (or solid solubility) of N in bulk GaAs is exceedingly low ([N] $< 10^{14}$ cm^{-3} at $T = 650°C$) [1–3] due to the formation of a fully relaxed, secondary GaN phase. On the other hand, single-phase epitaxial films grown at $T = 400–650°C$ with [N] as high as $\approx 10\%$ have been reported [4–9]. To explain the approximately eight-orders-of-magnitude difference in the nitrogen solubility in GaAs:N, surface-reconstruction-induced sub-surface strain has been considered [3]. While the calculated N solubility within this model is significantly larger than the previous calculations, it is still four orders of magnitude too small compared with experiments, and it is not clear how the surface energetics during the growth would affect defect formation and impurity incorporation deep inside the bulk.

Recently, Zhang and Wei [10] suggested that the formation of the secondary GaN phase could be suppressed during epitaxial growth. Indeed, a key factor in fabricating high [N] homogeneous GaAs:N films is to eliminate the formation of the GaN precipitates [11]. As such, there could exist a new region of the atomic chemical potentials (μ_{Ga}, μ_{As}, μ_{N}) available for epitaxial growth, but not for melt growth. The chemical potentials are the energies of individual atoms in the reservoirs (such as in the gas sources) that affect impurity substitutional energy [12]. First, let us consider *equilibrium N solubility.* The absolute formation energy of a charge-neutral defect or impurity is defined [12] as

$$\Delta H_f = E_{tot}(\text{Defect}) - E_{tot}(\text{Host}) + n_{Ga}\mu_{Ga} + n_{As}\mu_{As} + n_N\mu_N \quad (8.1)$$

where $E_{tot}(\text{Host})$ is the total energy of a supercell host used in the calculation, $E_{tot}(\text{Defect})$ is the total energy for the same supercell but with a defect, and n ($= n_{Ga}, n_{As}, n_N$) is the number of particles being removed to form the defect from the host to a reservoir of chemical potentials μ ($= \mu_{Ga}, \mu_{As}, \mu_N$). If one sets the energies of bulk Ga, bulk As, and N_2 gas as the reference zeros, then the chemical potentials satisfy

$$\mu_{Ga} \leq 0,\ \mu_{As} \leq 0,\ \text{and}\ \mu_N \leq 0 \quad (8.2)$$

This happens because if $\mu > 0$, an elemental solid (or gas phase) will spontaneously form that hinders any further increase of μ. For GaAs to be thermodynamically stable, it also requires that $\mu_{Ga} + \mu_{As} = \mu_{GaAs} = \Delta H(\text{GaAs}) = -0.62$ eV, which is the calculated formation enthalpy of GaAs. Thus, defect formation energies in GaAs:N are functions of only two independent variables, (μ_{As}, μ_N), satisfying (see Figure 8.1)

$$\Delta H(\text{GaAs}) \leq \mu_{As} \leq 0,\ \text{and}\ \mu_N \leq 0 \quad (8.3)$$

Physically, less negative μ_{As} (or μ_N) corresponds to more As (or N)-rich growth conditions, and vice versa. Spontaneous formation of secondary bulk GaN phase puts a further restriction on the chemical potentials, namely

$$\mu_{Ga} + \mu_N \leq \mu_{GaN} \quad (8.4)$$

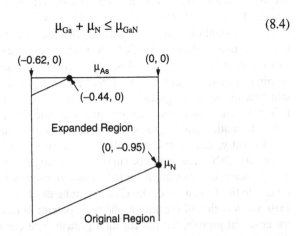

Figure 8.1 The physically accessible region of the chemical potentials, (μ_{As}, μ_N). The "original region" is defined by Equations 8.2 and 8.4, while the "expanded + original" region is defined by Equations 8.2 and 8.6. (From Ref. [10].)

Because $\mu_{GaN} = \Delta H(GaN) = -1.57$ eV, the upper limit for μ_N is not $\mu_N^{max} = 0$ in Equation 8.3, but $\mu_N^{max} = -1.57 + 0.62 = -0.95$ eV at the As-rich limit ($\mu_{As} = 0$). This defines the "original region" in Figure 8.1. Nitrogen substitution is a special case of Equation 8.1:

$$\Delta E_{sub} = \Delta E_{tot} - \mu_N + \mu_{As} \qquad (8.5)$$

where $\Delta E_{tot} = E_{tot}(N_{As}) - E_{tot}(GaAs)$. The higher the μ_N^{max} (and the lower the μ_{As}^{min}), the lower is the minimum ΔE_{sub}^{min}. The calculated ΔE_{sub}^{min} is 1.64 eV, which accounts for the extremely low [N] in melt-grown GaAs:N.

Second, let us consider *surface-enhanced N solubility*. In the epitaxial growth, a relaxed GaN bulk phase could form if both of the following conditions are met:

1. N accumulates and precipitates at the surface to form a GaN layer. This implies the spontaneous formation of a GaN surface layer, which, at low temperature, is equivalent to having a single nitrogen-substitution energy $\Delta E_{sub}^{Surf} = 0$.
2. The layer thickness exceeds the critical thickness for dislocation formation.

Condition 1 precedes condition 2 and thus sets an upper bound for μ_N and a lower bound for μ_{As} for epitaxial GaAs:N (cf. Equation 8.5), that is,

$$\Delta E_{sub}^{Surf}\left(\mu_{As}^{min}, \mu_N^{max}\right) = \Delta E_{tot}^{Surf} - \mu_N^{max} + \mu_{As}^{min} = 0 \qquad (8.6)$$

where

$$\Delta E_{tot}^{Surf} = E_{tot}^{Surf}\left(N_{As}\right) - E_{tot}^{Surf}(GaAs) \qquad (8.7)$$

is the total energy difference of a N substituting an As on the surface. If one neglects the energy due to nitrogen-nitrogen interaction at the surface, ΔE_{tot}^{Surf} would be independent of the surface N concentration. Therefore, one can solve Equation 8.6 to find ($\mu_{As}^{min}, \mu_N^{max}$) and then plug ($\mu_{As}^{min}, \mu_N^{max}$) into Equation 8.5 to determine ΔE_{sub}^{min}, from which [N] can be calculated using standard Boltzmann statistics.

Nitrogen incorporates into GaAs via an N-As exchange mechanism at the topmost surface layer [13–16]. A typical GaAs (0,0,1) surface is an

As-terminated 2×4 surface shown in the inset of Figure 8.2 [17], where the $n = 1$ and $n = 3$ layers contain six and eight As atoms per 2×4 cell, respectively. Figure 8. 2 shows ΔE_{sub} for the surface site 1 and 3D and for the bulk site as a function of the atomic chemical potentials along the N- and As-rich boundaries in Figure 8.1. We see that:

1. $\Delta E_{sub}(3D)$ and $\Delta E_{sub}(bulk)$ are only moderately higher than $\Delta E_{sub}(1)$ by 0.30, and 0.24 eV, respectively. Thus, ΔE_{sub} is relatively insensitive to where the N atom is at the surface layer, or even inside the bulk.

2. Using the lower ΔE_{sub}^{Surf} value $= \Delta E_{sub}(1)$, Zhang and Wei [10] determined that $(\mu_{As}^{min}, \mu_{N}^{max}) = (-0.44 \text{ eV}, 0.0)$, at which $\Delta E_{sub}^{min}(bulk)$ is significantly reduced from the original 1.64 eV to 0.24 eV. This gives rise to $[N] \approx 4\%$ at 650°C. The corresponding accessible (μ_{As}, μ_{N}) is the (expanded + original) region in Figure 8.1.

In essence, the physics of surface-enhanced N solubility resides in the difference between the formation energy of GaN on the GaAs surface and that of relaxed bulk GaN. The difficulty of forming a GaN at the GaAs surface leads to a higher achievable μ_{N} and, thus, a higher epitaxial N solubility.

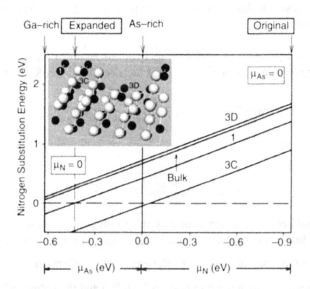

Figure 8.2 N substitutional energies ΔE_{sub} as functions of (μ_{As}, μ_{N}). The inset is the GaAs (2×4) surface, indicating the various substitution sites. The anion is denoted by a dark circle, while Ga is denoted by a light circle. (From Ref. [10].)

The above discussion does not consider kinetic effects on the nucleation processes of the secondary GaN phase. Consequently, N concentration is expected to increases with temperature T. In reality, however, at high T and high N concentration, N nucleation can occur even before ΔE_{tot}^{Surf} approaches zero, because the diffusion of the surface nitrogen atoms enhances the possibility of a surface N finding a low-energy nucleation site. This can lower the achievable μ_N and reduce N solubility. A number of experiments [5, 9, 18] indeed showed that [N] in high-concentration samples is nearly unchanged or decreases as T increases in the range between 400 and 650°C.

8.3   NITROGEN AND NATIVE DEFECT COMPLEXES

The expanded region of the chemical potentials discussed above forms a new basis for the study of defects and impurities in epitaxially grown dilute III-V nitrides. At high μ_N, thus high N concentration, defect physics changes qualitatively from that at low N concentration in at least two fundamental ways: (1) the dominant defects are no longer the same, and (2) the electronic properties of the leading defects can be qualitatively different from the corresponding defects in the absence of N. To study minority-carrier traps, Zhang and Wei [10] have considered the charge-neutral defects in GaAs shown in Figure 8.3.

1. *The N-N split interstitial:* Here, an N_2 molecule replaces one host As atom. Each N is threefold coordinated. The split interstitial is important because of the small size of the N atom and the exceptionally strong N-N bond. The calculated N-N bond length is 1.39 Å, compared with 1.10 Å for an N_2 molecule.

2. *The N-As split interstitial:* The N-As pair is a variation of the N-N split interstitial in the previous item. Here, the calculated N-As bond length is 1.85 Å.

3. *The $(As_{Ga}\text{-}N_{As})_{nn}$ pair* (where nn stands for the nearest neighbor): The N is very electronegative. Therefore, it always attracts donor defects such as As_{Ga}. Moreover, N_{As} is associated with compressive strain due to the small size of N, while As_{Ga} is associated with tensile strain due to two extra electrons in the nonbonding orbital. The $(As_{Ga}\text{-}N_{As})$ nn-pair has a 0.5-eV binding energy. Thus, isolated As_{Ga} is rare at high N concentration. The neutral N-As separation in the nn-pair is 2.86 Å, 47% larger than the sum of their atomic radii. Hence, the nn-pair, where both N and As are threefold coordinated, is qualitatively different from any other distant $As_{Ga}\text{-}N_{As}$ pairs. Note, however, that recent calculations showed that the atomic configurations of an $(A^V)_{B^{III}} - N_{A^V}$ pair (where A^V = As, P and B^{III} = Ga,In) depend sensitively on the charge states

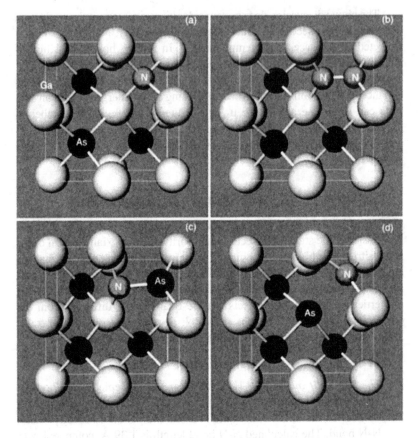

Figure 8.3 Calculated atomic positions for (a) substitutional N, (b) N-N, (c) N-As split inter-stitials, and (d) $(As_{Ga}\text{-}N_{As})_{nn}$ complexes. As is the dark circle, Ga is the light circle, and N is labeled. (From Ref. [10].)

of the complex. For example, for doubly positively charge pairs, the N-(gr V) separation is considerably smaller than the neutral ones and is, in fact, comparable with the sum of atomic radii.

4. *The $(V_{Ga}\text{-}N_{As})_{nn}$ pair:* The neutral pair has a relatively small binding energy of <0.05 eV. However, due to a level repulsion between the N_{As} and V_{Ga} gap states (see below), the binding increases with the charge state to about 0.43 eV for q = 3⁻.

Figure 8.4a shows the defect formation energy ΔH_f as a function of the atomic chemical potentials. At the expanded N solubility limit, the dominant defect in GaAs:N is the $(N\text{-}N)_{spl}$ split-interstitial, not the As_{Ga} antisite observed in As-rich GaAs. The calculated $\Delta H_f (N\text{-}N)_{spl} = 1.52$ eV is considerably smaller than either As_{Ga} (2.24 eV), $(As_{Ga}\text{-}N_{As})_{nn}$ (1.96 eV), or $(N\text{-}As)_{spl}$ (2.49 eV). The calculated $(N\text{-}N)_{spl}$ concentration at $T = 650°C$

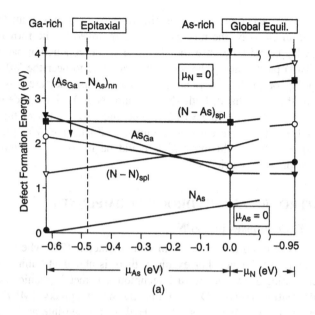

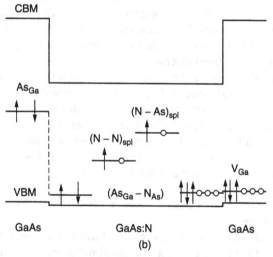

Figure 8.4 (a) Defect formation energies. The ordinates and legends are the same as in Figure 8.2. (b) Calculated single-particle defect energy levels with electron occupation indicated by arrows. The band diagrams are those of GaAs (ε_g = 1.57 eV) and GaAs:N (ε_g = 1.04 eV). (From Ref. [10].)

is [C] = 1×10^{14} cm^{-3}, which, however, further increases when the complex is charged. Figure 8.4b shows the calculated single-particle gap states, along with the band gaps of bulk GaAs and GaAs:N at an N concentration x = 3.125%. In bulk GaAs, As$_{Ga}$ has a mid-gap double donor state, whereas

V_{Ga} has several acceptor states near the valence band maximum (VBM). In contrast in GaAs:N, the mid-gap state is removed by the formation of the $(As_{Ga}-N_{As})_{nn}$ pairs because the N_{As} state at the conduction-band minimum (CBM) pushes down the As_{Ga} mid-gap state to near the VBM. The effects of N_{As} on the V_{Ga} states are negligible because they are close to the VBM. On the other hand, both $(N-N)_{spl}$ and $(N-As)_{spl}$ have a deep level (0.38 and 0.62 eV above the VBM or 0.66 and 0.42 eV below the CBM, respectively) with *single* electron occupancy. Zhang and Wei [10] have suggested that these split interstitials could be important recombination centers for minority carriers 0.

8.4   NITROGEN AND HYDROGEN COMPLEXES

8.4.1 H_2^* Complexes in GaPN

Hydrogen is an important impurity in semiconductors [20], whether placed there intentionally or not. For example, there is abundant "unintentional" H present during growth using such techniques as metal-organic chemical-vapor deposition (MOCVD) or molecular-beam epitaxy (MBE) when hydrides are used as the gas sources. Hydrogen often interacts with host materials [21, 22], as well as with other impurities/defects, causing changes in the electronic properties [23]. Being a fast diffuser [24], atomic hydrogen can also be used to probe either electronically or structurally the various defect properties in semiconductors.

Until recently, however, the interaction between hydrogen and nitrogen in III-V semiconductors remained very much unexplored [25]. An infrared study of GaPN:H [26] recently revealed three distinct local vibrational frequencies: 288.5, 2054.1, and 1049.8 cm^{-1}. To explain the data, an H-N-H dihydride model of trigonal symmetry was proposed [26], where two H atoms are bonded directly to an N atom, as shown in Figure 8.5a. There are, however, several difficulties with this model:

1.   The N is fivefold coordinated, which has never been observed before in any nitrides.
2.   The two H stretching frequencies are drastically different by a large amount, 830 cm^{-1}. In contrast, the calculated difference in the stretching frequencies for Si-H$_2$ dihydride on a silicon [0,0,1] surface is more than one order of magnitude smaller, only 50 cm^{-1} 0.
3.   The isotope shifts of the above two modes are distinctly different: 5.4 and 1.7 cm^{-1}.

From a more fundamental point of view, no diatomic H$_2$ complexes other than H$_2$ molecules [28, 29] have been predicted to be stable [30], nor have they been observed in any III-V semiconductors, even though an

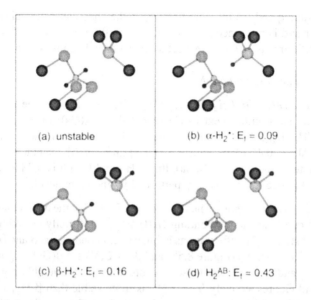

(a) unstable

(b) α-H_2^*: $E_f = 0.09$

(c) β-H_2^*: $E_f = 0.16$

(d) H_2^{AB}: $E_f = 0.43$

Figure 8.5 Atomic structures for the various diatomic H and N complexes in GaP:N, along with the calculated formation energies (in eV/H). The filled atoms are P, the gray atoms are Ga, and the small gray atoms are N. (From Ref. [33].)

H_2^* complex in Si was not only comparable in energy to H_2 molecule [31] and observed [32], but was also found to play an essential role in the nucleation and growth of the technologically important H platelets [22].

Using first-principle total energy calculations, Janotti, Zhang, and Wei [33] have investigated the various H structural configurations in GaP either associated with or without N. They found that nitrogen plays a pivotal role in the relative stability between the H_2^* complexes and H_2 molecules. Without N, H_2^* is 0.3 eV per H *higher* in energy than H_2. With one N, however, H_2^* is 0.26 eV per H *lower* than H_2. At the typical experimental condition listed by Clerjaud et al. [26], $\varepsilon_F \geq 0.5$ eV, H_2^* is also stable against other forms of hydrogen complexes, including against dissociating into two isolated H-N complexes. The calculated local frequencies and isotope shifts using the H_2^* model were in good agreement with experiment. In contrast, the H-N-H model (involving one dihydride) is unstable against spontaneous transformation into the H_2^* complex (involving, instead, two monohydrides).

The formation energy per hydrogen is defined as,

$$\Delta H_f \frac{1}{n}\left[E_{tot}\left(\text{host} + nH, q\right) - E_{tot}\left(\text{host}\right) - n\mu_H + q\varepsilon_F \right] \qquad (8.8)$$

where $E_{tot}(\text{host} + nH, q)$ and $E_{tot}(\text{host})$ are, respectively, the total energies for (a) a configuration with n H atoms in the host (GaP:N or GaP) at a charge-state q and (b) the pure host; μ_H is the chemical potential of the H reservoir and is referenced to H_2 molecule in the free space at $T = 0$; and ε_F is the Fermi energy with respect to the valence-band maximum (VBM).

8.4.1.1   The case without N

8.4.1.1.1   Interstitial H_2   The H_2 molecules are located at the tetrahedral interstitial sites, either next to Ga [denoted as H_2(Ga)] or P [denoted as H_2(P)]. The H_2(Ga) is more stable than H_2(P). The H_2(Ga) has a formation energy $\Delta H_f = 0.46$ eV/H with the [1,1,1] and [1,0,0] orientations almost degenerate in energy. For H_2(Ga), the H-H bond length is 0.78 Å, and the GaP host lattice is only weakly perturbed by the H_2 molecule.

8.4.1.1.2 H_2^* Complex   The H_2^* consists of one H at the bond center (BC) site and one H at the antibonding (AB) site [21]. Strictly speaking, there are two distinct H_2^* configurations in binary semiconductors, α and β, where $\alpha = BC(Ga) + AB(V)$ (Figure 8.5b) and $\beta = BC(V) + AB(Ga)$ (Figure 8.5c) where V and Ga in parentheses denote the nearest-neighbor group V anion and Ga, respectively. The α-H_2^* is more stable than β-H_2^*, but even α-H_2^* is 0.3 eV/H higher in energy than H_2(Ga). The calculated H(AB)-P and H(BC)-Ga bond lengths are 1.46 and 1.56 Å, respectively.

8.4.1.2   The case with N

Janotti, Zhang, and Wei [33] have studied the changes in α-H_2^* and β-H_2^* in GaPN, where α-H_2^* remains more stable than β-H_2^*. For α-H_2^*, the H(AB)-N bond length is 1.05 Å, and the H(BC)-Ga bond length is 1.53 Å. The distance between H(BC) and N is, however, 1.97 Å, which is 81% longer than the sum of the H-N radii of 1.09 Å. This is a strong indication that no chemical bond forms between the two. To be more convincing, Janotti et al. plotted the charge-density distributions for individual H states as well as for the total valence charge density (see Figure 8.6); the plot shows no trace of any H(BC)-N bond. The formation energy of the α-H_2^* is $\Delta H_f = 0.09$ eV/H. In contrast, the H-N-H model in Figure 8.5a is not only unstable, it *spontaneously* transforms into the α-H_2^* configuration without any energy barrier. Janotti, Zhang, and Wei [33] also studied H_2^{AB}, where both H are in the antibonding positions, one close to N and the other close to Ga (see Figure 8.5d). In this case, the H-N and H-Ga bond lengths are 1.05 and 1.61 Å, respectively. Both the Ga and N atoms are displaced along the [111] and [111] directions to form planar structures. It, however, has a much higher formation energy, $\Delta H_f = 0.43$ eV/H, and is, therefore, unlikely to form.

8.4.1.3   The effects of N on the H pairs

Figure 8.7 shows how N affects the energy of the various H pairs. First, regarding the H_2 molecules, nitrogen has little effect on H_2(V) but lowers

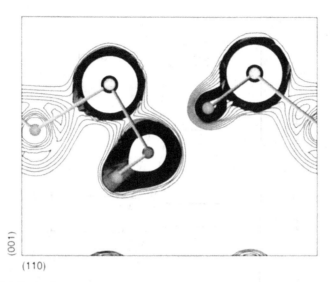

(001)

(110)

Figure 8.6 Total valence charge-density plot showing the lack of the H(BC)-N bond. The contours are in units of 1.13×10^{-4} electrons/cell. (From Ref. [33].)

the energy of $H_2(Ga)$ by 0.2 eV/H. The reason is that N is more electronegative than P, so it takes more electrons away from the vicinity of the Ga than P does. Because interstitial H_2 prefers low-charge-density regions, the net result of the charge transfer is that $H_2(Ga)$ becomes more stable than $H_2(N)$. Second, nitrogen drastically reduces the energies of both H_2^* complexes by about 0.7 eV/H. It also lowers the energy of H_2^{AB} in Figure 8.5d by a similar 0.6 eV/H. One can quantitatively understand this universal reduction by comparing the H-N and H-P bond strengths in ammonia (NH_3) and phosphine (PH_3). The difference in the cohesive energy between the two is 2.03 eV per molecule or 0.68 eV per H, in remarkable agreement with the above calculation. This somewhat unexpected effect places the H_2^* complexes significantly lower in energy than the H_2 molecule, which has never been the case in conventional III-V semiconductors.

8.4.1.4 H_2^* dissociation

The unusually strong H-N bond raises the question whether the H pairs are stable against dissociation into two individual H-N complexes, especially when the N concentration is greater than H. This would maximize the number of H-N bonds. Janotti, Zhang, and Wei [33] have calculated the α-H_2^* binding energy from distant H-N pairs of various charge states. It was found that for $\varepsilon_F \leq 0.55$ eV, two distant $[H(BC)-N]^+$ plus two electrons have lower energy than the charge-neutral α-H_2^*. For $\varepsilon_F \geq 0.55$ eV, however, the opposite is true. The reason is because H(AB) has a low defect level near the VBM. Even though the H-Ga bond in H_2^* is weaker than the H-N bond,

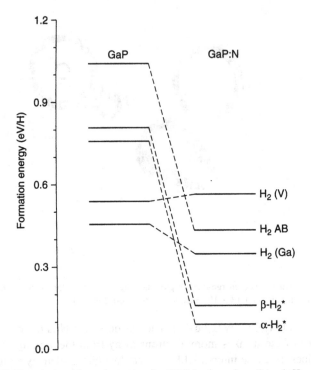

Figure 8.7 The calculated formation energy (in eV/H) for the various diatomic H complexes: (a) without N and (b) with N. (From Ref. [33].)

the electronic energy gain by a two-electron transfer from a relatively high ε_F to the defect level of H(AB) is enough to overcome the bond energy loss. The experiments by Clerjaud et al. [26] were done with $\varepsilon_F \geq 0.5$ eV. Hence, regardless of the relative H/N ratio, H_2^* is always the most stable, and thus the most abundant, H centers in GaP:N.

8.4.1.5 H vibrational frequencies

Janotti, Zhang, and Wei [33] also investigated the local vibration modes of H in GaP:N. The vibrational frequencies are calculated by evaluating the force-constant matrix K [24, 34], where the matrix elements, K_{ij}, were calculated via the Hellman-Feynman force induced on the ith atom by the displacement of the jth atom. The results are shown in Table 8.1.

The highest-frequency mode-1 at 3081.2 cm^{-1} is associated with the stretching of the H(AB)-N bond. The calculated isotopic shift is 6.3 cm^{-1}. Both are in reasonable agreement with experiment [26]: 2885.5 and 5.8 cm^{-1}, respectively. A 7% error here is typical for the local density approximation (LDA) calculations. As expected from the H_2^* model, this H-N mode is little affected by the ^{69}Ga-^{71}Ga isotopes. The middle-range-fre-

Table 8.1 Local Vibrational Frequencies and Isotope Shifts for the H_2^* Complex in GaP:N (cm^{-1})

	Mode^a		
	1 (shift)	2 (shift)	3 (shift)
Calculated	3081.2 (6.3)	2052.3 (0.4)	968.3 (1.2)
Experimental ^{b}	2885.4 (5.4)	2054.1 (1.7)	1049.8
Difference	195.7 (0.9)	−1.8 (−1.3)	−81.5

^{a} Mode 1 and 2 are the stretching modes, whereas mode 3 is a doubly degenerate wagging mode.
^{b} The experimental results are from Clerjaud, B. et al., *Phys. Rev. Lett.* 77: 4930–4933 (1996).

quency mode-2 at 2051.3 cm^{-1} is associated with the H(BC)-Ga bond stretching. It changes by 0.4 cm^{-1} if one replaces ^{14}N with ^{15}N and ^{69}Ga with ^{71}Ga. Experimentally, the isotopic shift for this mode is 1.7 cm^{-1}. Because H(BC) is not directly bonded to N, a qualitatively smaller N isotope shift for mode-2 can be expected. Also, isotope effect due to the heavier Ga atoms should be small. Interestingly, the highest H-N mode in experiment (2885.5 cm^{-1}) is significantly smaller than the normal mode in ammonia of 3444 cm^{-1}. A 16% reduction here is quite difficult to understand based on the H-N-H model. However, it has been shown [23, 32] in Si that significant reduction between 11 to 21% is possible for the H(AB)-Si stretching mode, due presumably to the large local strain that weakens the H(AB)-host atom bond [23]. Finally, the lowest-frequency mode at 968.3 cm^{-1} was found to be doubly degenerate, corresponding to the wagging of the H-N bond, again in good agreement with experiment. The 1.2-cm^{-1} isotope shift in this case is caused by replacing ^{14}N with ^{15}N.

8.4.2 H_2^* Complexes in GaAsN

Recent experimental results by Xin et al. [35, 36] showed that hydrogen incorporation increases with nitrogen concentration [N] in InGaAsN alloys grown by gas-source MBE. Based on Hall measurements, Xin et al. suggested that H acts as an isolated donor in InGaAsN, and makes the as-grown undoped samples slightly n-type. Annealing above 700°C reduces the hydrogen concentration [H] and renders the samples p-type. This behavior differs from the effect of hydrogen in other semiconductors where H acts as a passivation agent, but not as a source of doping in its own right. A second puzzling result emerged from experiments by Baldassarri et al. [37] and Polimeni et al. [38], which revealed that postgrowth hydrogenation of InGaAsN alloys can lead to a complete reversal of the drastic band-gap reduction caused by N, observed for various N and In concentrations in InGaAsN alloys. These results suggest strong interactions between hydrogen and nitrogen in III-V-N alloys. Baldassarri et al. [37] and Polimeni et al. [38] attributed the InGaAs band-gap restoration to the binding of H to the N atom in a passivation process. However, there is no microscopic theory on the nature of these interactions, i.e., neither the atomic nor the electronic structures of the hydrogen donors and of the N-H complex responsible for the band-gap opening in GaAsN and InGaAsN alloys have been studied.

Using first-principles total-energy calculations, Janotti et al. [39] studied the atomic structure and stability of the various hydrogen-related configurations (see Figure 8.8) along with their effects on the electronic properties of GaAsN alloys. They found that monatomic H in dilute GaAsN acts as a donor for all Fermi-level positions in the band gap. This is quite surprising because, in conventional semiconductors, monatomic H can exist in either donor or acceptor charge states, depending on the Fermi energy position [40]. For complexes that involve two H atoms, it was found that a $H_2^*(N)$ complex is more stable than an interstitial H_2 molecule because of the strong bonding between hydrogen and nitrogen, similar to H_2^* in GaP. The formation of the $H_2^*(N)$ complexes leads to a complete removal of the band-gap reduction in GaAsN with respect to GaAs. These findings also apply to InGaAsN alloys, thus providing a qualitative explanation for the experimental observations.

First, consider the donor (+), neutral (0), and acceptor (−) states of a monatomic H in the various monohydride configurations: in the bond-center sites next to N (BC_N) and far away from N (BC_{As}), in the N and Ga antibonding sites (AB_N and AB_{Ga}), and in the tetrahedral interstitial sites. The results showed that BC_N is the lowest-energy configuration for all charge states. Other configurations are at least 1 eV higher in energy except for AB_N. In the positively charged BC_N^+ configuration (Figure 8.8a), the H atom strongly binds to the N atom with a H-N bond length of 1.05 Å. The Ga atom is displaced along the [111] direction to the basal plane formed by its three nearest-neighbor As atoms. The Ga-H distance is 2.41 Å, 53% longer than the 1.58 Å expected from their respective atomic radii. The N atom is displaced along the opposite [$\bar{1}\bar{1}\bar{1}$] direction. In the positively charged AB_N^+ configuration (Figure 8.8b), the H atom at the antibonding site strongly binds to the N atom with a N-H bond length of 1.05 Å. In this case, similar to the BC_N^+ case, both N and Ga are displaced away from each other to their respective basal planes.

The AB_N^+ configuration has a formation energy that is 0.37 eV higher than BC_N^+, mainly because the electron charge density of the host at the BC_N^+ site is higher than that at the AB_N^+ site. Thus, the coulomb binding energy for H^+ at the BC_N site is larger. This is in contrast to pure GaN, where H^+ prefers the AB_N site 0. The Ga-N bond length is 2.05 Å in GaAsN,

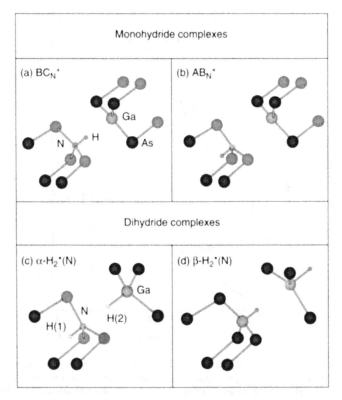

Figure 8.8 Ball-and-stick models for the hydrogen and nitrogen complexes in GaAsN. Configurations involving one H include (a) the bond-center site next to N (BC_N) and (b) the antibonding site (AB_N). Complexes involving two H include (c) α-H_2^*(N) and (d) β-H_2^*(N). (From Ref. [39].)

which is 7% longer than that in GaN of 1.92 Å. Due to the longer host Ga-N bond, the BC-site H^+ is less strained in GaAsN. Reduced strain energies are also responsible for the stabilization of the BC_N^0 and BC_N^- configurations in GaAsN.

The lowest formation energies for each charge state are plotted in Figure 8.9 as a function of ε_F. The results show that as long as [H] is less than [N] (typically a few atomic percent for dilute alloys), monatomic H in GaAsN exists only in the donor charge state, with the (+/−) level above the CBM. These results are in clear contrast to those in GaN and GaAs, the parent compounds, where H is an amphoteric impurity, positively charged in p-type samples but negatively charged in n-type samples [30, 40, 41]. The difference can be explained by the exceptionally large bowing effect of N that lowers the CBM of GaAsN [2, 42, 43] below both the CBM of GaAs and GaN, as well as below the H (+/−) level. In the case of InGaAsN, the presence of indium lowers the CBM even further below that

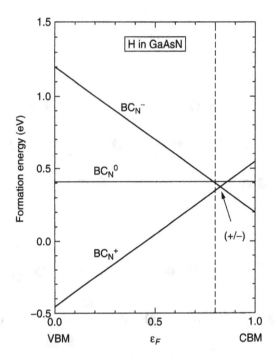

Figure 8.9 Formation energies of monatomic H in GaAsN as a function of the Fermi energy ε_F. Only the lowest-energy BC_N configuration is shown for each charge state: (+), (0), and (−). The vertical dashed line indicates the calculated band gap of GaAsN. (From Ref. [39].)

of GaAsN. Hence, it is expected that the hydrogen (+/−) level will be even further above the CBM. In general, when the host CBM level is very low [44], monatomic H can behave exclusively as a donor, as has been proposed for ZnO [45] and InN [46]. However, a similar behavior due solely to an alloying effect has not been suggested before.

The experiments described by Xin et al. [35, 36] also showed that the total hydrogen concentration [H] is higher than the free-electron concentration in as-grown unintentionally n-doped InGaAsN. This implies that hydrogen is also present in inactive states such as interstitial H_2 molecules or H_2^* complexes [21]. Furthermore, our calculations show that the formation of $H(BC_N^+)$ has only a minor effect on the band gap. Hence, a defect other than active H^+ has to be responsible for the large band-gap opening (of several tenths of an eV) observed in the post-growth hydrogenation experiments [37, 38]. To identify this defect, Janotti et al. [39] have studied a number of $H_2^*(N)$ complexes that involve two hydrogen atoms adjacent to a nitrogen. Table 8.2 shows that the α-$H_2^*(N)$ complex (Figure 8.8c) has the lowest formation energy, $\Delta H_f = -0.07$ eV/H. In this complex, H(1) is at the antibonding site, similar to $H(AB_N^+)$ in Figure 8b, with a N-H bond length of 1.05 Å. H(2) is at the bond-center site with a Ga-H bond length

Table 8.2 Formation Energies of the Double-H Complexes in GaAsN and GaAs (eV per H)

	Configuration				
	α-H_2^*	β-H_2^*	$H_2(Ga)$ [a]	$H_2(V)$ [b]	H_2^{AB} [c]
GaAsN	−0.07	0.01	0.23	0.46	0.26
GaAs	0.72	0.67	0.28	0.38	0.92

[a] $H_2(Ga)$ refers to H_2 molecules at the tetrahedral interstitial site next to Ga.

[b] $H_2(V)$ refer to H_2 molecules at the tetrahedral interstitial site next to anions (one of which is N for GaAsN).

[c] H_2^{AB} refers to a complex where both H atoms are at the antibonding sites.

of 1.54 Å. No chemical bond is formed between H(2) and N, as the N-H separation of 2.06 Å is 93% longer than the 1.07 Å expected from their respective atomic radii. The β-H_2^*(N) complex (Figure 8.8d; $\Delta H_f = 0.01$ eV/H) is slightly higher in energy than the α-H_2^*(N) complex by about 0.1 eV/H. The N-H(1) bond length is 1.05 Å, and the Ga-H(2) bond length is 1.60 Å. Here, both the N and Ga atoms assume the planar configuration, raising the strain energy of the β structure with respect to the α structure. The α-H_2^*(N) complex is also strongly favored by as much as 0.3 eV/H over the interstitial H_2 molecule ($\Delta H_f = 0.23$ eV/H). In contrast, the interstitial H_2 molecule is considerably more stable in GaAs.

The formation energy of H(BC_N^+) increases as the Fermi energy ε_F increases (Figure 8.9). A study of detailed balance between H^+ and H_2^* suggests that the relative concentration $[H^+]/[H]$ decreases with [H], whereas $[H_2^*]/[H]$ increases. Interestingly, however, the absolute $[H^+]$ increases with [H] instead of decreasing. As a result, the Fermi energy also increases with [H]. At $[H] = 10^{19}$ cm^{-3}, the calculated $[H^+]$ of 10^{16} cm^{-3} agrees with the experimental free-electron concentration of 7×1015 cm^{-3} [36]. This suggests that a majority of the single H complexes are ionized, so the (+/−) level is indeed shallow. At higher [H], comparable to [N], however, the role of H is shifted toward restoring the GaAs band gap by H_2^*. An analysis of the density of states near the band edges suggests that while N incorporation lowers the CBM of GaAs to 0.6 eV for 3.125% N, H_2^* formation pushes the CBM back up completely — an observation that holds for every [N] being studied: 1.5625, 3.125, and 6.25%. This remarkable effect can be qualitatively understood via a three-step process schematically shown in Figure 8.10.

First, the bonding of hydrogen to N leads to large atomic displacements along the <1,1,1> direction, breaking one of the Ga-N bonds. This eliminates

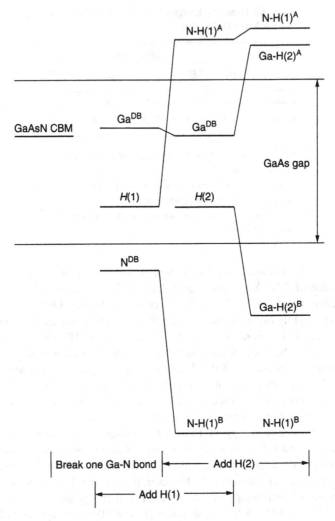

Figure 8.10 A schematic plot of the effect of the α-H$_2$*(N) complexes on the band gap of GaAsN. In step one, one of the Ga-N bonds is broken. A Ga DB-state and an N DB-state emerge at the expense of one GaAsN CBM state. In step two, H(1) is added to the AB side of N, forming the N-H(1)^B and N-H(1)^A states. In step three, H(2) is added next to the Ga, forming the Ga-H(2)^B and Ga-H(2)^A states. (From Ref. [39].)

one of the nitrogen-derived GaAsN CBM states [2], creating a N dangling bond (DB)-like state in the valence band and a Ga DB-like state near the GaAsN CBM. Second, the binding of H(1) to the N DB-like state creates a N-H(1)^B bonding state deep in the valence band and a N-H(1)^A antibonding state in the GaAs conduction band. Due to their spatial proximity, some interaction between the N-H(1)^A state and the reactive Ga DB-like state can be expected. Third, the binding of H(2) to the Ga DB-like state creates a bonding Ga-H(2)^B state below the VBM and an antibonding Ga-H(2)^A state inside the conduction band of GaAs. The net result of the process depicted in Figure 8.10 is that one GaAsN CBM state is completely removed by the formation of one $H_2^*(N)$ complex.

As H is gradually introduced during postgrowth hydrogenation [37, 38], the concentration of $[H_2^*]$ increases, whereas the concentration of nonhydrogenated N decreases. The band gap also increases because the N-induced gap reduction is proportional to the concentration of nonhydrogenated N. This process approaches completion when [H] is larger than [N], which leads to the exposure of the original GaAs band gap. The same conclusion can also be straightforwardly applied to InGaAsN alloys, which have an even lower CBM than GaAsN. A similar conclusion was also reached by Bonapasta et al. [47].

8.5 NITROGEN, HYDROGEN, AND CATION VACANCY COMPLEXES

Several recent experimental studies have implied the effect of N on the presence of gallium vacancies in InGaAsN. Li et al. [48] described positron annihilation measurements of gallium vacancies in InGaAsN grown by gas-source molecular-beam epitaxy (MBE). Toivonen et al. [49], also using positron annihilation spectroscopy, reported an increase in defect complexes containing Ga vacancies up to 10^{18} cm^{-3} when the N concentration is increased (up to 5%). They found that the concentration of gallium vacancies, $[V_{Ga}]$, decreases and the PL lifetime and intensity improve upon annealing at 700°C. Moreover, InGaAsN grown without the intentional addition of dopants is frequently observed to have a significant ($\approx 10^{17}$ cm^{-3}) acceptor concentration [11, 50]. Although these acceptors have sometimes been ascribed to unintentional carbon doping, the acceptor concentration often exceeds the observed carbon concentration, implying that intrinsic defects may be acting as acceptors [50]. These experiments provide strong motivation for developing a microscopic understanding of vacancy physics in the dilute-nitride alloys.

Recently, Janotti et al. [51] examined the important roles of N and H in the formation of gallium vacancies (V_{Ga}). It was found that:

1. The binding energy of V_{Ga}^{3-} and N is 0.43 eV, although N is an isovalent impurity.
2. Hydrogen strongly binds to V_{Ga} in GaAs, with binding energies as high as 1.27 eV.

3. In the presence of N, the binding energy between H and N-V_{Ga} increases to 2.15 eV.
4. In the presence of hydrogen, the binding energy between N and H-V_{Ga} increases to 1.31 eV.

These results showed that V_{Ga} binds strongly with both N and H, which lowers the formation energy of V_{Ga} considerably and raises [V_{Ga}] well beyond the level expected for GaAs. The presence of In in an InGaAsN alloy would not directly affect the interactions between N (or H) and V_{Ga}.

Similar to Equation 8.1, the formation energy of H- and V_{Ga}-related defects α with charge-state q is defined as

$$\Delta H_f(\alpha^q) = \Delta E_{tot}(\alpha^q) + n_{Ga}\, \mu_{Ga} + n_H\, \mu_H + q\, \varepsilon_F \qquad (8.9)$$

Here, $\Delta E_{tot}(\alpha^q) = E_{tot}(\alpha^q) - E_{tot}(\text{host}) + n_{Ga}\, \mu_{Ga}^{solid} + n_H\, \mu_H^{H_2} + q\, \varepsilon_{VBM}$. $E_{tot}(\alpha^q)$ is the total energy of the defect, and $E_{tot}(\text{host})$ is the total energy of the host (GaAs or GaAsN). The variables n_{Ga} and n_H represent, respectively, the number of Ga and H atoms removed from the host to form the defect. For example, for (N-H-V_{Ga})$^{2-}$ in GaAsN, q = -2, n_{Ga} = 1, and n_H = -1. The variable μ_H is the chemical potential of hydrogen referenced to H_2 molecule in vacuum at $T = 0$, and μ_{Ga} is the Ga chemical potential referenced to bulk Ga crystal, which obeys the equilibrium condition $\mu_{Ga} + \mu_{As} = \Delta H_f(\text{GaAs})$. In the following, we will discuss the results at the Ga- and H-rich limits, so $\mu_{Ga} = 0$ and $\mu_H = 0$. The formation energy at other chemical potentials can be easily obtained from Equation 8.9.

First, consider the effect of nitrogen on the formation energy of V_{Ga} in dilute GaAsN. It was found that isolated V_{Ga} in GaAs is an acceptor that is predominantly stable in the (3-) charge state (V_{Ga}^{3-}). The transition levels to other charge states (-2, -1, and 0) are at less than 0.2 eV above the VBM. (The [3-/2-], [2-/-], and [-/0] transition levels are 180, 150, and 130 meV above the valence band maximum, respectively.) The formation energy ΔH_f of V_{Ga}^{3-} is 3.60 eV for $\varepsilon_F = 0$. When the V_{Ga} is associated with N, i.e., when one vacancy is surrounded by 3 As and 1 N, the resulting complex is still an acceptor, predominantly in the 3- charge state [(N-V_{Ga})$^{3-}$], but the corresponding formation energy is 0.43 eV lower than that of isolated V_{Ga}^{3-}, as shown in Figure 8.11. (In the case of N-V_{Ga}, the [3-/2-] and [2-/-] transition energies are 110 and 30 above the valence band maximum [VBM], respectively, while the [-/0] is 10 meV below the VBM.)

Figure 8.11 also shows the effect of hydrogen on the formation energy of the Ga vacancy. Consider two cases: (1) isolated V_{Ga} in GaAs, and (2) V_{Ga} with an N nearest neighbor in GaAsN. For isolated V_{Ga}, H bonds to one of the As neighbors, saturating its dangling bond, and forms a V_{Ga}-H complex occurring predominantly in the 2- charge state [(H-V_{Ga})$^{2-}$]. The

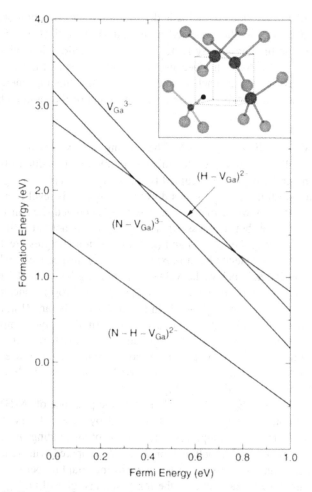

Figure 8.11 Formation energy vs. Fermi energy for V_{Ga}^{3-} and $(H\text{-}V_{Ga})^{2-}$ in GaAs and for the same defects bonded to N, $(N\text{-}V_{Ga})^{3-}$ and $(N\text{-}H\text{-}V_{Ga})^{2-}$, in dilute GaAsN alloys under Ga-rich and H-rich conditions. The inset shows the local atomic configuration for the $(N\text{-}H\text{-}V_{Ga})^{2-}$ complex. (From Ref. [51].)

formation energy of $(H\text{-}V_{Ga})^{2-}$ is 0.78 eV lower than the isolated V_{Ga}^{3-}. In the case of $(N\text{-}V_{Ga})^{3-}$, H forms a strong chemical bond with N, and the formation energy is 1.66 eV lower than that of $(N\text{-}V_{Ga})^{3-}$. In this case, the energy reduction of 1.66 eV compared with 0.78 eV in the absence of nitrogen, chiefly reflecting the stronger chemical bond between N and H compared with As-H. Because lower formation energies translate into higher concentrations, our results clearly show that in the presence of N, the V_{Ga} concentration will increase, and that the presence of H will further strongly enhance the effect of nitrogen on $[V_{Ga}]$ in dilute GaAsN alloys.

In order to provide a systematic description of the reduction of the V_{Ga} formation energy in the presence of N and H, Janotti et al. [51] have calculated the binding energy E_b for a series of reactions involving N, H, and V_{Ga} in GaAs. Note that because the charge states and the number of atoms are conserved in each of the reactions, the binding energies are independent of the position of the Fermi level and of the atomic chemical potentials.

1. $(N\text{-}V_{Ga})^{3-} + E_b \leftrightarrow V_{Ga}^{3-} + N$: The binding energy for this reaction is $E_b = 0.43$ eV. The reason for this energy reduction is due to the level repulsion between the occupied $t_{2v}(V_{Ga})$ and the empty $t_{2c}(N_{As})$ states, which push the $t_{2v}(V_{Ga})$ states down in energy. This level repulsion is most efficient when the V_{Ga} state is fully occupied, i.e., in the (3–) charged state, but it is less efficient when V_{Ga} is neutral [10]. (In the case of $N\text{-}V_{Ga}$, the (3–/2–) and (2–/–) transition energies are 110 and 30 above the valence band maximum (VBM), respectively, while the (–/0) is 10 meV below the VBM.) The binding between N and V_{Ga}^{3-} leads us to expect an increase in $[V_{Ga}]$ with the nitrogen concentration.
2. $(H\text{-}V_{Ga})^{2-} + E_b \leftrightarrow V_{Ga}^{3-} + H^+$: In the absence of N, an H^+ at a bond-center site could be attracted to a V_{Ga}^{3-} to form the stable complex $(H\text{-}V_{Ga})^{2-}$ with $E_b = 1.27$ eV. The large binding energy here indicates that $[V_{Ga}]$ in GaAs will increase with H concentration. Note, however, that when ε_F moves closer to the CBM, H^+ becomes less stable than H^-, resulting in isolated H^- and V_{Ga}^{3-}.
3. $(N\text{-}H\text{-}V_{Ga})^{2-} + E_b \leftrightarrow V_{Ga}^{3-} + (N\text{-}H)^+$: In the presence of $(N\text{-}H)^+$ pairs, the formation energy of V_{Ga}^{3-} is lowered by $E_b = 1.48$ eV, forming stable $(N\text{-}H\text{-}V_{Ga})^{2-}$ complexes. The increase of the binding energy (1.05 eV compared with reaction 1 or 0.21 eV compared with reaction 2) suggests that the presence of H enhances the coupling between N and V_{Ga} and vice versa, reducing the formation energies of all V_{Ga}-related complexes.

Alternatively, the formation of the $(N\text{-}H\text{-}V_{Ga})^{2-}$ complex in reaction 3 can be considered as a product of the reaction $(N\text{-}H\text{-}V_{Ga})^{2-} + E_b \leftrightarrow (N\text{-}V_{Ga})^{3-} + H^+$. The presence of H^+ significantly lowers the formation energy of the $(N\text{-}V_{Ga})^{3-}$ complex, with $E_b = 2.15$ eV. The formation of the $(N\text{-}H\text{-}V_{Ga})^{2-}$ complex in reaction 3 can also be considered as a product of the reaction $(N\text{-}H\text{-}V_{Ga})^{2-} + E_b \leftrightarrow (H\text{-}V_{Ga})^{2-} + N$. This reaction shows that in the presence of N, the formation energy of $(H\text{-}V_{Ga})^{2-}$ is lowered by $E_b = 1.31$ eV. These large binding energies (compared with reactions 1 and 2 above) are due mainly to the formation of the strong N-H chemical bond in the $(N\text{-}H\text{-}V_{Ga})^{2-}$ complex.

Janotti et al. [51] also studied the $N\text{-}V_{Ga}$ defects with multiple H atoms: $N\text{-}nH\text{-}V_{Ga}$ with $n = 2, 3, 4$. The calculated formation energies are shown

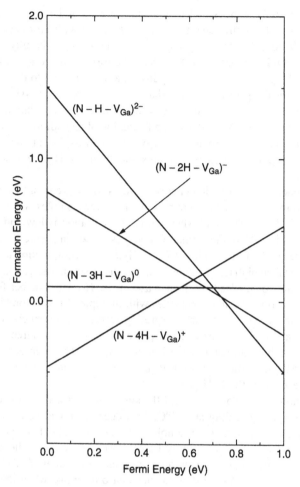

Figure 8.12 Formation energy vs. Fermi energy for $(N\text{-}H\text{-}V_{Ga})^{2-}$, $(N\text{-}2H\text{-}V_{Ga})^{-}$, $(N\text{-}3H\text{-}V_{Ga})^{0}$, and $(N\text{-}4H\text{-}V_{Ga})^{+}$ complexes in dilute GaAsN alloys. Ga-rich conditions and equilibrium with H_2 at $T = 0$ are assumed. (From Ref. [51].)

in Figure 8.12, where $(N\text{-}H\text{-}V_{Ga})^{2-}$ has a formation energy of 1.51 eV at ε_F = 0. An $N\text{-}2H\text{-}V_{Ga}$ with a second hydrogen atom bonding to one of the As neighboring atoms could be stable in the (−) and 0 charge states with formation energies of 0.77 and 0.70 eV, respectively. The lowering of the formation energy of $(N\text{-}2H\text{-}V_{Ga})^{-}$ by 0.74 eV with respect to that of $(N\text{-}H\text{-}V_{Ga})^{2-}$ is mostly due to the formation of an As-H chemical bond. Note that this value of 0.74 eV is smaller than the value of 1.66 eV for the addition of the first H^+. This is because the first H is attached to N, and the N-H bond is strong, whereas the second H is bonded to As, and the As-H bond is weaker. Furthermore, the second H also experiences H-H repulsion.

When a third H atom is added, it bonds to a second As atom, which is stable only in the neutral-charge state, $(N-3H-V_{Ga})^0$, with a formation energy of 0.10 eV. In this case, the $t_2(V_{Ga})$ state in the band gap is fully occupied. Again, we find a reduction of 0.67 eV in the formation energy with respect to $(N-2H-V_{Ga})^-$. Finally, a fourth hydrogen atom can bond to the remaining As atom, resulting in a complex that acts as a donor: $(N-4H-V_{Ga})^+$. The formation energy of −0.46 eV at $\varepsilon_F = 0$ again reflects a reduction of 0.56 eV relative to $(N-3H-V_{Ga})^0$ due to the formation of an additional As-H bond. The reduction of the formation energy decreases as the number of added H atoms in the vacancy increases, presumably due to H-H repulsion in the complex.

In general, the formation energy of the $N-nH-V_{Ga}$ complexes decreases with n up to $n = 4$. For entropic reasons, however, complexes with a larger number of H are less likely when the H concentration is low and the temperature is high. Also, the formation energies shown in Figure 8.12 assume equilibrium with H_2 molecules at $T = 0$. Hydrogen incorporation takes place during growth at high temperatures, which cause a strong decrease in μ_H [52]. Lower values of μ_H lead to an increase in the H-related defect formation energy, rendering vacancy complexes with multiple H less favorable.

The above results help in interpreting several recent experiments. The finding that $[V_{Ga}]$ increases in the presence of nitrogen agrees with the positron annihilation studies of Li et al. [48] and Toivonen et al. [49]. It shows clearly that the formation of the $(N-H-V_{Ga})$ complexes is enhanced in the presence of the N-H pair.

In addition, Toivonen et al. [49] found that the vacancy concentration decreases upon annealing at 700°C. This decrease could be explained by the diffusion of V_{Ga} out of the sample. Janotti et al. [51] noted that in GaAs, the migration barrier of V_{Ga} is only 1.5 eV [53]. In pure GaAs, the vacancies would therefore be mobile at temperatures well below 700°C. But in MOCVD-grown GaAsN, the vacancies can only move when their bonds with N and H are broken. The reaction that is most appropriate to describe this process is therefore reaction 3 above, with a binding energy of 1.48 eV. This binding energy needs to be added to the migration barrier of V_{Ga} in pure GaAs, resulting in a total barrier of $1.48 + 1.5 \approx 3.0$ eV. With a reasonable prefactor, a barrier of this magnitude would render Ga vacancies mobile above 650°C, in good agreement with Toivonen et al. [49].

Janotti et al. [51] pointed out that their results about the V_{Ga} mobility also nicely explain the observed interdiffusion on the group III sublattice in quantum-well structures, as reported by Albrecht et al. [54] and Pang et al. [55]. Interdiffusion of Ga and In is mediated by the Ga vacancies. As pointed out earlier, the interaction with N would render formation of V_{Ga} more likely than in pure GaAs, enhancing the concentration of vacancies and hence the interdiffusion. Furthermore, Albrecht et al. [54] observed a PL band centered at around 0.82 eV in the as-grown samples (about 0.1

eV below the band edge), which was significantly reduced upon annealing. It was speculated that this band might arise from the N-V_{Ga} complexes, which have transition levels about 0.1 eV above the VBM. (In the case of N-V_{Ga}, the (3–/2–) and (2–/–) transition energies are 110 and 30 above the valence band maximum (VBM), respectively, while the (–/0) is 10 meV below the VBM.)

Finally, the calculations suggested that N-H-V_{Ga} might act as a shallow acceptor. Although the error bar on the calculated transition levels is too large to allow an unambiguous conclusion, N-H-V_{Ga} acting as a shallow acceptor could plausibly explain the reported background acceptor concentration in as-grown InGaAsN [11, 50]. Moreover, growth of InGaAsN by solid-source MBE has been shown to result in much lower acceptor concentrations [56], consistent with the lower likelihood of V_{Ga} formation in this relatively H-free growth environment. Annealing at high temperature will cause both H and V_{Ga} to become mobile. If the hydrogen, which is bonded to an anion and behaves as a donor, is lost faster than V_{Ga}, then the acceptor concentration may increase, as was observed for InGaAsN annealed under nitrogen [11]. However, when H is depleted more slowly than gallium vacancies, $[V_{Ga}]$ will decrease, and the observed conversion from p-type to n-type GaAsN could be a result of the transformation from $(N$-H-$V_{Ga})^{2-}$ to $(N$-$H)^{+}$ [57].

8.6. SUMMARY

In summary, we have reviewed some of the recent developments in the theoretical understanding of defects/impurities in dilute III-V nitrides. Key in the development that has enabled the defect studies is the extension of the atomic chemical potentials to the so-called epitaxial regime. Qualitatively different defect physics from that of conventional III-V compounds was obtained. For example, the nitrogen solubility could be many orders of magnitude higher than that under bulk equilibrium. It was also demonstrated that nitrogen split interstitials, negligible in conventional III-V compounds, could play a very important role in this family of materials. We also reviewed the role of hydrogen in the dilute nitrides. It was shown that the H_2^{*} complexes, while being only metastable in conventional semiconductors, have exceptional stability due to the involvement of nitrogen. This allows for the identification of H-related IR spectra in the dilute nitrides, as well as a whole new range of H-related electronic properties, including the exclusive-donor behavior of H and the complete reversal of the band-gap reduction by the nitrogen atoms. Finally, we reviewed the effects of nitrogen and hydrogen on the formation energy and the electronic properties of the Ga vacancies in GaAsN. Both N and H have the effect of further stabilizing vacancies in GaAsN than in pure binaries. A number of recent puzzling experimental observations were elucidated.

Acknowledgments

We would like to offer our special thanks to Anderson Janotti (Oak Ridge National Laboratories) for much of the work being discussed here, as well as to Chris Van de Walle and Sarah Kurtz (National Renewable Energy Laboratory), who also made invaluable contributions at the various stages of the studies. This work was supported by the U.S. DOE-SC-BES under contract No. DE-AC36-99GO10337.

References

1. Ho, I.H., and Stringfellow, G.B., "Solubility of Nitrogen in Binary III-V Systems," *J. Crystal Growth* 178: 1–7 (1997).
2. Wei, S.H., and Zunger, A., "Giant and Composition-Dependent Optical Bowing Coefficient in GaAsN Alloys," *Phys. Rev. Lett.* 76: 664–667 (1996).
3. Zhang, S.B., and Zunger, A., "Surface-Reconstruction-Enhanced Solubility of N, P, As, and Sb in III-V Semiconductors," *Appl. Phys. Lett.* 71: 677–679 (1997).
4. Wolford, D.J., Bradley, J.A., Fry, K., and Thompson, J., "The Nitrogen Isoelectronic Trap in GaAs," in *Proceedings of the 17th International Conference on the Physics of Semiconductors*, Ed. J.D. Chadi and W.A. Harrison (New York: Springer, 1984), 627.
5. Weyers, M.M., and Sato, M., "Growth of GaAsN Alloys by Low-Pressure Metalorganic Chemical Vapor Deposition Using Plasma-Cracked NH_3," *Appl. Phys. Lett.* 62: 1396–1398 (1993).
6. Kondow, M., Uomi, K., Kitatani, T., Watahiki, S., and Yazawa, Y., "Extremely Large N Content (up to 10%) in GaNAs Grown by Gas-Source Molecular Beam Epitaxy," *J. Crystal Growth* 164: 175–179 (1996).
7. Qiu, Y., Nikishin, S.A., Temkin, H., Faleev, N.N., and Kudriavtsev, Y.A., "Growth of Single Phase $GaAs_{1-x}N_x$ with High Nitrogen Concentration by Metal-Organic Molecular Beam Epitaxy," *Appl. Phys. Lett.* 70: 3242–3244 (1997).
8. Bi, W.G., and Tu, C.W., "Bowing Parameter of the Band-Gap Energy of GaNAs," *Appl. Phys. Lett.* 70: 1608–1610 (1997).
9. Uesugi, K., and Suemune, I., "Metalorganic Molecular Beam Epitaxy of GaNAs Alloys on (0 0 1) GaAs," *J. Crystal Growth* 189/190: 490–495 (1998).
10. Zhang, S.B., and Wei, S.H., "Nitrogen Solubility and Induced Defect Complexes in Epitaxial GaAsN," *Phys. Rev. Lett.* 86: 1789–1792 (2001).
11. Geisz, J.F., Friedman, D.J., Olson, J.M., Kurtz, S.R., and Keyes, B.M., "1-eV Solar Cells with GaInNAs Active Layer," *J. Crystal Growth* 195: 409–415 (1998); and private communications.
12. Zhang, S.B., and Northrup, J.E., "Chemical Potential Dependence of Defect Formation Energies in GaAs: Application to Ga Self-Diffusion," *Phys. Rev. Lett.* 67: 2339–2342 (1991).
13. Hauenstein, R.J., Collins, D.A., Cai, X.P., O'Steen, M.L., and McGill, T.C., "Reflection High Energy Electron Diffraction Study of Nitrogen Plasma Interactions with a GaAs (100) Surface," *Appl. Phys. Lett.* 66: 2861–2863 (1995).
14. Gwo, S., Tokumoto, H., and Miwa, S., "Atomic-Scale Nature of the (3×3)-Ordered GaAs(001):N Surface Prepared by Plasma-Assisted Molecular-Beam Epitaxy," *Appl. Phys. Lett.* 71: 362–364 (1997).
15. Aksenov, I., Iwai, H., Nakada, Y., and Okumura, H., "Nitridation of GaAs(001) Surface: Auger Electron Spectroscopy and Reflection High-Energy Electron Diffraction," *J. Vac. Sci. Technol. B* 17: 1525–1539 (1999).

16. Sato, M., "Plasma-Assisted MOCVD Growth of GaAs/GaN/GaAs Thin-Layer Structures by N-As Replacement Using N-Radicals," *Jpn. J. Appl. Phys. (Part 1)* 34: 1080–1084 (1995).

17. Northrup, J.E., and Froyen, S., "Structure of GaAs(001) Surfaces: The role of Electrostatic Interactions," *Phys. Rev. B* 50: 2015–2018 (1994).

18. Moto, A., Tanaka, S., Ikoma, N., Tanabe, T., Takagishi, S., Takahashi, M., and Katsuyama, T., "Metalorganic Vapor Phase Epitaxial Growth of GaNAs Using Tertiarybutylarsine and Dimethylhydrazine (DMHy)." *Jpn. J. Appl. Phys.* 38 (Part 1): 1015–1018 (1999).

19. See, S., and Sze, M., *Semiconductor Devices: Physics and Technology* (New York: John Wiley & Sons, 1985), 48.

20. Pankove, J.I., and Johnson, N.M., "Introduction to Hydrogen Semiconductors," in *Semiconductors and Semimetals*, vol. 34, ed. R.K. Willardson and A.C. Beer (Boston: Academic Press, 1991).

21. Chang, K.J., and Chadi, D.J., "Diatomic-Hydrogen-Complex Diffusion and Self-Trapping in Crystalline Silicon," *Phys. Rev. Lett.* 62: 937–940 (1989); "Vibrational Properties of Metastable Diatomic Hydrogen Complexes in Crystalline Silicon," *Phys. Rev. B* 42: 7651–7654 (1990;).

22. Zhang, S.B., and Jackson, W.B., "Formation of Extended Hydrogen Complexes in Silicon," *Phys. Rev. B* 43: 12142–12145 (1991).

23. Zhang, S.B., and Chadi, D.J., "Microscopic Structure of Hydrogen–Shallow-Donor Complexes in Crystalline Silicon," *Phys. Rev. B* 41: 3882–3884 (1990).

24. Van de Walle, C.G., Denteneer, P.J.H., Bar-Yam, Y., and Pantelides, S.T., "Theory of Hydrogen Diffusion and Reactions in Crystalline Silicon," *Phys. Rev. B* 39: 10791–10808 (1989).

25. Shan, W., Yu, K.M., Walukiewicz, W., Ager III, J.W., Haller, E.E., and Ridgway, M.C., "Reduction of Band-Gap Energy in GaNAs and AlGaNAs Synthesized by N^+ Implantation," *Appl. Phys. Lett.* 75: 1410–1412 (1999).

26. Clerjaud, B., Côte, D., Hahn, W.S., Lebkiri, A., Ulrici, W., and Wasik, D., "Nitrogen-Dihydrogen Complex in GaP," *Phys. Rev. Lett.* 77: 4930–4933 (1996).

27. Tagami, K., Tsuchida, E., and Tsukada, M., "First-Principles Study of Vibrational Spectra on Dihydride-Terminated Si(001)/H Surfaces," *Surf. Sci.* 446: L108–L112 (2000).

28. Murakami, K., Fukata, N., Sasaki, S., Ishioka, K., Kitajima, M., Fujimura, S., Kikuchi, J., and Haneda, H., "Hydrogen Molecules in Crystalline Silicon Treated with Atomic Hydrogen," *Phys. Rev. Lett.* 77: 3161–3164 (1996).

29. Vetterhoffer, J., Wagner, J., and Weber, J., "Isolated Hydrogen Molecules in GaAs," *Phys. Rev. Lett.* 77: 5409–5412 (1996).

30. Pavesi, L., and Giannozzi, P., "Atomic and Molecular Hydrogen in Gallium Arsenide: A Theoretical Study," *Phys. Rev. B* 46: 4621–4629 (1992).

31. Zhang, S.B., and Branz, H.M., "Hydrogen above Saturation at Silicon Vacancies: H-Pair Reservoirs and Metastability Sites," *Phys. Rev. Lett.* 87: 105503/1-4 (2001).

32. Holbech, J.D., Bech Nielsen, B., Jones, R., Sitch, P., and Oberg, S., "H^*_2 Defect in Crystalline Silicon," *Phys. Rev. Lett.* 71: 875–878 (1993).

33. Janotti, A., Zhang, S.B., and Wei, S.H., "Hydrogen Vibration Modes in GaP:N: The Pivotal Role of Nitrogen in Stabilizing the H_2^* Complex," *Phys. Rev. Lett.* 88: 125506/1-4 (2002).

34. Northrup, J.E., "Surface Phonon Frequencies and Eigenvectors on Si(111) $\sqrt{3} \times \sqrt{3}$: Al," *Phys. Rev. B* 39: 1434–1437 (1989).

35. Xin, H.P., Tu, C.W., and Geva, M., "Annealing Behavior of *p*-Type $Ga_{0.892}In_{0.108}N_xAs_{1-x}$ ($0 \leq X \leq 0.024$) Grown by Gas-Source Molecular Beam Epitaxy," *Appl. Phys. Lett.* 75: 1416–1418 (1999).

36. Xin, H.P., Tu, C.W., and Geva, M., "Effects of Hydrogen on Doping of GaInNAs Grown by Gas-Source Molecular Beam Epitaxy," *J. Vac. Sci. Technol. B* 18: 1476–1479 (2000).

252 S.B. ZHANG and SU-HUAI WEI

37. Baldassarri Höger von Högersthal, G., Bissiri, M., Polimeni, A., Capizzi, M., Fischer, M., Reinhardt, M., and Forchel, A., "Hydrogen Induced Band-Gap Tuning of (InGa)(AsN/GaAs) Single Quantum Wells," *Appl. Phys. Lett.* 78: 3472–3474 (2001).
38. Polimeni, A., Baldassarri Höger von Högersthal, G., Bissiri, M., Capizzi, M., Fischer, M., Reinhardt, M., and Forchel, A., "Effect of Hydrogen on the Electronic Properties of $In_xGa_{1-x}As_{1-y}N_y$/GaAs Quantum Wells," *Phys. Rev. B* 63: 201304/1-4 (2001).
39. Janotti, A., Zhang, S.B., Wei, S.H., and Van de Walle, C.G., "Effects of Hydrogen on the Electronic Properties of Dilute GaAsN Alloys," *Phys. Rev. Lett.* 89: 086403/1-4 (2002).
40. Estreicher, S.K., "Hydrogen-Related Defects in Crystalline Semiconductors: A Theorist's Perspective," *Mat. Sci. Eng. Rep.* 14: 319–412 (1995).
41. Neugebauer, J., and Van de Walle, C.G., "Hydrogen in GaN: Novel Aspects of a Common Impurity," *Phys. Rev. Lett.* 75: 4452–4455 (1995).
42. Weyers, M., Sato, M., and Ando, H., "Red Shift of Photoluminescence and Absorption in Dilute GaAsN Alloy Layers," *Jap. J. Appl. Phys.* 31: L853–L855 (1992).
43. Neugebauer, J., and Van de Walle, C.G., "Electronic Structure and Phase Stability of $GaAs_{1-x}N_x$ Alloys," *Phys. Rev. B* 51: 10568–10571 (1995).
44. Wei, S.H., and Zunger, A., "Calculated Natural Band Offsets of All II-VI and III-V Semiconductors: Chemical Trends and the Role of Cation d Orbitals," *Appl. Phys. Lett.* 72: 2011–2013 (1998).
45. Van de Walle, C.G., "Hydrogen as a Cause of Doping in Zinc Oxide," *Phys. Rev. Lett.* 85: 1012–1015 (2000).
46. Limpijumnong, S., and Van de Walle, C.G., "Passivation and Doping due to Hydrogen in III-Nitrides," *Phys. Stat. Solidi B* 228: 303–307 (2001).
47. Amore Bonapasta, A., Filippone, F., Giannozzi, P., Capizzi, M., and Polimeni, A., "Structure and Passivation Effects of Mono- and Di-Hydrogen Complexes in $GaAs_yN_{1-y}$ Alloys," *Phys. Rev. Lett.* 89: 216401/1-4 (2002).
48. Li, W., Pessa, M., Ahlgren, T., and Decker, J., "Origin of Improved Luminescence Efficiency after Annealing of Ga(In)NAs Materials Grown by Molecular-Beam Epitaxy," *Appl. Phys. Lett.* 79: 1094–1096 (2001).
49. Toivonen, J., Hakkarainen, T., Sopanen, M., Lipsanen, H., Oila, J., and Saarinen, K., "Observation of Defect Complexes Containing Ga Vacancies in GaAsN," *Appl. Phys. Lett.* 82: 40–42 (2003).
50. Kurtz, S., Reedy, R., Keyes, B., Barber, G.D., Geisz, J.F., Friedman, D.J., McMahon, W.E., and Olson, J.M., "Evaluation of NF_3 versus Dimethylhydrazine as N Sources for GaAsN," *J. Crystal Growth* 234: 323–326 (2002).
51. Janotti, A., Wei, S.H., Zhang, S.B., Kurtz, S., Van de Walle, C.G., "Interactions between Nitrogen, Hydrogen, and Gallium Vacancies in GaAsN Alloys," *Phys. Rev. B* 67: 161201/1-4 (2001).
52. Van de Walle, C.G., and Neugebauer, J., "First-Principles Surface Phase Diagram for Hydrogen on GaN Surfaces," *Phys. Rev. Lett.* 88: 066103/1-4 (2002).
53. Dabrowski, J., and Northrup, J.E., "Microscopic Theory of Diffusion on the Ga Sublattice of GaAs: Vacancy-Assisted Diffusion of Si and Ga," *Phys. Rev. B* 49: 14286–14289 (1994).
54. Albrecht, M., Grillo, V., Remmele, T., Strunk, H.P., Egorov, A.Y., Dumitras, G., Riechert, H., Kaschner, A., Heitz, R., and Hoffmann, A., "Effect of Annealing on the In and N Distribution in InGaAsN Quantum Wells," *Appl. Phys. Lett.* 81: 2719–2721 (2002).
55. Peng, C.S., Pavelescu, E.M., Jouhti, T., Konttinen, J., Fodchuk, I.M., Kyslovsky, Y., and Pessa, M., "Suppression of Interfacial Atomic Diffusion in InGaNAs/GaAs Heterostructures Grown by Molecular-Beam Epitaxy," *Appl. Phys. Lett.* 80: 4720–4722 (2002).
56. Ptak, A.J., Johnston, S.W., Kurtz, S., Friedman, D.J., and Metzger, W.K., "A Comparison of MBE- and MOCVD-Grown GaInNAs," *J. Crystal Growth* 251: 392–398 (2003).

57. Kurtz, S., Geisz, J.F., Friedan, D.J., Olason, J.M., Duda, A., Karam, N.H., King, R.R., Ermer, J.H., and Jolsin, D.E., "Modeling of Electron Diffusion Length in GaInAsN Solar Cells," in *Proceedings of 28th IEEE Photovoltaic Specialists Conference* (New York: IEEE, 2000), 1210–1213.

CHAPTER 9

Recombination Processes in Dilute Nitrides

I.A. BUYANOVA and W.M. CHEN

*Department of Physics and Measurement Technology,
Linköping University, Sweden*

1-591-69019-6/04/$0.00+$1.50
© 2004 by CRC Press LLC

9.1   INTRODUCTION

Dilute nitrides form a novel material system with fascinating physical properties [1] desirable for various applications in optoelectronics and photonics. In combination with the possibility of lattice-matching to Si and GaAs, a giant bowing in the band-gap energy in dilute nitrides provides an increased flexibility in tuning material properties for applications as highly efficient light emitters, operating within the fiber-optic communications window (1.3–1.55 μm), and in multijunction solar cells with improved efficiency. This has led to considerable research efforts to understand recombination processes in the materials. The aim of this chapter is to provide a brief account of the present status of this issue.

The chapter is organized as follows. In Section 9.2, we show the nature of photoluminescence (PL) in Ga(In)NAs as the result of a recombination of excitons localized by potential fluctuations in band edges, and we discuss an origin of tail states. In Section 9.3, we discuss the nature of radiative recombination in GaNP as well as effects of the band crossover from an indirect band-gap in GaP to a direct band-gap in GaNP. Experimental findings obtained on chemical identities and formation mechanisms of nonradiative defects in Ga(In)NAs and Ga(Al)NP alloys are reviewed in Sections 9.4 and 9.5, respectively.

9.2   RADIATIVE RECOMBINATION IN Ga(In)NAs ALLOYS

Among the primary motivations for the development of Ga(In)NAs alloys is their application as efficient light emitters operating within the spectral window of 1.3–1.55 μm for fiber-optic communications. Naturally, numerous optical studies have been carried out to address the origin of radiative recombination in the alloys. These efforts have led to a rather satisfactory understanding of this issue. They have also provided important insights into some key material-related problems, such as effects related to nonequilibrium growth conditions and strain, as well as into the fundamental physical processes underlying the band-structure changes.

9.2.1   Mechanism for Radiative Recombination

PL spectra of Ga(In)NAs alloys in the near-band-edge spectral region are dominated by a rather broad structureless emission that experiences a strong redshift with increasing N composition [2–7]. The observed shift of the PL peak position correlates with the reduction in the alloy band-gap energy, deduced, for example, from PL excitation (PLE) and absorption measurements and thus was interpreted as representing the band-gap bowing effect. Compositional dependence of the PL and PLE spectra is demonstrated in Figure 9.1, taking as an example GaNAs epilayer structures.

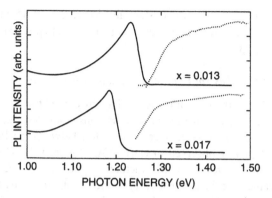

Figure 9.1 Typical PL spectra (solid lines) of the near-band-edge emission in GaNAs alloys measured from the GaNAs epilayers with two nitrogen compositions. The corresponding PLE spectra are shown by the dashed lines and demonstrate a clear redshift of the GaNAs band-gap with increasing N content. The spectra are shifted relatively in the vertical direction, for clarity.

The PL emissions in bulk Ga(In)NAs epilayers and Ga(In)NAs/GaAs quantum-well (QW) structures exhibit the following characteristic properties, which have allowed identification of the corresponding optical transitions as being due to a recombination of localized excitons (LE) [7–10] trapped within band tail states.

9.2.1.1 Asymmetric line shape

The near-band-gap emission in Ga(In)NAs is typically very asymmetric (Figure 9.1), with a tail at low energies and a sharp cutoff at high energies that experiences a blueshift with increasing excitation power. The exponential dependence of the PL intensity within the low-energy tail suggests that the PL transitions occur within the tail states with an exponential density distribution, typical for random fluctuations of alloy compositions. The slope of the PL tail,

$$E_0 = \left(\frac{d \ln I(hv)}{d(hv)} \right)^{-1} \tag{9.1}$$

provides some qualitative measure of the fluctuation potential responsible for the exciton localization. (Possible contributions to the low-energy PL tail from the excitonic emissions within the deep N-related cluster states [11] can somewhat affect the quantitative values for the fluctuation potential deduced by this approach.) The value of E_0 was shown to increase with increasing N content and depends critically on the growth conditions. The estimated values range from 35 to 60 meV for the as-grown Ga(In)NAs epilayers [7, 10, 12], but are substantially lower, i.e., 9–15 meV, for the alloys grown either at high temperatures or subjected to postgrowth thermal annealing [12, 13].

9.2.1.2 Large stokes shift between PL and PLE spectra

Since the PLE spectra are mainly determined by optical transitions between extended states, whereas the radiative transitions at low temperatures occur within the localized states, a large Stokes shift between the PL and PLE spectra was typically observed in Ga(In)NAs (Figure 9.1). The energy of the Stokes shift scales with the fluctuation potential in the alloy. The estimated values are in reasonable agreement with the ones determined from the PL line-shape analysis discussed in the previous section [7].

9.2.1.3 Spectral dependence of PL decay

The PL decay of the near-band-gap emission measured at a certain emission energy is predominantly single exponential, with a PL decay time being a strong function of the emission energy (Figure 9.2). A shortening of the PL decay time is known to occur at the high-energy side of the LE PL spectrum and is attributed to exciton transfer to the lower-energy tail states [14] or to competing recombination channels. Within this model, the spectral dependence of the PL decay time is described by the function

$$\tau(h\nu) = \frac{\tau_{rad}}{1 + E_0^{-1} \cdot \exp(h\nu - E_m)} \tag{9.2}$$

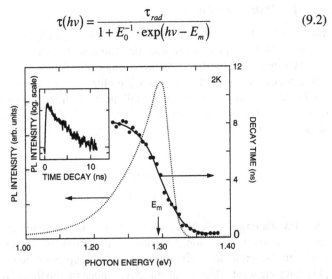

Figure 9.2 Spectral dependence of the PL decay time (dots) measured at 2 K from the GaN$_{0.011}$As$_{0.989}$/GaAs multiple QW structure. The solid line is a fit by using Equation 9.2. The PL spectrum of the sample is shown by the dotted line, for easy reference. The insert shows PL decay detected at 1.23 eV.)

where τ_{rad} is the radiative lifetime and E_m represents the "mobility edge," i.e., the energy [14] at which the probability of the exciton transfer to deeper localized states becomes equal to the probability of the radiative recombination. Such spectral dependence of the PL decay time was indeed observed in Ga(In)NAs alloys (Figure 9.2). Fitting of the experimental data by Equation 9.2 provided the value of $\tau_{rad} \approx$ 5–8 ns for the GaNAs epilayers and quantum-well structures [7], i.e., comparable with the radiative lifetime of 3 ns in the parental GaAs. The somewhat smaller τ_{rad} value of 0.35 ns has been deduced for the GaInNAs epilayers [8]. The PL transient measurements were also used as a complementary approach to characterize potential fluctuations in the alloy, providing similar values of the E_0.

9.2.1.4 S-shape temperature dependence of the PL maximum

The dependence of the PL maximum position on measurement temperature exhibits an apparent S-shape [7, 9] (Figure 9.3). At low temperatures, a strong redshift of the PL maximum with increasing temperature T is observed and reflects thermal depopulation of the localized states, which starts from the shallowest localization potential. At elevated temperatures, the appearance of free exciton (FE) recombination (Figure 9.3) causes a blueshift of the PL maximum position. As expected, further increase in measurement temperature is accompanied by a redshift of the FE transition due to temperature-induced reduction in the band-gap energy.

The FE transition dominates in the radiative recombination in Ga(In)NAs alloys at elevated temperatures, typically above 100 K [15,16].

9.2.2 Origin of Potential Fluctuations

A direct experimental proof confirming the origin of the low-temperature PL emission in Ga(In)NAs has most recently been obtained from the near-field scanning optical microscopy (NSOM) [17, 18]. Sharp PL peaks corresponding to individual LE transitions have been resolved within the broad profile of the LE PL band observed in the macro-PL measurements (Figure 9.4). The PL transitions were concluded to occur within confined regions acting as quantum dots (QDs), due to compositional fluctuations in N content (clustering of N atoms). They involve delocalized bandlike states, based on the observation of a diamagnetic shift for the individual LE transitions [18]. The average lateral dimension of the QDs was estimated to be around 20 nm in $In_{0.08}Ga_{0.92}As_{0.97}N_{0.03}$ layers whereas the average value of compositional fluctuation in the cluster was around 0.5%. These relatively low deviations in N content have a profound effect on the conduction-band (CB) edge of the alloy due to the giant bowing and, thus, easily create confining potentials for free

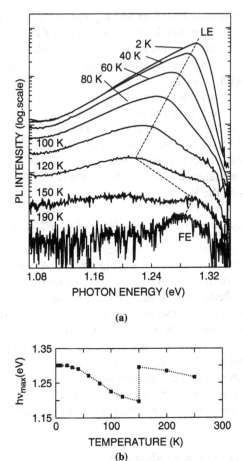

Figure 9.3 (a) Temperature dependence of the PL spectra measured from the $GaN_{0.011}As_{0.989}$/ GaAs multiple QW structure. (b) The PL maximum position as a function of measurement temperature.

carriers. On the other hand, band-tail states seem to contain also some contribution from impuritylike states related to various N clusters, as revealed by magneto-tunneling spectroscopy [19].

Formation of N-rich clusters with lateral sizes of 10–30 nm was independently confirmed by cross-sectional transmission electron microscopy (TEM) measurements, in accordance with previously reported structural studies [9, 20]. An increase in cluster density of more than ten orders of magnitude greater than that expected for a random nitrogen distribution was concluded [18], indicative of their spontaneous formation. This result seems to contradict previous studies by scanning tunneling microscopy (STM) [21], where a predominantly random distribution of N atoms in GaNAs layers with a modest (≤50%) enhancement in the number of the

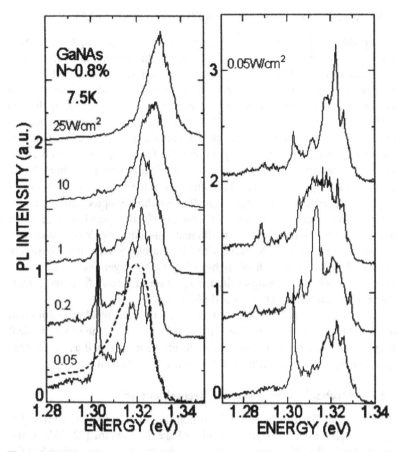

Figure 9.4 Left panel: Near-field PL spectra with various excitation power densities at 7.5 K. The dotted line represents the macro-PL spectrum. Right panel: Near-field PL spectra measured at different spatial locations under low excitation conditions. (From Ref. [17].)

nearest-neighbor pairs was found. In GaInNAs, the fluctuations in N and In compositions have been suggested to have a predominantly bimodal character [22–25] which has a decisive effect on the band-gap energy.

The observed nonuniform distribution of N atoms is not surprising and, in fact, is expected to be promoted under nonequilibrium growth conditions, such as low-temperature growth by molecular-beam epitaxy (MBE), commonly used to ensure high N incorporation. Supporting this assumption, an increase in growth temperature [12], postgrowth annealing [13, 26], or postgrowth hydrogenation [27] of GaNAs has been found to suppress the potential fluctuations, likely due to a randomization of the nitrogen distribution [13, 26]. Rearrangements in local N environment upon annealing are more dramatic in GaInNAs, alloys [22–24]. A detailed description of these phenomena can be found in Chapters 5 and 14 of this book.

9.3   RADIATIVE RECOMBINATION IN GaNP ALLOYS

Incorporation of nitrogen in GaP is predicted [11, 28, 29] to have even more exciting effects on the band structure of the forming alloy than those observed in Ga(In)NAs, leading not only to a huge bowing in band-gap energy but also to the N-induced crossover from an indirect band-gap in GaP to a direct band-gap in GaNP alloys with N composition as low as 0.5% (see also Chapters 1 and 2 of this book). The first experimental observations indicating this crossover include the appearance of the photomodulation signal at the absorption edge of GaN_xP_{1-x} [29], as well as the spectral dependence of the absorption coefficient typical for a direct band-gap semiconductor [30]. The transformation to the direct band-gap should improve the radiative efficiency of GaNP, desirable for optoelectronic applications. Moreover, the possibility of growing GaNP layers lattice-matched to Si substrates provides hope for efficient light-emitting devices and multijunction solar cells compatible with Si technology. Recent encouraging results [31] of the growth of high-quality GaNP layers on Si substrates opens the door for the long-desired integration of mature Si-based microelectronics and optoelectronics and photonics.

A strong effect of the N incorporation on the near-band-edge emission of GaNP has, in fact, been observed [32–35]. However, the exact physical origin of the observed increase in the PL efficiency, as well as the mechanism for light emission in the alloy, remain controversial.

9.3.1   Mechanism for Radiative Recombination

Similar to the Ga(In)NAs alloys, incorporation of N in GaP transforms PL spectra and causes an overall redshift of the PL spectra [32–35]. In the doping limit, the PL is dominated by a series of sharp lines, related to the recombination of excitons bound to isoelectronic centers formed by N pairs (NN pairs) [36]. These lines disappear with increasing N content above about 0.2% and are replaced by some broader features within the 2.0–2.15-eV spectral range, attributed to N clusters [35]. A much larger line width of the corresponding optical transitions has been suggested to reflect a gradual formation of impurity bands from the overlapping N-related states [35], which will form the CB edge of GaNP at the later stages of alloy formation at higher N compositions. For even higher N compositions ($x >$ 2%), only a broad, structureless PL band is detected (Figure 9.5). This emission has been attributed to the tail states formed after merging of the impurity bands and-gap band edge [33], based on the large Stokes shift between PL and PL-excitation spectra. Alternatively, the broad PL emission has been suggested to represent the band-to-band recombination of GaNP, where high radiative efficiency is ensured by the band crossover [29].

We have recently pointed out [37] that all broader PL emissions, which become dominant in the PL spectra from GaNP with higher N compositions,

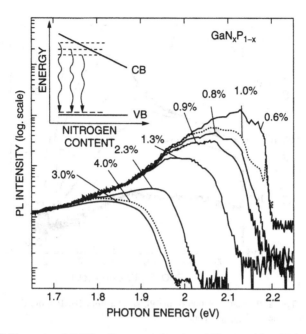

Figure 9.5 PL spectra of GaNP epilayers as a function of N composition. PL intensities in all spectra are displayed in the logarithmic scale and normalized at the low energy. The PL spectra of the GaN_xP_{1-x}/GaP multiple QW structures are also shown by the dotted lines, to demonstrate the effect of band-gap reopening due to quantum confinement. The insert shows the disappearance of the high-energy PL components when the corresponding energy levels start to be resonant with the conduction band, due to the N-induced downshift of the CB edge. (From Ref. [16].)

do not appear during the CB transformation, as they are already seen in the PL spectra of samples with N as low as 0.24%. Moreover, the spectral position and shape of each PL feature are not affected by an increase in N content, indicating that the feature is related to some deep level, with its absolute energy position insensitive to the alloy formation, and thus does not participate in the formation of the new CB edge, in agreement with pseudopotential calculations [11] (see Chapter 1). Thus we have attributed the broader PL lines to PL transitions within deep defects, most likely N-related clusters. The commonly observed redshift of the PL emission with increasing N content is then a consequence of the downshift of the CB edge due to the giant bowing effect, which causes a subsequent disappearance of the high-energy PL components (see insert in Figure 9.5). Vice versa, a reopening of the band gap, caused, e.g., by quantum confinement in GaNP/GaP multiple quantum wells (MQW), leads to a reappearance (or an increase in intensity) of the high-energy PL components, depicted by the dotted lines in Figure 9.5 supporting the proposed model.

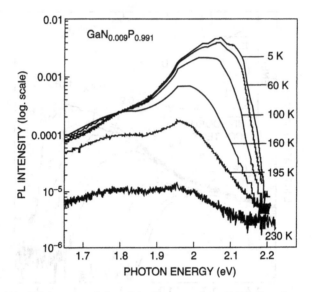

Figure 9.6 Typical thermal evolution of the PL spectra from GaNP alloys demonstrated by the example of the $GaN_{0.009}P_{0.991}$ epilayer.

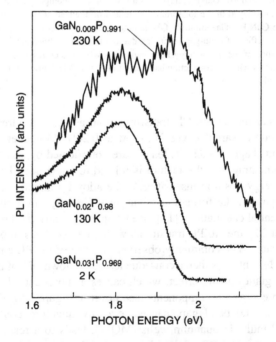

Figure 9.7 A comparison of several PL spectra measured at different temperatures from the GaNP alloys with several N compositions, to demonstrate the existence of the same optical transitions at 1.81 eV.

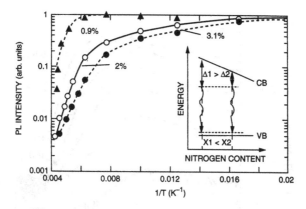

Figure 9.8 Arrhenius plots of the intensity of the 1.81-eV PL emission measured from the GaNP epilayers with the N compositions of 0.9% (solid triangles), 2.0% (open dots), and 3.1% (solid dots). The lines are the fitting curves with the activation energies of 300, 200, and 140 meV for the N compositions of 0.9%, 2.0%, and 3.1%, respectively. The insert shows a decrease in the thermal-activation energy of a certain optical transition with increasing N content, due to the N-induced downshift of the CB edge. (From Ref. [16].)

An increase in the measurement temperature causes one-by-one thermal ionization of the high-energy PL components, again indicating that they are related to several deep levels in the band-gap (Figure 9.6). In fact, the temperature increase has a very similar effect on the PL spectra as increasing N compositions. This is clearly seen in Figure 9.7, where the identical PL emission peaking at 1.81 eV becomes dominant either due to an increase in N content or by raising the measurement temperature.

Moreover, assuming that the energy levels of the deep N-related states remain almost constant with respect to the valence-band edge [11], they should become closer to the CB edge with increasing x content due to the downshift of the CB energy (see insert in Figure 9.8). This should result in a corresponding decrease of the activation energy describing thermal quenching of a specific optical transition, as indeed observed in the experiments (Figure 9.8). By fitting the Arrhenius plots of the 1.81-eV PL intensity, the activation energies for the N compositions of 0.9%, 2.0%, and 3.1% can be estimated to be 300, 200, and 140 meV, respectively. The obtained values are consistent with the reduction of the band-gap energy from 2.1 eV (for $x = 0.9\%$) to 1.92 eV (for $x = 3.1\%$), determined from the absorption measurements.

The observed temperature behavior of the broad PL band in the GaN$_x$P$_{1-x}$ with $x > 2\%$ (Figure 9.9a) is very similar to that with a lower x value (Figure 9.6), except for a lack of resolved features from individual N clusters. Specifically, an increase in temperature causes a strong redshift of the PL maximum, which is much larger than the temperature-induced shift of the GaNP band-gap energy (shown by the open dots in Figure 9.9b).

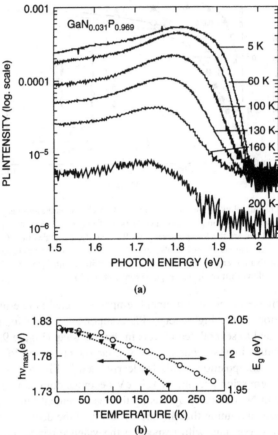

(a)

(b)

Figure 9.9 (a) Typical thermal evolution of the PL spectra from GaNP alloys with relatively high N content, demonstrated by the example of the GaN$_{0.031}$P$_{0.969}$ epilayer. (b) Comparison between the temperature dependence of the PL maximum position (solid triangles) and temperature-induced shift of the band-gap energy (open dots) deduced for the same structure from optical absorption measurements.

This rules out the possible band-to-band transitions as the origin of the broad emissions in the alloy and suggests that the PL arises from optical transitions within some deep centers, e.g., the N-related clusters that contain a larger number of N atoms. Increasing local variations due to, for example, the variety of such clusters, local strain, and composition makes it impossible to resolve the individual features. If the cluster states interact with the CB edge, the PL transitions can be considered as a localized exciton recombination, similar to the Ga(In)NAs alloy. However, in Ga(In)NAs, thermal activation of carriers from the localized states makes the band-to-band recombination dominant at elevated temperatures, apparent from the strong blueshift of the PL maximum. Such a blueshift was not observed for the GaNP samples (see Figure 9.6 and Figure 9.9), suggesting that the band-to-band recombination in the alloy remains mainly nonradiative.

The aforementioned results provide clear evidence that the near-band-gap emission in the studied GaN_xP_{1-x} alloy with $x \leq 4\%$ originates from the optical transitions within the N-related localized defects that do not participate in the formation of the alloy CB edge. The band-to-band recombination in GaNP remains predominantly nonradiative and thus is not seen in the PL experiments, probably due to a still stronger effect of competing recombination channels.

9.3.2 Band Structure Enhancement Effect

N-induced admixture of the Γ_c component to the CB states of GaNP, either due to the Γ-X-L mixing [11, 28] (see also Chapter 1) or because of the anticrossing between localized nitrogen states and conduction-band states [29] (see also Chapter 2), makes partially allowed the transitions between the CB minimum and the valence-band maximum. Gradual changes of the character of the band edge leading to a directlike band-gap [29, 30, 38] should strongly enhance the oscillator strength of the band-to-band optical transitions, providing hope for improved radiative efficiency of GaNP. Indeed, the PL intensity seems to increase with increasing N content.

The physical mechanism for this improvement has most recently been determined from the transient PL measurements [39]. The abrupt decrease in the PL lifetime was observed for N compositions of around 0.4%, followed by a minor reduction with further increase in the N content (Figure 9.10). The observed drastic decrease occurs over the same range of N compositions as the appearance of the band-edge signal in the photomodulated absorption spectra of GaNP [29], implying that the change in the character of the extended conduction-band states affects the oscillator strength of the optical transitions at the N-related localized states. The effect has been attributed to an enhancement in the oscillator strength of the optical transitions at the N-related isoelectronic centers due to the band crossover to the direct band gap, i.e., the so-called band-structure-enhancement (BSE) effect, well known from previous studies of excitons bound to the isolated N centers in GaAsP alloys [40, 41].

The observed reduction in the PL lifetime in GaNP is, however, much less pronounced than that observed for the conventional GaAsP:N alloys, where a two-order-of-magnitude decrease in the radiative lifetime (down to 1 ns) of the N-bound excitons [41] was observed after the crossover. In fact, the radiative lifetime in GaNP even after the crossover remains much longer than that typically observed for a direct band-gap semiconductor. This is consistent with the predicted [28] unusual properties of the conduction-band states in GaNP after the crossover, which are formed not only

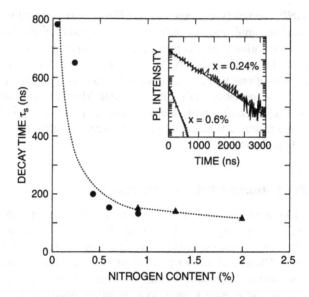

Figure 9.10 Compositional dependence of the radiative lifetimes, measured at 2 K for two emission energies of 2.07 eV and 1.95 eV. The insert shows PL decay curves measured for the same emission energy of 2.067 eV from the GaNP alloys with N content of 0.24% and 0.6%, to demonstrate the abrupt shortening of the PL lifetime due to the band crossover. In the insert, the straight solid lines are fitting curves assuming radiative lifetime of 650 ns and 150 ns for the N compositions of 0.24% and 0.6%, respectively.

from the Γ_c states (as in conventional alloy systems) but also from many other states.

9.4   NONRADIATIVE RECOMBINATION IN Ga(In)NAs

The incorporation of nitrogen strongly degrades radiative efficiency of the Ga(In)NAs alloys. This degradation of the optical quality is commonly attributed to formation of some competing nonradiative (NR) defects induced by nonequilibrium growth and/or promoted by the nitrogen incorporation. The importance of this issue was recently underlined when about 50% of the threshold current in the state-of-the-art GaInNAs-based lasers was concluded to have been caused by the NR recombination [42] (see also Chapter 12). The NR defects have also been found responsible for a degraded minority-carrier diffusion length in GaInNAs solar cells, causing a reduction in the internal quantum efficiency (see Chapter 13).

Despite the recognition of the importance of the NR recombination, only limited information is available so far on the origin of the NR centers as well as their formation mechanism. The main body of our present knowledge on NR centers was obtained by employing several experimental techniques, such as electrical measurements [43–48], nuclear reaction analysis (NRA) combined with Rutherford backscattering (RBS) channeling technique [49–51], magnetic resonance measurements [52–54], and positron annihilation spectroscopy [50, 55]. The cross-links between the experimental results obtained by these techniques are currently not well established. Therefore, the following discussion of the NR centers will be organized according to the employed experimental approaches.

9.4.1   Electrical Measurements

Information on the electronic properties of recombination centers in Ga(In)NAs alloys has been primarily obtained through the use of deep-level transient spectroscopy (DLTS). While a number of electron and hole traps of both intrinsic and extrinsic origin have been detected, only some of them were suggested to act as important recombination centers controlling carrier lifetime, and these will be reviewed here.

In n-type GaNAs, incorporation of nitrogen has been shown [43] to promote formation of dominant electron traps with energy levels fixed at E_V + 0.8 eV and E_V + 1.1 eV independent of N composition. From depth profiling measurements, the E_V + 0.8-eV trap is primarily formed near the surface, where its concentration can reach as high as 10^{18} cm^{-3} in as-grown GaNAs layers, but it is suppressed by more than two orders of magnitude after annealing. On the contrary, the E_V + 1.1-eV traps are homogeneously distributed throughout the GaNAs layers. Concentration of this defect increases drastically with increasing N content, reaching about 1×10^{17} cm^{-3} for 0.5% N. The large capture cross-section for electrons on the order of 10^{-15}–10^{-14} cm^{-2} places it among the primary candidates for efficient non-radiative recombination that, unfortunately, cannot be removed by thermal annealing. The defect was tentatively attributed to As-N antisite, based on comparison with theoretical predictions [56].

In p-type GaNAs layers, two dominant extrinsic defect levels at E_v + 0.35 eV and E_v + 0.45 eV (denoted as HK3 and HK4, respectively [44]) were observed. Their large capture cross-section of 10^{-15}–10^{-14} cm^{-2} results in a hole lifetime shorter than 1 ns, i.e., comparable with the PL decay time in GaNAs. Hence, these defects have been suggested to be, at least partly, responsible for the degradation of the optical quality of as-grown GaNAs layers, but they can be efficiently removed after rapid thermal annealing (RTA). The defects were tentatively attributed to residual contamination by Cu and Fe impurities during MBE growth, based on the similarity of their properties with those of well-known Cu_{Ga} and Fe_{Ga} defects in GaAs.

Several electron and hole traps of unknown origin were also found in GaInNAs layers [45, 46]. Most of them can be removed by postgrowth annealing accompanied by a substantial increase of the band-edge photoluminescence intensity and an increase in minority-carrier diffusion length [45]. This correlation indicates that thermally unstable deep levels largely determine the quality of as-grown layers, but these can be efficiently removed from the materials by annealing and, thus, are less harmful in terms of device applications. On the other hand, thermally stable traps, i.e., E_v + 0.48-eV hole trap in the metal-organic chemical-vapor deposition (MOCVD)-grown p-type $In_{0.07}Ga_{0.93}As_{0.98}N_{0.02}$ layers may be more important for future InGaAsN devices.

Thermal annealing was also found to introduce a new defect (denoted as E2/H1 in [47]) in GaInNAs layers, with an energy level at E_c – 0.34 eV, which has been suggested to be the primary contributor to reverse-bias generation current in InGaAsN diodes.

Intentional doping of GaInNAs with oxygen leads to the appearance of an electron trap at E_c – 0.59 eV and a hole trap at E_v + 0.59 eV, attributed to the same oxygen-related defect [48]. The midgap positions of these levels and their large capture cross-sections suggest that the defect should act as an efficient recombination center.

9.4.2 Nuclear Reaction Analysis and Rutherford Backscattering Channeling Studies

The deviation from the Vegard's low for high N compositions (e.g., above 1.5% [57] or above 2.9% [49]) suggests that a significant fraction of N atoms can be incorporated during growth at other locations than the group V lattice sites. The combined NRA and RBS channeling studies have provided strong evidence that a large amount of N atoms in MBE-grown GaNAs layers reside at nonsubstitutional positions [49–51]. The ratio between the on-site and off-site N atoms seems to depend critically on growth conditions. For example, a very high (2.3×10^{19} cm^{-3}) concentration of N interstitials in GaNAs alloys with only 0.7% of N, which remains constant with further increase of N content up to 3%, was detected in epilayers grown by gas-source molecular-beam epitaxy (MBE) [50, 51]. In contrast, the formation of N interstitials was apparent only for relatively high N compositions (exceeding 2.9%) in GaNAs epilayers grown by solid-source MBE [49].

Concentration of N interstitial (N_i) atoms can be substantially suppressed upon annealing [49, 50] by up to one order of magnitude, accompanied by a narrowing of the high-resolution X-ray diffraction peak and improved luminescence efficiency. This might suggest that nitrogen interstitials are among important nonradiative centers responsible for the low PL efficiency of as-grown Ga(In)NAs. This assumption is supported by theoretical studies [56], which show that the formation of various interstitial defects becomes energetically favorable during epitaxial growth of dilute nitrides. One of the defects, i.e., N-N split interstitial, is predicted to have a midgap energy level and, thus, to act as an efficient recombination center (see also Chapter 8). Unfortunately, this prediction has not yet been verified experimentally, since the NRA and RBS channeling studies are unable to determine the exact microscopic configuration or the electronic structure of the corresponding defects.

9.4.3   Optically Detected Magnetic Resonance Spectroscopy

Reliable identification of nonradiative (NR) defects was achieved by the use of optically detected magnetic-resonance (ODMR) measurements [52–54]. Two ODMR signals were found in both GaNAs and GaInNAs alloys grown by gas-source molecular-beam epitaxy (GS-MBE) (Figure 9.11). The signals correspond to a spin-resonance-induced decrease in the PL intensity, and they could be detected via both the near-band-gap and defect-related emissions in Ga(In)NAs. Thus they have been attributed to competing NR channels. (NR defects can be monitored using the ODMR technique, since a magnetic-resonance-induced increase in the efficiency of the dominant NR channels can lead to a corresponding decrease in the free-carrier concentration available for radiative recombination and, thus, the PL intensity.) Based on the magnetic-field position of the signals, the g-value for the corresponding defects is found to be around 2, which is typical for deep-level centers in GaAs-based materials known to act as efficient NR channels. This conclusion regarding the role of corresponding defects as efficient NR channels was further supported by a strong anticorrelation observed between the PL and ODMR intensities.

One of the signals (labeled "1" in Figure 9.11) was identified as a complex involving the As_{Ga} antisite based on its resolved hyperfine structure [52]. This signal consists of a group of four ODMR lines, characteristic of the hyperfine interaction between an unpaired electron spin $S = 1/2$ and a nuclear spin $I = 3/2$ of an ^{75}As atom (Figure 9.11). Unfortunately, no chemical identification could be obtained for the other NR defect (labeled "2" in Figure 9.11) due to a lack of resolved hyperfine structure. This defect has an effective electronic spin $S = 1/2$ and gives rise to an isotropic ODMR signal.

The intensities of both ODMR signals were found to strongly enhance with increasing N composition up to 3.3% [53, 54], indicating increasing concentrations of the corresponding defects (Figure 9.12). Such behavior is not surprising, at least in the case of the As_{Ga} antisite complex defect. Defects related to As_{Ga} antisites are commonly observed in GaAs grown at low temperatures, which can be further promoted by an increase in the amount of N atoms substituting As. On the other hand, the effect of In incorporation was found to be rather marginal for the range of In compositions studied, i.e., below 3% (Figure 9.12).

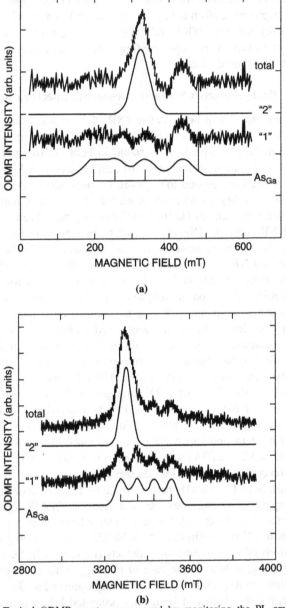

Figure 9.11 Typical ODMR spectrum measured by monitoring the PL emission from Ga(In)NAs, obtained at microwave frequencies of (a) 9.28 GHz and (b) 95 GHz. The numbers "1" and "2" denote two deconvoluted ODMR signals. Signal 1 exhibits a four-line hyperfine structure, indicating the involvement of a defect with an electronic spin S = 1/2 and a nuclear spin I = 3/2 (m_I = −3/2, −1/2, +1/2, +3/2). The lowest curves are simulated ODMR spectra for an As_{Ga} antisite complex. The ODMR signal 2 arises from a deep-level defect of unknown origin with S = 1/2 and a g-factor of 2.03.

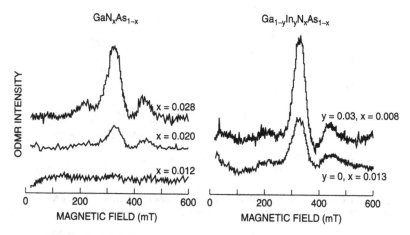

Figure 9.12 Effects of nitrogen and indium incorporation on the ODMR spectra from Ga(In)NAs alloys.

Increasing the growth temperature causes a strong reduction in the intensity of the ODMR signal, suggesting that the studied NR defects are primarily introduced during growth at low temperatures [53]. The observed increasing presence of the As_{Ga}-antisites during the low-temperature growth is likely related to a lower adatom mobility during such growth conditions and is also in agreement with the earlier findings of As-rich conditions during the low-temperature growth of GaAs. Moreover, a dramatic suppression of the ODMR intensity was also found after RTA in the Ga(In)NAs alloys, accompanied by a significant improvement in the radiative efficiency of the alloys (Figures 9.13 and 9.14). The observed anticorrelation between the ODMR and PL intensities indicates the importance of the studied defects, identifying them as being among the dominant NR channels degrading the radiative efficiency of the alloy.

9.4.4   Positron Annihilation Spectroscopy

Vacancy-type defects in Ga(In)NAs were detected via positron annihilation spectroscopy [50, 55]. Based on a comparison of line-shape parameters measured in the annihilation spectra with line shapes estimated for Ga and As vacancies in GaAs, the observed defects were attributed to Ga vacancies. Since isolated Ga vacancies in GaAs are known to have low thermal stability, a complex involving V_{Ga} has been suggested.

The concentration of the defect increases with increasing N content in the alloy, reaching a value of 10^{18} cm^{-3} for $x = 5\%$, but it can be suppressed after annealing. The concentration of vacancy defects anticorrelates with integrated PL intensity, possibly indicating the importance of this defect as a competing recombination channel.

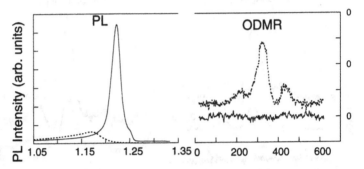

Figure 9.13 Comparison of the ODMR and PL spectra measured from the $GaN_{0.028}As_{0.972}$/GaAs multiple QW structure before (dashed curves) and after (solid curves) RTA.

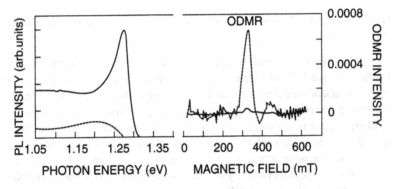

Figure 9.14 Comparison of the ODMR and PL spectra measured from the $Ga_{0.97}In_{0.03}N_{0.008}As_{0.992}$ epilayer before (dashed curves) and after (solid curves) RTA.

Defect formation was suggested to be related to the damage induced by energetic N ions from the radio frequency (RF) plasma during MBE growth and by low adatom mobility due to the low growth temperature of 450°C [50]. However, even higher vacancy concentration was observed in GaNAs layers grown by MOVPE at temperatures higher than 500°C [55]. The most recent first-principle calculations [58] have showed that the low formation energy of Ga vacancies in Ga(In)NAs — and thus high defect concentration — is primarily attributed to binding with N or H atoms (see also Chapter 8).

9.5 NONRADIATIVE RECOMBINATION IN Ga(Al)NP

Similar to Ga(In)NAs alloys, the optical properties of GaNP layers severely deteriorate when N composition exceeds a few percent. This is largely caused by the formation of misfit dislocations/cracks that are generated to relieve large internal stresses in GaNP epilayers caused by the large lattice mismatch with GaP substrates, as these defects can act as nonradiative

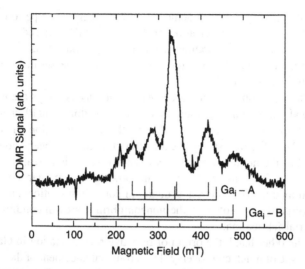

Figure 9.15 ODMR spectrum observed at 5 K from the GaAlNP alloy, taken as an example of the sample with N composition of 1% and Al composition of 2%. The expected positions of the ODMR lines for the Ga_i-A and Ga_i-B defects derived from the spin Hamiltonian analysis are also indicated. Two sets of four lines for each defect are due to the hyperfine interactions of the electron spin with the nuclear spin (I = 3/2) of the two gallium isotopes ^{69}Ga and ^{71}Ga.

centers [59]. In addition, N incorporation may also promote formation of various point defects that participate in nonradiative recombination. The only available information on the properties of these defects has recently been obtained using ODMR measurements, which have identified several point defects, specifically complexes involving Ga interstitials in GaAlNP [60]. These are discussed below.

In GaAlNP alloys, ODMR spectra are dominated by two paramagnetic defect centers, denoted as Ga_i-A and Ga_i-B [60], with an effective electron spin S = 1/2 (Figure 9.15). Both defects exhibit the characteristic hyperfine (HF) structure of the two Ga isotopes, providing clear evidence for the involvement of one Ga atom in each defect. The large and nearly isotropic hyperfine interaction implies an A_1 symmetry of the electron wavefunction characteristic of Ga_i self-interstitials. Thus, a Ga interstitial was concluded to form the core of the two defects, which were tentatively attributed to two Ga_i complexes.

9.6 CONCLUSIONS

In summary, dilute nitrides represent a novel material system that has many exciting physical properties and a great potential for applications in optoelectronics and photonics. By far the most studied dilute nitrides are the Ga(In)NAs alloys, where impressive progress in material growth and device fabrication have been achieved within a very short period of time, leading to commercial fabrication of the first GaInNAs-based laser structures. To secure full exploration of the alloy potential for device applications, comprehensive studies of the recombination processes in the materials have been undertaken.

Radiative recombination processes in Ga(In)NAs alloys are now well understood. The common consensus regarding the dominant PL mechanism at low temperatures is that it is due to the recombination of localized excitons. The observed strong localization potential has been shown to be attributed to the nonuniformity of N distribution, which creates confining potentials for free carriers due to giant bowing in band-gap energy, and these confined regions thus act as quantum dots. Compositional nonuniformities are partly promoted by the nonequilibrium growth conditions used for material fabrication.

On the other hand, the origin of radiative recombination in GaNP has been an issue of much controversy. The strong enhancement of the radiative efficiency of the alloy due to N incorporation is likely attributed to the band crossover from an indirect band-gap in GaP to a direct band-gap in GaNP alloys with a surprisingly low amount of nitrogen of about 0.4%. However, the question of efficient band-to-band recombination in these materials, beneficial for device applications, still remains rather contentious. We believe that contrary to Ga(In)NAs alloys, the PL emission in GaNP remains of bound exciton origin, even at room temperature, whereas the band-to-band recombination in the alloy remains predominantly nonradiative and thus is not seen in the PL experiments, probably due to a still stronger effect of competing recombination channels.

Among the most challenging issues in the development of dilute nitrides is learning how to control nonradiative recombination. Several intrinsic defects have been identified as efficient recombination channels likely controlling carrier lifetime and reducing the radiative efficiency of the alloy. In many cases, RTA has been shown to be efficient in removing these unwanted defects from the material. However, a high-threshold current in the state-of-the-art GaInNAs-based lasers, a short diffusion length of minority carriers, and a large dark current in GaInNAs-based solar cells are undoubtedly caused by efficient nonradiative recombination via unknown defects. Removal of these defects represents a major challenge in further developments of optoelectronic and photonic devices based on dilute nitrides, and this will require further comprehensive research efforts.

References

1. Buyanova, I.A., Chen, W.M., and Monemar, B., "Electronic Properties of Ga(In)NAs Alloys," *MRS Internet J. Nitride Semicond. Res.* 6: 1–19 (2001).

2. Weyers, M., Sato, M., and Ando, H., "Red Shift of Photoluminescence and Absorption in Dilute GaAsN Alloy Layers," *Jap. J. Appl. Phys.* 31: L853–L855 (1992).

3. Kondow, M., Uomi, K., Niwa, A., Kitani, T., Watahiki, S., and Yazawa, Y., "Gas-Source Molecular Beam Epitaxy of GaNAs Using N Radical as the N Source," *Jap. J. Appl. Phys.* 33: L1056–L1058 (1994).

4. Quagazzaden, A., Le Bellego, Y., Rao, E.V.K., Juhel, M., Leprince, L., and Patriarche, G., "MOVPE Growth of GaAsN on GaAs Using Dimethylhydrazine and Tertiarybuty-larsine," *Appl. Phys. Lett.* 70: 2681–2683 (1997).

5. Francoeur, S., Sivaraman, G., Qiu, Y., Nikishin, S., and Temkin, H., "Luminescence of As-Grown and Thermally Annealed GaAsN/GaAs," *Appl. Phys. Lett.* 72: 1857–1859 (1998).

6. Grüning, H., Chen, L., Hartmann, T., Klar, P.J., Heimbrodt, W., Höhnsdorf, F., Koch, J., and Stolz, W., "Optical Spectroscopic Studies of N-Related Bands in GaNAs," *Phys. Stat. Solidi B* 215: 39–45 (1999).

7. Buyanova, I.A., Chen, W.M., Pozina, G., Bergman, J.P., Monemar, B., Xin, H.P., and Tu, C.W., "Mechanism for Low-Temperature Photoluminescence in GaNAs/GaAs Structures Grown by Molecular-Beam Epitaxy," *Appl. Phys. Lett.* 75: 501–503 (1999).

8. Mair, R.A., Lin, J.Y., Jiang, H.X., Jones, E.D., Allerman, A.A., and Kurtz, S.R., "Time-Resolved Photoluminescence Studies of InGaAsN," *Appl. Phys. Lett.* 76: 188–190 (2000).

9. Grenouillet, L., Bru-Chevallier, C., Guillot, G., Gilet, P., Duvaut, P., Vannuffel, S., Millon, A., and Chenevas-Paule, A., "Evidence of Strong Carrier Localization below 100 K in GaInNAs/GaAs Single Quantum Well," *Appl. Phys. Lett.* 76: 2241–2243 (2000).

10. Polimeni, A., Cappizzi, M., Geddo, M., Fischer, M., Reinhardt, M., and Forchel, A., "Effect of Temperature on the Optical Properties of (InGa)(AsN)/GaAs Single Quantum Wells," *Appl. Phys. Lett.* 77: 2870–2872 (2000).

11. Kent, P.R.C., and Zunger, A., "Theory of Electronic Structure Evolution in GaAsN and GaPN Alloys," *Phys. Rev.* 64: 115208/1-23 (2001).

12. Buyanova, I.A., Chen, W.M., Monemar, B., Xin, H.P., and Tu, C.W., "Effect of Growth Temperature on Photoluminescence of GaNAs/GaAs Quantum Well Structures," *Appl. Phys. Lett.* 75: 3781–3783 (1999).

13. Buyanova, I.A., Pozina, G., Hai, P.N., Thinh, N.Q., Bergman, J.P., Chen, W.M., Xin, H.P., and Tu, C.W., "Mechanism for Rapid Thermal Annealing Improvements in GaNAs/GaAs Structures Grown by MBE," *Appl. Phys. Lett.* 77: 2325–2327 (2000).

14. Queslati, M., Zouaghi, M., Pistol, M.E., Samuelson, L., Grimmeiss, H.G., and Balkanski, M., "Photoluminescence Studies of Localization Effects Induced by the Fluctuating Random Alloy Potential in Indirect Band-Gap GaAsP," *Phys. Rev. B* 32: 8220–8227 (1985).

15. Buyanova, I.A., Chen, W.M., and Tu, C.W., "Magneto-Optical and Light Emission Properties of III-As-N Alloys," *Semicond. Sci. Technol.* 17: 815–822 (2002).

16. Buyanova, I.A., Chen, W.M., and Tu, C.W., "Recombination Processes in N-Containing III-V Ternary Alloys," *Solid State Electron.* 47: 467–475 (2003).

17. Matsuda, K., Sakai, T., Takahashi, T., Moto, A., and Takagishi, S., "Near Field Photo-luminescence Study of GaNAs Alloy Epilayer at Room and Cryogenic Temperature," *Appl. Phys. Lett.* 78: 1508–1510 (2001).

18. Mintairov, A.M., Kosel, T.H., Merz, J.L., Blaganov, P.A., Vlasov, A.S., Ustinov, V.M., and Cook, R.E., "Near-Field Magnetophotoluminescence Spectroscopy of Composition Fluctuations in InGaAsN," *Phys. Rev. Lett.* 87: 277401/1-4 (2001).

19. Neumann, A., Patane, A., Eaves, L., Belyaev, A.E., Gollub, D., Forchel, A., and Kamp, M., "Magneto-Tunneling Spectroscopy of Nitrogen Clusters in GaAsN Alloys," *IEE Proc.-Optoelectron.* 150: 49–51 (2003).

20. Xin, H.P., Kavanagh, K.L., Zhu, Z.Q., and Tu, C.W., "Observation of Quantum Dotlike Behavior of GaInNAs in GaInNAs/GaAs Quantum Wells," *Appl. Phys. Lett.* 74: 2337–2339 (1999).
21. McKay, H.A., Feenstra, R.M., Schmidtling, T., and Pohl, U.W., "Arrangement of Nitrogen Atoms in GaNAs Alloys Determined by Scanning Tunneling Microscopy," *Appl. Phys. Lett.* 78: 82–84 (2001).
22. Kurtz, S., Webb, J., Gedvilas, L., Friedman, D., Geisz, J., Olson, J., Ring, K., Joslin, D., and Karam, N., "Structural Changes during Annealing of GaInNAs," *Appl. Phys. Lett.* 78: 748–750 (2001).
23. Klar, P.J., Grüning, H., Koch, J., Schäfer, S., Volz, K., Heimbrodt, M., Kalm Saadi, A.M., Lindsay, A., and O'Reilly, E.P., "(Ga,In)(N,As) Fine Structure of the Band-gap due to Nearest-Neighbour Configurations of the Isovalent Nitrogen," *Phys. Rev. B* 64: 121203/1-4 (2001).
24. Lordi, V., Gambin, V., Friedrich, S., Funk, T., Takizawa, T., Uno, K., and Harris, J.S., "Nearest-Neighbor Configuration in (GaIn)(NAs) Probed by X-ray Absorption Spectroscopy," *Phys. Rev. Lett.* 90: 145505/1-4 (2003).
25. Kim, K., and Zunger, A., "Spatial Correlation in GaInNAsN Alloys and Their Effects on Band-Gap Enhancement and Electronic Localization, *Phys. Rev. Lett.* 86: 2609–2612 (2001).
26. Grenouillet, L., Bru-Chevallier, C., Guillot, G., Gilet, P., Ballet, P., Duvaut, P., Rolland, G., and Million, A., "Rapid Thermal Annealing in GaNAs/GaAs Structures: Effects of Nitrogen Reorganization on Optical Properties," *J. Appl. Phys.* 91: 5902–5908 (2002).
27. Buyanova, I.A., Izadifard, M., Chen, W.M., Polimeni, A., Capizzi, M., Xin, H.P., and Tu, C.W., "Hydrogen-Induced Improvements in Optical Quality of GaNAs Alloys," *Appl. Phys. Lett.* 82: 3663–3665 (2003).
28. Bellaiche, L., Wei, S.H., and Zunger, A., "Composition Dependence of Interband Transition Intensities in GaPN, GaAsN, and GaPAs Alloys," *Phys. Rev. B* 56: 10233–10240 (1997).
29. Shan, W., Walukiewicz, W., Yu, K.M., Wu, J., Ager, J.W., Haller, E.E., Xin, H.P., and Tu, C.W., "Nature of the Fundamental Band-gap in GaN_xP_{1-x} Alloys," *Appl. Phys. Lett.* 76: 3251–3253 (2000).
30. Xin, H.P., Tu, C.W., Zhang, Y., and Mascarenhas, A., "Effect of Nitrogen on the Band Structure of GaNP Alloys," *Appl. Phys. Lett.* 76: 1267–1269 (2001).
31. Yonezu, H., "Control of Structural Defects in Group III-V-N Alloys Grown on Si," *Semicond. Sci. Technol.* 17: 762–768 (2002).
32. Liu, X., Bishop, S.G., Baillargeon, J.N., and Cheng, K.Y., "Band-gap Bowing in GaPN Alloys," *Appl. Phys. Lett.* 63: 208–210 (1993).
33. Onabe, K., "MOVPE Growth and Optical Characterization of GaPN Metastable Alloy Semiconductor," in *MRS Symposium Proceedings*, ed. F.A. Ponce, T.D. Moustakas, I. Akasaki, and B. Monemar (Pittsburgh: Material Research Society, 1997), 449: 23–34.
34. Kuroiwa, R., Asahi, H., Iwata, K., Tampo, H., Asami, K., and Gonda, S., "Observation of Quantum-Dot-Like Properties in the Phase-Separated GaN-Rich GaNP," *Phys. Stat. Solidi B* 216: 461–464 (1999).
35. Zhang, Y., Fluegel, B., Mascarenhas, A., Xin, H.P., and Tu, C.W., "Optical Transitions in the Isoelectronically Doped Semiconductor GaP:N: An Evolution from Isolated Centers, Pairs and Clusters to an Impurity Band," *Phys. Rev. B* 62: 4493–4500 (2000).
36. Thomas, D.G., and Hopfield, J.J., "Isoelectronic Traps due to Nitrogen in GaP," *Phys. Rev.* 150: 680–689 (1966).
37. Buyanova, I.A., Rudko, G.Y., Chen, W.M., Xin, H.P., and Tu, C.W., "Radiative Recombination Mechanism in GaNP Alloys," *Appl. Phys. Lett.* 80: 1740–1742 (2002).
38. Rudko, G.Y., Buyanova, I.A., Chen, W.M., Xin, H.P., and Tu, C.W., "Temperature Dependence of GaNP Bandgap in N Composition Range of Indirect-Direct Bandgap Crossover," *Appl. Phys. Lett.* 81: 3984–3986 (2002).

39. Buyanova, I.A., Pozina, G., Bergan, J.P., Chen, W.M., Xin, H.P., and Tu, C.W., "Time-Resolved Studies of Photoluminescence in GaNP Alloys: Evidence for Indirect-Direct Band Crossover," *Appl. Phys. Lett.* 81: 52–54 (2002).
40. Chevallier, J., Mariette, H., Diguet, D., and Poiblaud, G., "Direct Experimental Observation of Band-Structure Effects in GaPAs:N Alloys by Radiative Lifetime Measurements," *Appl. Phys. Lett.* 28: 375–377 (1976).
41. Kash, J.A., Collet, J.H., Wolford, D.J., and Thompson, J., "Luminescence Decays of N-Bound Excitons in GaAsP," *Phys. Rev. B* 27: 2294–2300 (1983).
42. Fehse, R., Jin, S., Sweeney, S.J., Adams, A.R., O'Reilly, E.P., Riechert, H., Illek, S., and Egorov, A.Y., "Evidence for Large Monomolecular Recombination Contribution to Threshold Current in 1.3-μm GaInNAs Semiconductor Lasers," *Electron. Lett.* 37: 1518–1520 (2001).
43. Krispin, P., Gambin, V., Harris, J.S., and Ploog, K.H., "Nitrogen-Related Electron Traps in Ga(As,N) Layers (≤3% N)," *J. Appl. Phys.* 93: 6095–6099 (2003).
44. Krispin, P., Spruytte, S.G., Harris, J.S., and Ploog, K.H., "Origin and Annealing of Deep-Level Defects in p-Type GaAs/Ga(As,N)/GaAs Heterostructure Grown by Molecular Beam Epitaxy," *J. Appl. Phys.* 89: 6294–6301 (2001).
45. Kwon, D., Kaplar, R.J., Ringel, S.A., Allerman, A.A., Kurtz, S.R., and Jones, E.D., "Deep Levels in p-Type InGaAsN Lattice Matched to GaAs," *Appl. Phys. Lett.* 74: 2830–2832 (1999).
46. Kurtz, S.R., Allerman, A.A., Jones, E.D., Gee, J.M., Bans, J.J., and Hammons, B.E., "InGaAsN Solar Cells with 1.0-eV Bandgap, Lattice Matched to GaAs," *Appl. Phys. Lett.* 74: 729–731 (1999).
47. Kaplar, R.J., Arenhart, A.R., Ringel, S.A., Allerman, A.A., Sieg, R.M., and Kurtz, S.R., "Deep Levels and Their Impact on Generation Current in Sn-Doped InGaAsN," *J. Appl. Phys.* 90: 3405–3408 (2001).
48. Balcioglu, A., Ahrenkiel, R.K., and Friedman, D.J., "Evidence of an Oxygen Recombination Center in p⁺-n GaInNAs Solar Cells," *Appl. Phys. Lett.* 76: 2397–2399 (2000).
49. Spruytte, S.G., Colden, C.W., Harris, J.S., Wampler, W., Krispin, P., Ploog, K., and Larson, M.C., "Incorporation of Nitrogen in Nitride-Arsenides: Origin of Improved Luminescence Efficiency after Anneal," *J. Appl. Phys.* 89: 4401–4406 (2001).
50. Li, W., Pessa, M., Ahlgren, T., and Decker, J., "Origin of Improved Luminescence Efficiency after Annealing of Ga(In)NAs Materials Grown by Molecular-Beam Epitaxy," *Appl. Phys. Lett.* 79: 1094–1096 (2001).
51. Ahlgren, T., Vainonen-Ahlgren, E., Likonen, J., Li, W., and Pessa, M., "Concentration of Interstitial and Substitutional Nitrogen in GaNAs," *Appl. Phys. Lett.* 80: 2314–2316 (2002).
52. Thinh, N.Q., Buyanova, I.A., Hai, P.N., Chen, W.M., Xin, H.P., and Tu, C.W., "Signature of an Intrinsic Point Defect in GaNAs," *Phys. Rev. B* 63: 033203/1-4 (2001).
53. Thinh, N.Q., Buyanova, I.A., Chen, W.M., Xin, H.P., and Tu, C.W., "Formation of Nonradiative Defects in MBE Epitaxial GaNAs Studied by Optically Detected Magnetic Resonance," *Appl. Phys. Lett.* 79: 3089–3091 (2001).
54. Chen, W.M., Thinh, N.Q., Buyanova, I.A., Xin, H.P., and Tu, C.W., "Nature and Formation of Nonradiative Defects in GaNAs and InGaNAs," in *MRS 2001 Fall Meeting Proceedings* (Pittsburgh: Material Research Society, 2001), 67–72.
55. Toivonen, J., Hakkarainen, T., Sopanen, M., Lipsanen, H., Oila, J., and Saarinen, K., "Observation of Defect Complexes Containing Ga Vacancies in GaAsN," *Appl. Phys. Lett.* 82: 40–42 (2003).
56. Zhang, S.B., and Wei, S.H., "Nitrogen Solubility and Induced Defect Complexes in Epitaxial GaAsN," *Phys. Rev. Lett.* 86: 1789–1792 (2001).
57. Li, W., Pessa, M., and Likonen, J., "Lattice Parameter in GaNAs Epilayers on GaAs: Deviation from Vegard's Law," *Appl. Phys. Lett.* 78: 2864–2866 (2001).

58. Jannoti, A., Wei, S.H., Zhang, S.B., Kurtz, S., and Van de Walle, C.G., "Interactions between Nitrogen, Hydrogen, and Gallium Vacancies in GaAsN Alloys," *Phys. Rev. B* 67: 161201/1–4 (2001).
59. Buyanova, I.A., Chen, W.M., Goldys, E.M., Phillips, M.R., Xin, H.P., and Tu, C.W., "Strain Relaxation in GaNP: Effect on Optical Quality," *Physica B* 308–310: 106–109 (2001).
60. Vorona, I.P., Thinh, N.Q., Buyanova, I.A., Chen, W.M., Hong, Y.G., and Tu, C.W., "Identification of Ga Interstitials in GaAlNP," *Physica B* 340–342:466–469 (2003).

CHAPTER 10

Growth, Characterization, and Band-Gap Engineering of Dilute Nitrides

CHARLES W. TU

Department of Electrical and Computer Engineering,
University of California, San Diego

10.1 INTRODUCTION

Interest in III-N-V compound semiconductors has increased recently because of the discovery that only a small amount of nitrogen incorporation ($\approx 2\%$) in conventional GaAs- and InP-based III-V compounds results in very large band-gap bowing [1, 2]. Hence, these materials are also called "dilute nitrides" or "low-band-gap nitrides." As described in Chapter 2, for GaN_xAs_{1-x}, the band anticrossing model describes the repulsion of the GaAs host conduction band and the nitrogen localized states, which are in resonance with the conduction band. Thus, the band-gap bowing is a result of the downward movement of the conduction band [3]. Band-gap lowering allows 1.3-μm light emission at room temperature from $Ga_{1-y}In_yN_xAs_{1-x}$/ GaAs quantum wells (QWs), with y usually about 0.3. Chapters 10 and 13 describe the latest results in GaInNAs 1.3-μm edge-emitting and vertical-cavity surface-emitting lasers (VCSELs). In this chapter, we show that GaInNAs and GaInNP, both grown on GaAs (001) substrates, can also be beneficial to heterojunction bipolar transistors (HBTs).

Because of the large difference in the atomic size between N and As or N and P atoms, the large local strain, when N is in an As or P sublattice site, results in a very low solubility limit of N and a very wide miscibility gap [4]. To incorporate N into the lattice, therefore, the growth temperature has to be lower than optimal, resulting in many point defects, which degrade the optical and transport properties of the material [5, 6]. Therefore, usually postgrowth thermal annealing is employed to improve material quality. After summarizing growth conditions in Section 10.2, we describe in Section 10.3 the effects of postgrowth thermal annealing on the optical and transport properties of GaInNAs and GaInNP.

In Section 10.4 we report on various band-gap engineering possibilities with N incorporation. We investigate $Ga_{0.3}In_{0.7}N_xAs_{1-x}$ quantum dots (QDs) for longer wavelengths, toward 1.55 μm, and show that low-band-gap GaInNAs can also be used as the base of an HBT to obtain a low turn-on voltage for low-power applications. Another effect of N incorporation is on the band offset. If N is incorporated in a quantum well, the conduction-band discontinuity is increased, providing better electron confinement, as in InNAsP/GaInAsP quantum-well microdisk lasers. On the other hand, if N is incorporated in the barrier as in GaInNP/GaAs, the conduction-band discontinuity is reduced. This should be ideal for *npn* HBTs. Finally, nitrogen incorporation can also change the band structure of the host material, e.g., from indirect band-gap to direct band gap, as in the case of GaNP/GaP, from which light-emitting diodes (LEDs) were fabricated.

10.2 GROWTH OF DILUTE NITRIDES BY MBE AND THEIR CHARACTERIZATIONS

All samples reported in this chapter are grown by gas-source molecular-beam epitaxy (MBE) with a radio frequency (RF) plasma nitrogen radical beam source, elemental group III sources, and cracked arsine and phosphine. Be is the *p*-type dopant, and Si the *n*-type. GaInNAs has also been grown by metal-organic chemical-vapor deposition (MOCVD) using dimethylhydrazine as the nitrogen source because it has a lower decomposition temperature than ammonia [7] (see also Chapter 13). Other precursors such as hydrazine and t-butylamine have also been used [8]. For MBE growth of GaAs buffer layers, the growth temperature is 600°C. For the GaInNAs layer and the GaAs cap layer, the growth temperature is decreased to 420°C in order to incorporate In and N into the GaInNAs layers. Photoluminescence (PL) measurement is carried out with the 514.5-nm line of an Ar^+ laser as the excitation source. A thermoelectrically cooled Ge photodiode was used to detect the signal at the exit of a 50-cm monochromator through an amplifier. Secondary-ion mass spectroscopy (SIMS) measurements were performed using CAMECA IMS-4*f* SIMS. The compositions were determined by X-ray rocking curve (XRC) measurements and simulations based on the dynamic theory.

10.3 EFFECT OF ANNEALING

In order to incorporate nitrogen in the films and avoid phase separation, the growth temperature is relatively low, resulting in point defects. Hence, the PL intensity, in general, decreases with increasing nitrogen incorporation [9]. To anneal out the point defects and improve the material properties, thermal annealing usually is applied postgrowth *in situ* under an arsenic overpressure or *ex situ* placed faced down on a GaAs substrate to minimize loss of arsenic at elevated temperatures. We use rapid thermal annealing (RTA) with a halogen lamp for 10 s and the sample in ambient nitrogen. Chapter 5 compares annealing of MBE and MOCVD samples of GaInNAs.

10.3.1 Optical Properties

The PL spectra of seven-period $Ga_{0.7}In_{0.3}N_xAs_{1-x}$/GaAs QW structures with 0, 1, and 2% of nitrogen are compared. Figures 10.1a and 10.1b show the relative PL intensity and blueshift, respectively, as a function T_a for GaInNAs/GaAs QW and different N concentrations [10].

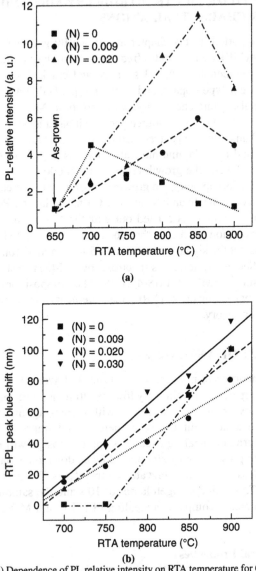

Figure 10.1 (a) Dependence of PL relative intensity on RTA temperature for $Ga_{0.7}In_{0.3}N_xAs_{1-x}$/GaAs MQWs with different N concentrations. The data point at 650°C corresponds to the sample without annealing. (b) Dependence of PL blueshift on RTA temperature. (From Ref. [10].)

The PL intensity of the $Ga_{0.7}In_{0.3}As$/GaAs QW sample increases by a factor of three after 700°C RTA due to the reduction of nonradiative centers. For annealing temperature $T_a > 750$°C, the PL intensity decreases and the peak blueshifts. Since Rao et al. [11] reported only a negligibly small blueshift in the bulk GaNAs samples, even with higher N content ($\approx$4%), our large blueshift after 700°C RTA is most likely caused by Ga and In interdiffusion, which would yield a graded GaInAs/GaAs interface. This would cause the quantum-confined state to shift upward in energy. The decrease of PL intensity after higher-temperature RTA may be caused by the thermal generation of new nonradiative defects. For N-containing seven-period QWs, the effects of RTA are similar, but the PL intensity reaches its maximum at higher T_a, and the peak starts to blueshift at lower T_a (700°C).

With $T_a > 750$°C, the blueshift is larger for the N-free sample than for the N-containing sample, indicating that the latter is more stable due to the smaller average strain. The higher the N concentration, the larger is the blueshift and the larger is the increase in the PL intensity, suggesting more nonradiative centers and increased defect-assisted diffusion with N incorporation.

10.3.2 Transport Properties of p-GaInNAs

The transport properties of unintentionally p-type (background carrier concentration of 10^{17} cm^{-3}) $Ga_{0.92}In_{0.08}N_{0.03}As_{0.97}$ grown by MOCVD, lattice-matched to GaAs substrates, have been reported recently [12] with the room-temperature (RT) hole mobility ranging from 20 to 80 cm^2/V/s. Here we report on highly p-type Be-doped $Ga_{0.892}In_{0.108}N_xAs_{1-x}$ layers, which are to be used in HBTs, on semi-insulating GaAs substrates. The p-type $Ga_{0.892}In_{0.108}N_xAs_{1-x}$ (1100 Å) samples with a 100-Å-thick GaAs cap layer and different N concentrations were grown on a GaAs buffer layer on semi-insulating (1,0,0) GaAs substrates. We first grew a reference $Ga_{0.892}In_{0.108}As$ bulk layer and then $Ga_{0.892}In_{0.108}N_xAs_{1-x}$ samples, changing only the N_2 flow rate, with all other growth conditions unchanged. Therefore, the Ga and In compositions in all the samples should be the same.

Figures 10.2a and 10.2b show XRCs of as-grown and annealed $Ga_{0.892}In_{0.108}As$ and $Ga_{0.892}In_{0.108}N_{0.017}As_{0.983}$ samples, respectively [13]. The dashed lines correspond to simulation results. The $Ga_{0.892}In_{0.108}N_{0.017}As_{0.983}$ peak shifts closer to the GaAs substrate peak, demonstrating that adding N into the $Ga_{0.892}In_{0.108}As$ layer does reduce the net compressive strain of the system. For N-containing samples, there are clear Pendelloesung fringes, indicating high crystalline quality and uniformity of the film, and smooth interfaces between GaAs and GaInNAs. After RTA at 700°C for 10 s, the $Ga_{0.892}In_{0.108}As$ peak broadens, indicating strain relaxation. For $Ga_{0.892}In_{0.108}N_{0.017}As_{0.983}$, however, the Pendelloesung fringes after RTA at 700°C are sharper than those of the as-grown sample, indicating improved structural quality. After annealing at 850°C, the XRC peaks are still well resolved, indicating that the $Ga_{0.892}In_{0.108}N_{0.017}As_{0.983}$ layer remains pseudomorphically strained and has better thermal stability than $Ga_{0.892}In_{0.108}As$ due to reduced lattice mismatch (0.37% vs. 0.7%).

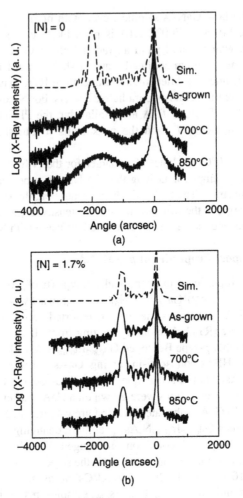

Figure 10.2 [(400) rocking curves of $Ga_{0.892}In_{0.108}N_xAs_{1-x}$ samples annealed at 700°C. (a) $x =$ 0, (b) $x = 0.017$. θ_B and ϕ are the Bragg angle and the angle between (1,0,0) and {511} diffraction planes, respectively. (From Ref. [13].)

SIMS was done on the samples to check Be diffusion after RTA. It was found that no H is incorporated in $Ga_{0.892}In_{0.108}As$, but H is incorporated alongside N in as-grown GaInNAs samples, possibly due to the very large electronegativity of N atoms (N, 3.0; As, 1.57). The higher the N concentration, the greater is the H concentration. There is no detectable Be diffusion at 700°C RTA. At 850°C, some diffusion could be detected, and the free-carrier concentration is also decreased to 7.2 $\times 10^{18}$ cm^{-3}, compared to 1.1 $\times 10^{19}$ cm^{-3} for the sample annealed at 700°C. Chapter 6 discusses in great detail the role of hydrogen in dilute nitrides.

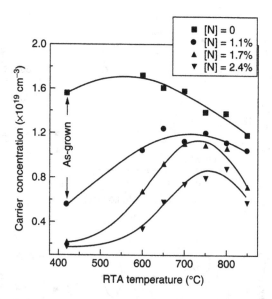

Figure 10.3 Free-hole concentration as a function of RTA temperature for as-grown and annealed Be-doped $Ga_{0.892}In_{0.108}N_xAs_{1-x}$ samples. The lines are drawn to guide the eye. (From Ref. [13].)

Figure 10.3 shows the free-carrier concentrations as a function of RTA temperature for as-grown and annealed GaInNAs samples. With the N concentration increasing from 0 to 0.024, the carrier concentration of the as-grown samples decreases by one order of magnitude, and the Hall mobility decreases from 60 to 30–45 $cm^2/V/s$. After RTA at 700°C, the carrier concentration of N-containing samples is increased to half that of GaInAs, and the Hall mobility also increases to ≈50 $cm^2/V/s$. Therefore, the product of carrier concentration and hole mobility is increased to half that of the GaInAs sample, which, with reduced band gap, may still be acceptable for HBTs.

10.3.3 Transport Properties of *n*-GaInNP and *n*-GaInNAs

In Section 10.4.4 we show that the conduction-band discontinuity between $Ga_{0.48}In_{0.52}N_yP_{1-y}$, lattice-matched to GaAs, and GaAs can be made to be nearly zero, so that GaInNP could be an ideal material for the emitter or collector of an *npn* HBT. Therefore, we need to investigate the electron transport properties of GaInNP doped with Si. Hall measurements show decreased electron concentration and mobility with increasing N concentration for both $Ga_{0.48}In_{0.52}N_yP_{1-y}$ and $Ga_{0.92}In_{0.08}N_yP_{1-y}$, as shown in Figure 10.4 [14]. Incorporation of 0.5% nitrogen decreases electron mobility from 406 $cm^2/V/s$ to 51 $cm^2/V/s$. The rapid decrease of electron mobility in $Ga_{0.48}In_{0.52}N_yP_{1-y}$ is very similar to that of GaInNAs, which could be caused by intrinsic defects or scattering due to the effects of local strain associated

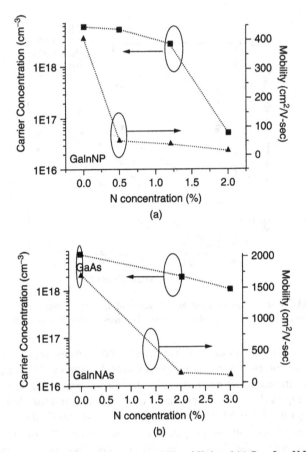

Figure 10.4 Free-electron concentration and mobility of Si-doped (a) $Ga_{0.48}In_{0.52}N_yP_{1-y}$ and (b) $Ga_{0.92}In_{0.08}N_yAs_{1-y}$ as a function of N concentration. (From Ref. [14].)

with nitrogen incorporation [15]. Furthermore, the electron concentration of $Ga_{0.48}In_{0.52}N_yP_{1-y}$ decreases with increasing N content, which indicates increasing N-related electron traps. Scattering by such isoelectronic traps is less temperature-dependent.

In order to improve the sample quality, we annealed the samples for 10 s under N_2 ambient at different RTA temperatures. For the n-type $Ga_{0.48}In_{0.52}P$ sample, the free-electron concentration decreased gradually from 5.1×10^{18} to 1.1×10^{18} cm^{-3} as the RTA temperature increased to 800°C (see Figure 10.5), which could be due to the amphoteric behavior of Si. At higher temperatures, an increasing number of Si atoms would occupy P sites. On the other hand, the electron concentration of $Ga_{0.48}In_{0.52}N_yP_{1-y}$ decreases much more rapidly, from 4.4×10^{18} to 8.0×10^{16} cm^{-3}, although no structural degradation was observed by X-ray measurements even for the sample annealed at 800°C.

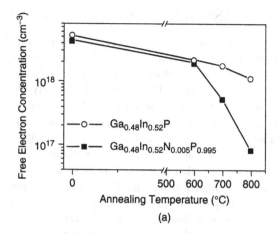

(a)

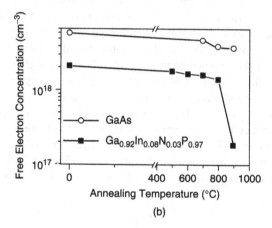

(b)

Figure 10.5 Free-electron concentration of Si-doped (a) $Ga_{0.48}In_{0.52}N_yP_{1-y}$ and (b) $Ga_{0.92}In_{0.08}N_yAs_{1-y}$ as a function of RTA temperature. (From Ref. [14].)

The monotonic decrease with annealing temperature of Si-doped $Ga_{0.48}In_{0.52}N_yP_{1-y}$ is contrary to the expectation that annealing should anneal out point defects, as in the behavior of the hole concentration of Be-doped $Ga_{0.892}In_{0.108}N_{0.011}As_{0.989}$ in Figure 10.3. Recently, Yu et al. [16] proposed a model of mutual passivation between Si and N, that is, Si is passivated by N, and vice versa, through the formation of Si-N pairs. N is more electronegative than P (Pauling electronegativities of N and P being 3.04 and 2.19, respectively), and, thus, N has a tendency to bind the free valence electron of Si. When $Ga_{0.48}In_{0.52}N_yP_{1-y}$ is grown at a relatively low temperature ($\approx$420–450°C) by MBE, Si atoms are randomly distributed in the Ga sublattice sites. During RTA at high temperatures, Si atoms have enough thermal energy to diffuse to be near N to form Si-N pairs. This passivation process results in a significant drop in the electron concentration in N-containing $Ga_{0.48}In_{0.52}N_{0.005}P_{0.995}$.

10.4 BAND-GAP ENGINEERING OF DILUTE NITRIDES

Nitrogen has a large electronegativity [17], and according to the band anticrossing model, incorporating nitrogen into the conventional III-V materials pulls down the conduction band edge, resulting in a larger conduction-band offset. Figure 10.6 shows qualitatively the band edges for GaNAs (InNP) and GaInAs (InAsP) as a function of strain. Adding In (As) into GaAs (InP) increases the lattice constant, resulting in compressive strain in the plane parallel to the interface. The band-gap becomes smaller; the conduction band edge moves down; and the valence-band edge moves up. On the other hand, adding N to GaAs (InP) results in tensile strain and a smaller band gap. From linear interpolation of the large electronegativity of nitrogen, we expect the valence-band edge to move down with nitrogen incorporation. However, Buyanova et al. [18] provide direct experimental evidence that the band alignment in $GaN_x As_{1-x}$/GaAs heterojunction is actually Type I for N concentration, x, up to 3.3%. Furthermore, experimental evidence from cross-section scanning tunneling microscopy of the filled states of InP, InAsP, and InNAsP indicates that the valence-band edge of InNAsP is higher than that of InAsP [19]. The important point is that adding N to a compressive-strained layer reduces the strain from ε_1 to ε_2 and increases the conduction-band offset from ΔE_{c1} to ΔE_{c2}, resulting in better electron confinement. Such a band offset due to N incorporation can be used to increase the conduction-band offset if N is in a quantum well (Sections 10.4.2 and 10.4.3) and to decrease the band offset if N is in barriers (Section 10.4.4). Because the amount of nitrogen is small, the effect on the valence-band offset can

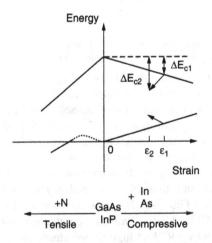

Figure 10.6 Conduction-band and valence-band edges for GaInAs (InAsP) and GaNAs (InNP).

be ignored. Chapter 5 also discusses conduction-band structure and Type I and Type II alignment.

10.4.1 Band-Gap Lowering: toward 1.5-μm Wavelength through GaInNAs Quantum Dots

Because adding N into GaInAs decreases PL intensity rapidly, one approach to achieve long wavelength is to use as large an In concentration as possible and as small an N concentration as possible without generating misfit dislocations. Thus, this would be in the regime of self-assembled QDs. Other chapters in this book describe the current worldwide effort in 1.3-μm GaInNAs QW VCSELs; here we report on efforts to extend the wavelength to 1.55 μm by the use of GaInNAs QDs [20].

The sample structure consists of a 150-nm-thick GaAs buffer layer grown at 580°C on a GaAs substrate, one layer of GaInNAs islands grown at different temperatures (discussed below), a 50-nm-thick GaAs barrier layer, and finally a second layer of GaInNAs islands using growth parameters similar to the first layer. A transition from two-dimensional to three-dimensional growth mode, i.e., Stranski-Krastanov growth mode, is confirmed by the transformation of the reflection high-energy electron diffraction pattern from streaks to chevrons at the nominal thickness of approximately three monolayers (MLs).

Figure 10.7 shows atomic force microscope (AFM) images of (a) $Ga_{0.3}In_{0.7}As$ QDs grown at 520°C, (b) $Ga_{0.3}In_{0.7}N_{0.02}As_{0.98}$ QDs grown at 520°C, (c) $Ga_{0.3}In_{0.7}As$ QDs grown at 450°C, and (d) $Ga_{0.3}In_{0.7}N_{0.02}As_{0.98}$ QDs grown at 450°C. The dot density decreases with increasing growth temperature due to the longer migration lengths of Ga and In. It also decreases with increasing nitrogen composition due to reduced strain in the wetting layer.

For samples grown at 520°C, the PL intensity decreases dramatically with increasing nitrogen composition. However, the PL intensity of $Ga_{0.3}In_{0.7}N_xAs_{1-x}$ QDs becomes comparable with that of $Ga_{0.3}In_{0.7}As$ QDs with a decrease in the growth temperature to 450°C, as shown in Figure 10.8. The full width at half maximum (FWHM) of the GaInNAs QDs also becomes as narrow as that of GaInAs QDs (34 meV). The AFM images of the GaInNAs QDs grown at different temperatures show that the dot-size uniformity also improves as the growth temperature decreases. The comparable PL intensity at lower growth temperature occurs because the difference in the density between $Ga_{0.3}In_{0.7}As$ and $Ga_{0.3}In_{0.7}N_xAs_{1-x}$ QDs becomes smaller when the QDs are grown at lower growth temperature. The reason is as follows.

Incorporating N into GaInAs QDs increases the critical wetting-layer thickness due to reduced strain and causes inhomogeneous local strain in the

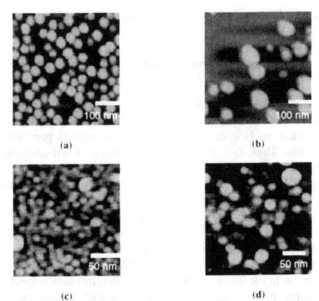

(a)   (b)

(c)   (d)

Figure 10.7 AFM images of (a) $Ga_{0.3}In_{0.7}As$ QDs grown at 520°C, (b) $Ga_{0.3}In_{0.7}N_{0.02}As_{0.98}$ QDs grown at 520°C, (c) $Ga_{0.3}In_{0.7}As$ QDs grown at 450°C, and (d) $Ga_{0.3}In_{0.7}N_{0.02}As_{0.98}$ QDs grown at 450°C. (From Ref. [20].)

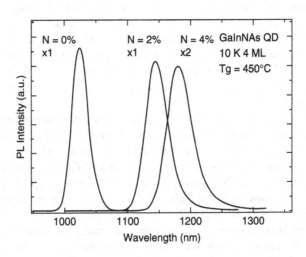

Figure 10.8 PL spectra (at 10 K) from GaInNAs QDs with different nitrogen compositions. (From Ref. [20].)

wetting layer due to composition fluctuation, so that the number of dots that form at the initial stage of GaInNAs QD growth is less than that of GaInAs QDs. After a few monolayer depositions, the dot density of QDs grown at high temperature is still low because Ga and In can migrate to those initial dots, while that of QDs grown at low temperature increases because Ga and In, which cannot reach those initial dots, form new dots. Thus, the difference of the dot density between GaInNAs and GaInAs QDs grown at 450°C becomes small. As a result, the emission peak of $Ga_{0.3}In_{0.7}N_xAs_{1-x}$ QDs shifts to longer wavelength because of higher nitrogen composition, and the PL intensity is comparable because of similar dot density.

Even though the PL intensity of $Ga_{0.3}In_{0.7}N_xAs_{1-x}$ QDs becomes comparable with that of $Ga_{0.3}In_{0.7}As$ QDs, the emission wavelength is still short because decreasing the growth temperature also decreases the dot size, resulting in a stronger quantum size effect. Therefore, in order to extend the emission wavelength further, the nominal thickness of the QD layer was increased to reduce the quantum-size effect. According to the AFM images of the QD surface and the PL spectra, the redshift of the PL peak is observed until the dot coalescence takes place. In our case, after increasing nominal thickness up to five monolayers, the emission wavelength of $Ga_{0.3}In_{0.7}N_{0.04}As_{0.96}$ QDs was extended to 1.42 μm at room temperature with high PL intensity.

To reduce the quantum-confinement effect further, we incorporated a small amount (2.6%) of N into the GaAs barriers. Compared with GaAs, GaNAs barrier layers reduce the barrier potential by diminishing the band-gap reduction due to the downward movement of the conduction band [3], thus reducing the quantum-confinement effect. From AFM images, the dot size and density of GaInNAs QDs grown on GaNAs are similar to those of GaInNAs QDs grown on GaAs. This result indicates that the effect of GaNAs barriers on emission wavelength can be discussed without considering the effect of dot size and density change. Figure 10.9 shows the PL spectra of the corresponding GaInNAs QDs at room temperature. The redshift of the emission peak is observed with the insertion of GaNAs barrier layers. The longest wavelength we have achieved at room temperature is 1.47 μm, with a PL intensity only 1/6 that of the GaInAs QDs. This intensity ratio is much better than that reported earlier ($\approx$1/20 at 1.21 μm, $\approx$1/200 at 1.52 μm) [21].

10.4.2 Band-Gap Lowering: GaInNAs-Base HBTs

GaAs-based HBTs have achieved widespread use in high-performance microwave and digital applications. They have, however, a large base-emitter turn-on voltage of approximately 1.4 V (at high current density), which is awkward for use in circuits with low power supply voltage. It is, therefore, important to develop techniques to reduce the value of the turn-on voltage for low-power applications. The turn-on voltage is related to the band-gap of the base. In this section, we describe our investigation using p-type GaInNAs as the base of an HBT [22].

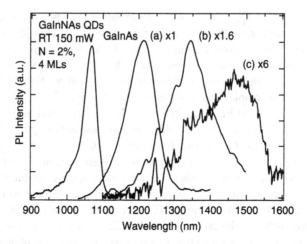

Figure 10.9 PL spectra of GaInAs and GaInNAs QDs measured at room temperature. The barrier layers for GaInNAs QDs are (a) GaAs, (b) GaAs on top and GaNAs on bottom, and (c) GaNAs on top and bottom. (From Ref. [20].)

Table 10.1 shows a schematic of the GaInNAs-base HBT structure. Since most of the band-gap lowering from nitrogen incorporation is a result of lowering of the conduction band edge [3], nitrogen incorporation results in an increase of the conduction-band discontinuity for GaInNAs/GaAs. Grading- and delta-doping layers (interface-doping dipole) have been designed to eliminate the barriers at the base-emitter and base-collector junctions. The grading between GaAs and $Ga_{0.89}In_{0.11}N_{0.02}As_{0.98}$ was done with a $Ga0.89In0.11N0.02As0.98$/GaAs chirped superlattice with a period of 1.1 nm to make sure that the electrons can tunnel through the barriers. In our experiments, the $Ga_{0.89}In_{0.11}N_{0.02}As_{0.98}$ base was 40-nm thick with a nominal doping of 8×10^{18} cm^{-3}.

The HBTs were fabricated using a self-aligning process. All measurements were done on a device with a 120×120-μm^2 emitter. The ideality factor for the base-collector junction was 1.05, indicating near-ideal collector current. The ideality factor for the base-emitter junction, however, was 1.59, indicating a substantial space-charge recombination current.

Figure 10.10 shows a comparison of the collector current for a $Ga_{0.89}In_{0.11}N_{0.02}As_{0.98}$ base HBT with a conventional GaAs HBT with a base doping of 4×10^{19} cm^{-3} and AlGaAs grading between the base-emitter junction. The large series resistance from the low base doping limits the $Ga_{0.89}In_{0.11}N_{0.02}As_{0.98}$ base HBT when the base-emitter bias is greater than 0.7 V. The $Ga_{0.89}In_{0.11}N_{0.02}As_{0.98}$ base HBT shows a 0.4-V reduction in the turn-on voltage compared with the GaAs base HBT. Other groups have also used p-GaInNAs in HBTs, but they used a smaller concentration of N and achieved a reduction of turn-on voltage of 0.2 V [23].

Table 10.1 A GaInNAs-Base HBT Structure

Layer	Material	Thickness (nm)	Doping (cm³)
Cap	GaAs	200	$n = 5e18$
Emitter	GaAs	200	$n = 5e17$
δ doping	GaAs	0.5	$n = 3e19$
Graded	GaAs ⇐ $Ga_{0.89}In_{0.11}N_{0.02}As_{0.98}$	30	$n = 3e17$
Spacer	$Ga_{0.89}In_{0.11}N_{0.02}As_{0.98}$	5	undoped
Base	$Ga_{0.89}In_{0.11}N_{0.02}As_{0.98}$	40	$p = 8e18$
Spacer	$Ga_{0.89}In_{0.11}N_{0.02}As_{0.98}$	5	undoped
Graded	$Ga_{0.89}In_{0.11}N_{0.02}As_{0.98}$ ⇐ GaAs	30	$n = 3e16$
δ doping	GaAs	5	$n = 1.5e18$
Collector	GaAs	400	$n = 3e16$
Subcollector	GaAs	700	$n = 5e18$
S.I. GaAs substrate			

Source: Welty, R.J. et al., *Solid-State Electronics* 46: 1–5 (2002).

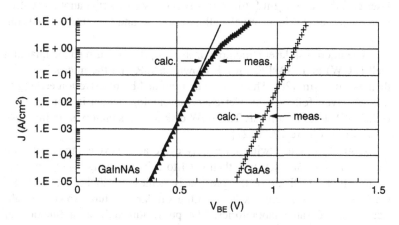

Figure 10.10 Comparison of the collector current for a GaAs/GaInNAs/GaAs HBT with an AlGaAs/GaAs/GaAs HBT. (From Ref. [22].)

Recently DeLuca et al. [24] reported improved performance using a graded-composition GaInNAs in an GaInP/GaInNAs HBT. Their structure was grown by MOCVD. Because of the degraded minority-carrier transport, as reported in Chapter 13, they used a graded-composition base to create a quasi-electric field that accelerates the electrons across the base. The In concentration of GaInNAs was about 5%, and the N concentration was about 0.3%. The GaInNAs band-gap varied approximately 40 meV over a 55-nm thickness, resulting in an electric field of $\approx$7 kV/cm. The increased electron velocity reduced the base transit time, resulting in higher (1.7 ×) current gain and better (35%) RF performance. A dc-current gain as high as 200 and cutoff frequencies of 70 GHz have been demonstrated [24]. The turn-on voltage was reduced by about 0.1 V.

10.4.3   Increased Conduction-Band Offset: GaInNAs and InNAsP Applied to Lasers

Long-wavelength lasers emitting at 1.3 and 1.55 μm are important for optical-fiber communications and have been intensively investigated. Such lasers are commonly realized with the GaInAsP-InP material system, but they have poor performance at high temperature (25°C to 85°C). The high-temperature performance is described by a characteristic temperature T_0, where the temperature dependence of the threshold current density is proportional to $\exp(T/T_0)$. Obviously, a higher T_0 is desirable. In the case of the GaInAsP-InP system, T_0 is about 60 K, due to the small conduction-band offset, resulting in poor electron confinement in the QWs [25]. (The hole confinement is less of a problem because of heavier effective hole masses.) In 1995 Kondow et al. reported a T_0 of 126 K for a GaInAs/GaAs laser emitting at 1.2 μm [26]. In this section, we describe another Al-free material system for long-wavelength lasers at 1.3 and 1.55 μm: InNAsP/InP [27].

The microdisk laser structure consists of just a 1.3-μm single quantum well (SQW) of InNAsP/GaInAsP grown on InP, and the microdisk has a diameter of 5 μm [28]. The mesa is first defined by reactive-ion etching to the InP layer, followed by wet chemical etching, which is isotropic and etches InP but not GaInAsP. The SQW disk thus sits on top of an InP post, as shown in Figure 10.11 [27].

Optical pumping, with a 1% duty cycle at a wavelength of 514 nm from an argon-ion laser, was then used to achieve lasing. Lasers operate up to 340 K, well above the previously achieved 220 K of a 1.45-μm GaInAs/GaInAsP microdisk laser with a similar structure grown and fabricated in the same laboratories. The pump threshold as a function of temperature indicates a T_0 of 97 K, which is also well above the 60 K achieved with a GaInAs/GaInAsP microdisk laser, as shown in Figure 10.12 [27].

Figure 10.13 shows the pump-light intensity plots for (a) InNAsP/GaInAsP and (b) GaInAs/GaInAsP microdisk lasers at 85 K [27]. At this temperature, the threshold pump power of (a) is about 20 μW, compared with the 36 μW of (b). Furthermore, the emission intensity saturates in (b) but not in (a). These results all indicate a larger conduction-band offset in the InNAsP/GaInAsP system.

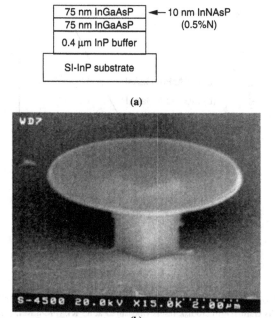

(a)

Figure 10.11 (a) Schematic layer structure of the InNAsP/GaInAsP microdisk laser; (b) InNAsP/GaInAsP microdisk laser. (From Ref. [28].)

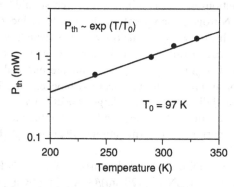

Figure 10.12 Laser-pump threshold as a function of temperature. (From Ref. [28].)

10.4.4 Decreased Conduction-Band Offset: GaInNP Applied to HBTs

Recently, $Ga_{0.52}In_{0.48}P$-grown structures that are lattice-matched to GaAs have received considerable attention because of their potential applications in optoelectronic and electronic devices, such as semiconductor lasers [29], HBTs [30], and high-efficiency tandem solar cells [31]. $Ga_{0.52}In_{0.48}P$/GaAs structures have significant advantages over AlGaAs/GaAs structures in that

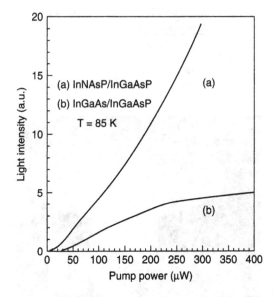

Figure 10.13 Output light intensity vs. pump power for 5-μm-diameter (a) InNAsP/GaInAsP and (b) GaInAs/GaInAsP microdisk lasers (measured at 85 K). (From Ref. [28].)

they exhibit larger valence-band discontinuity, better etch selectivity, and less oxidation effect.

The conduction-band discontinuity, however, should be eliminated for an *npn* HBT. Nitrogen incorporation drastically reduces the band-gap energy in $Ga_{1-x}In_xAs$, with the majority of the reduction resulting from lowering of the conduction band, and we expect a similar effect in $Ga_{1-x}In_xN_yP_{1-y}$. Thus, $Ga_{1-x}In_xN_yP_{1-y}$ may be a suitable material for the emitter and collector of an *npn* HBT, specifically a blocked-hole HBT [32]. The large valence-band discontinuity and large hole effective mass would block holes, while there would be no electron barriers at the base-collector junction. Here, we report on the determination of the $Ga_{1-x}In_xN_yP_{1-y}$/GaAs band alignment [33].

In order to determine the band alignment of $Ga_{1-x}In_xN_yP_{1-y}$/GaAs, a set of eight-period $Ga_{0.46}In_{0.54}N_yP_{1-y}$ (20 nm)/GaAs (5 nm)/$Ga_{0.46}In_{0.54}N_yP_{1-y}$ (20 nm) MQW (multiple quantum wells) samples are grown. There is no GaAs QW PL emission when the barrier has a higher nitrogen concentration, such as 1.2% and 2.4%. One possibility is that high nitrogen composition decreases the $Ga_{0.46}In_{0.54}N_yP_{1-y}$ conduction band too much and that $Ga_{0.46}In_{0.54}N_yP_{1-y}$/GaAs has a Type II alignment, such that the electron and hole separation results in much weaker PL. The PL spectrum of the $Ga_{1-x}In_xN_yP_{1-y}$/GaAs QW having a well thickness of $t_{QW} = 5$ nm taken at 10 K is shown in Figure 10.15. The PL peaks at 1.490 eV and 1.577 eV can be assigned as emissions from the bulk GaAs and the GaAs QW layer itself,

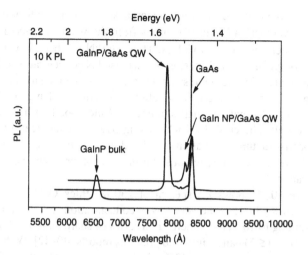

Figure 10.14 PL spectra (at 10 K) for eight-period $Ga_{1-x}In_xN_yP_{1-y}$/GaAs samples. The well thickness is 5 nm. (From Ref. [33].)

respectively. With 0.5% nitrogen incorporation into the $Ga_{1-x}In_xP$ barrier, the position of the GaAs QW PL peak shifts from 1.577 eV to 1.514 eV, as shown in Figure 10.14, due to a reduction of the barrier height.

Based on the experimental data, the $Ga_{1-x}In_xN_yP_{1-y}$/GaAs band lineup can be calculated using a finite-barrier quantum-well model, where the GaAs electron and hole effective mass at 10 K are 0.067 m_e and 0.51 m_e, respectively [34] (m_e is the free-electron mass). The conduction band (ΔE_c) and valence band (ΔE_v) discontinuity values for $Ga_{0.46}In_{0.54}P$/GaAs are determined to be (42 ± 3)% and (58 ± 3)% of the total band-gap difference (404 meV), respectively. This result is similar to previous reports based on the C-V profile method [35, 36]. With 0.5% nitrogen incorporation into the $Ga_{0.46}In_{0.54}P$ barrier, the ΔE_c and ΔE_v become (3 ± 1)% and (97 ± 1)% of total band-gap difference (204 meV), respectively. The band reduction mostly happens in the conduction band moving to lower energy, which is similar to nitrogen incorporation in GaInAs [37]. This situation of small ΔE_c but large ΔE_v is ideal for $Ga_{0.46}In_{0.54}N_{0.005}P_{0.995}$ to be the emitter or the collector of an HBT.

At present, we are fabricating GaInP/GaAs HBTs with a thin GaInNP hole-blocking layer in the collector.

10.4.5 Change in Band Structure: Ga(In)NP Applied to Light-Emitting Devices

It is well known that at very low concentrations, in the 10^{16} cm^{-3} range, isolated nitrogen introduces a highly localized state in GaP, where the energy level is located slightly below the conduction-band minimum. Even

though GaP is an indirect band-gap material, such spatial localization means
that the wavefunction has contributions from the whole k-space and leads
to quasi-direct transitions in GaP:N. Thus, GaP:N has been a widely used
material for green LEDs. With slightly increased N concentrations, N forms
NN pairs, which shift light emission wavelength from green to yellow. With
even higher N concentration, GaNP alloys are formed. Change in the band
structure of GaNP with a small amount of N incorporation has been predicted by Bellaiche et al. [38] who, using 512-atom supercell pseudopotential band-structure calculation, predicted the indirect-direct crossover at
3% N. As reported below, we observe this crossover at an even lower N
concentration.

Figure 10.15 shows PL spectra of 250-nm-thick GaN$_x$P$_{1-x}$ with $x \leq$
0.81% at 10 K [39]. For the very low N concentration sample ($x = 0.05\%$,
corresponding to 10^{19} cm^{-3}), there are a series of sharp emission lines from
different NN$_i$ ($i \leq 5$) pairs, similar to previous reports [40–42]. With increasing N concentration up to 0.43%, the sharp emissions from NN$_i$ pairs
disappear, and a broad PL peak with strong intensity from the GaNP alloy
appears. The PL peak redshifts, and the intensity also increases with increasing N concentration.

The room-temperature PL intensity of GaNP bulk layers increases with
increasing N concentration up to 1.3%, similar to Baillargeon et al.'s report

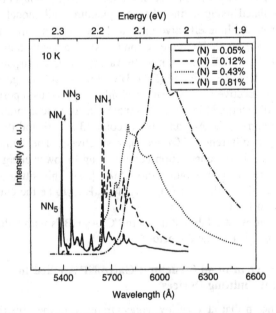

Figure 10.15 PL spectra (at 10 K) of 250-nm-thick GaN$_x$P$_{1-x}$ grown on GaP(1,0,0) with x
0.81%. (From Ref. [39].)

[43]. The PL intensity increases with higher N concentration, mainly due to the increased matrix element for the transition from the conduction band to the valence band. With a N concentration more than 1.3%, PL intensity decreases due to decreased sample quality, partly as a result of increased strain relaxation.

Our results qualitatively agree with the theoretical calculation of Bellaiche et al. [38], where they found a transition point from indirect to direct band-gap at a N concentration of 3% for GaNP. We attribute the large band-gap reduction to the formation of a nitrogen-related impurity band evolved from nitrogen-pair bound states in heavily doped GaP [44]. The intercenter interactions or the formation of large nitrogen clusters lead to band-edge states extending to the energy region below that of the lowest bound-state NN_1, and thus a further reduction of the band-gap of the alloy, inducing the formation of a direct band gap. Therefore, a very strong PL emission is expected in GaNP bulk layers. No room-temperature PL emission was reported previously in GaNP alloys with similar N compositions, likely due to poor sample quality and not the indirect nature of the band structure.

Figure 10.16 shows the square of the absorption coefficient of GaN_xP_{1-x} films as a function of photon energy [39]. Quite obviously, the absorption coefficient obeys a square law, indicating a direct band-gap behavior of GaNP. Furthermore, as the N concentration is increased, the band edge of GaNP shifts to lower energy, indicating a reduction of band-gap energy.

To confirm the direct-band-gap nature, we have the samples characterized by pressure-dependent PL, which is described in detail in Chapter 2. Figure 2.16 shows PL transition energies as a function of applied hydrostatic pressure for samples with three different alloy compositions [45]. Two extreme cases of the pressure dependence of the indirect band-gap of GaP are also shown. It is well known that the indirect band edge X_C shifts to lower energy with increasing pressure for all III-V semiconductors. The positive pressure coefficients of GaNP samples, on the other hand, confirm the direct band-gap nature [46].

We have fabricated LEDs from GaNP (1.1% N) *pn* homojunctions [47] as well as from GaP/GaNP/GaP double heterostructures (DH), shown in Figure 10.17 [48]. Electroluminescence at room temperature peaks at 670 nm, as shown in Figure 10.18, and the results are very encouraging. The DH LEDs are 20 times more intense than *pn* homojunction LEDs. Because the structures were grown on lightly doped *n*-type GaP substrates, instead of a heavily doped one, and there was no thick GaP window layer on top, current spreading is a problem. Although the quantum efficiency ($\approx$1–2%) is still an order of magnitude lower than that of high-brightness AlGaInP/GaAs LEDs, the structure and the active layer are not optimized. As GaNP is lattice-mismatched to GaP, the 500-nm-thick GaNP is about 15% strain-relaxed, as determined from asymmetric X-ray diffraction. Our

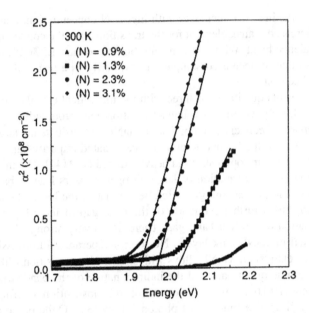

Figure 10.16 Square of the absorption coefficient of GaN_xP_{1-x} films as a function of photon energy. (From Ref. [39].)

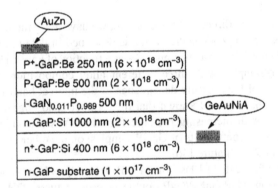

Figure 10.17 Schematic diagram of a GaNP/GaP LED structure. (From Ref. [48].)

recent results on incorporating In to GaInNP show lattice-matching and no strain relaxation, so better LED performance can be expected. The PL intensity of GaNP and GaInNP is comparable or higher than that of GaInP. Thus, GaInNP/GaP heterostructures are a viable candidate for red high-brightness LEDs, and a single epitaxial growth of LED structures on GaP substrates should be simpler and more cost-effective than the current two-step process of GaAs substrate removal and wafer-bonding to a transparent GaP substrate for high-brightness red LEDs [49].

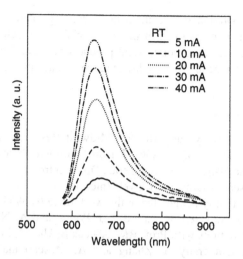

Figure 10.18 Room-temperature electroluminescence at different forward currents of an uncoated DH LED. (From Ref. [48].)

10.5 CONCLUSION AND OUTLOOK

Incorporation of a small amount of nitrogen in III-V compounds grown on GaAs, InP, or GaP substrates has dramatic effects on physical properties, such as band-gap narrowing, change in the conduction-band offset, and even the change of the band structure from indirect to direct band-gap (as in GaNP/GaP). These characteristics allow many interesting applications. Red GaNP/GaP LEDs are grown and fabricated on transparent GaP substrates. This is a simpler process than removing absorbing GaAs substrates and wafer-bonding to GaP substrates. The small-band-gap GaInNAs is used as the base of an HBT, which exhibits a lower turn-on voltage, suitable for low-power applications. The band-gap reduction occurs mainly in lowering of the conduction band edge. When N-containing material is in a quantum well, the conduction-band offset is increased, resulting in a deeper quantum well and better electron confinement, as in InNAsP/InP and GaIn-NAs/GaAs. The characteristic temperature of InNAsP/InP microdisk lasers is shown to be 97 K, higher than the conventional GaInAsP/InP lasers. On the other hand, when the N-containing material is in the larger band-gap material, the conduction-band offset is decreased, as in $Ga_{0.46}In_{0.54}N_{0.005}P_{0.995}$/GaAs, which has a nearly zero conduction-band offset and, therefore, could be an ideal material for *npn* HBTs.

The outlook for dilute nitrides is excellent in material science, device physics, and device applications. Many materials issues resulting from the large miscibility gap of nitrogen in III-V compounds offer a range of interesting topics to be studied, as reflected by the various topics in this book. The large shift in the conduction band edge with a small amount of nitrogen incorporation results in many interesting band-gap engineering possibilities, as demonstrated in this chapter. Finally, GaInNAs VCSELs and HBTs are becoming commercial products because they offer advantages over device counterparts fabricated from conventional materials.

Acknowledgments

This work is partially supported by Midwest Research Institute under subcontract No. AAD-9-18668-7 from National Renewable Energy Laboratory, the DARPA Heterogeneous Optoelectronic Technology Center, Army Research Office, Rockwell International, and the UC MICRO Program. I gratefully acknowledge the work done by W.G. Bi, H.P. Xin, Y.G. Hong, M. Sopanen, R. André, R.J. Welty, and A. Nishikawa, and fruitful collaborations with Prof. P.M. Asbeck of UCSD, Prof. S.T. Ho of Northwestern University, Y. Zhang and A. Mascarenhas of National Renewable Energy Laboratory, W. Shan and W. Walukiewicz of Lawrence Berkeley Laboratory, and W.M. Chen and I. Buyanova of Linköping University.

References

1. Weyers, M., Sato, M., and Ando, H., "Red Shift of Photoluminescence and Absorption in Dilute GaAsN Alloy Layers," *Jpn. J. Appl. Phys. (Part 2)* 31: L853–855 (1992).
2. Kondow, M., Uomi, K., Niwa, A., Kitatani, T., Watahiki, S., and Yazawa, Y., "GaInNAs: A Novel Material for Long-Wavelength-Range Laser Diodes with Excellent High-Temperature Performance," *Jpn. J. Appl. Phys. (Part 1)* 35: 1273–1275 (1996).
3. Shan, W., Walukiewicz, W., Ager, J.W., Haller, E.E., Geisz, J.F., Friedman, D.J., Olson, J.M., and Kurtz, S.R., "Band Anticrossing in GaInNAs Alloys," *Phys. Rev. Lett.* 82: 1221–1224 (1999).
4. Ho, I.H., and Stringfellow, G.B., "Solubility of Nitrogen in Binary III-V Systems," *J. Crystal Growth* 178: 1–7 (1997).
5. Kurtz, S.R., Klem, J.F., Allerman, A.A., Sieg, R.M., Seager, C.H., and Jones, E.D., "Minority Carrier Diffusion and Defects in InGaAsN Grown by Molecular Beam Epitaxy," *Appl. Phys. Lett.* 80: 1379–1381 (2002).
6. Bissiri, M., Gaspari, V., Polimeni, A., Baldassarri, G., von Högersthal, H., Capizzi, M., Frova, A., Fischer, M., Reinhardt, M., and Forchel, A., "High Temperature Photoluminescence Efficiency and Thermal Stability of (InGa)(AsN)/GaAs Quantum Wells," *Appl. Phys. Lett.* 79: 2585–2587 (2001).
7. Sato, S.I., Osawa, Y., and Saitoh, T., "Room-Temperature Operation of GaInNAs/GaInP Double-Heterostructure Laser Diodes Grown by Metalorganic Chemical Vapor Deposition," *Jpn. J. Appl. Phys. (Part 1)* 36: 2671–2675 (1997).
8. Friedman, D.J., Norman, A.G., Geisz, J.F., and Kurtz, S.R., "Comparison of Hydrazine, Dimethylhydrazine, and t-Butylamine Nitrogen Sources for MOVPE Growth of GaInNAs for Solar Cells," *J. Crystal Growth* 208: 11–17 (2000).
9. Xin, H.P., and Tu, C.W., "GaInNAs/GaAs Multiple Quantum Wells Grown by Gas-Source Molecular Beam Epitaxy," *Appl. Phys. Lett.* 72: 2442–2444 (1998).

10. Xin, H.P., Kavanagh, K.L., Kondow, M., and Tu, C.W., "Effects of Rapid Thermal Annealing on GaInNAs GaAs Multiple Quantum Wells," *J. Crystal Growth* 202: 419–422 (1999).

11. Rao, E.V.K., Ougazzaden, A., Lebellego, Y., and Juhel, M., "Optical Properties of Low Band-gap GaAsxNx Layers: Influence of Post-Growth Treatments," *Appl. Phys. Lett.* 72: 1409–1411 (1998).

12. Kurtz, S.R., Allerman, A.A., Seager, C.H., Sieg, R.M., and Jones, E.D., "Minority Carrier Diffusion, Defects, and Localization in InGaAsN, with 2% Nitrogen," *Appl. Phys. Lett.* 77: 400–402 (2000).

13. Xin, H.P., Tu, C.W., and Geva, M., "Annealing Behavior of p-Type $Ga_{0.892}In_{0.108}N_xAs_{1-x}$ (0 x 0.024) Grown by Gas-Source Molecular Beam Epitaxy," *Appl. Phys. Lett.* 75: 1416–1418 (1999).

14. Hong, Y.G., Nishikawa, A., and Tu, C.W., "Similarities between $Ga_{0.48}In_{0.52}N_yP_{1-y}$ and $Ga_{0.92}In_{0.08}N_yAs_{1-y}$ Grown on GaAs (001) Substrates," unpublished manuscript, 2003.

15. Geisz, J.F., Friedman, D.J., Olson, J.M., Kurtz, S.R., and Keyes, B.M., "Photocurrent of 1-eV GaInNAs Lattice Matched to GaAs," *J. Crystal Growth* 195: 401–408 (1998).

16. Yu, K.M., Walukiewicz, W., Wu, J., Mars, D.E., Chamberlin, D.R., Scarpulla, M.A., Dubon, O.D., and Geisz, J.F., "Mutual Passivation of Electrically Active and Isovalent Impurities," *Nature Materials* 1: 185–190 (2002).

17. Sakai, S., Ueta, Y., and Terauchi, Y., "Band-gap Energy and Band Lineup of III-V-Alloy Semiconductors Incorporating Nitrogen and Boron," *Jpn. J. Appl. Phys. (Part 1)* 32: 4413–4417 (1993).

18. Buyanova, I.A., Pozina, G., Hai, P.N., Chen, W.M., Xin, H.P., and Tu, C.W., "Type I Band Alignment in the GaN_xAs_{1-x}/GaAs Quantum Wells," *Phys. Rev. B* 63: 3303–3307 (2001).

19. Zuo, S.L., Bi, W.G., Tu, C.W., and Yu, E.T., "Atomic-Scale Compositional Structure of InAsP/InP and InNAsP/InP Heterostructures Grown by Molecular-Beam Epitaxy," *J. Vacuum Sci. Technol. B* 16: 2395–2398 (1998).

20. Nishikawa, A., Hong, Y.G., and Tu, C.W., "The Effect of Nitrogen on Self-Assembled GaInNAs Quantum Dots Grown on GaAs," *Phys. Stat. Solidi B* 240:310–313 (2003).

21. Sopanen, M., Xin, H.P., and Tu, C.W., "Self-Assembled GaInNAs Quantum Dots for 1.3- and 1.55-μm Emission on GaAs," *Appl. Phys. Lett.* 76: 994–996 (2000).

22. Welty, R.J., Xin, H.P., Mochizuki, K., Tu, C.W., and Asbeck, P.M., "GaAs/$Ga_{0.89}In_{0.11}N_{0.02}As_{0.98}$/GaAs NpN Double Heterojunction Bipolar Transistor with Low Turn-On Voltage," *Solid-State Electronics* 46: 1–5 (2002).

23. Chang, P.C., Baca, A.G., Li, N.Y., Xie, X.M., Hou, H.Q., and Armour, E., "InGaP/InGaAsN/GaAs NpN Double-Heterojunction Bipolar Transistor," *Appl. Phys. Lett.* 76: 2262–2264 (2000).

24. DeLuca, P.M., Lutz, C.R., Welser, R.E., Chi, T.Y., Huang, E.K., Welty, R.J., and Asbeck, P.M., "Implementation of Reduced Turn-On Voltage InGaP HBTs Using Graded GaIn-AsN Base Regions," *IEEE Electron. Device Lett.* 23: 582–584 (2002).

25. Pearsall, T.P., ed., *GaInAsP Alloy Semiconductors* (New York: Wiley, 1982).

26. Kondow, M., Nakatsuka, S., Kitatani, T., Yazawa, Y., and Okai, M., "Room-Temperature Pulsed Operation of GaInNAs Laser Diodes with Excellent High-Temperature Performance," *Jpn. J. Appl. Phys. (Part 1)* 35: 5711–5713 (1996).

27. Tu, C.W., Bi, W.G., Ma, Y., Zhang, J.P., Wang, L.W., and Ho, S.T., "A Novel Material for Long-Wavelength Lasers: InNAsP," *IEEE J. Selected Top. Quantum Electron.* 4: 510–513 (1998).

28. Bi, W.G., Ma, Y., Zhang, J.P., Wang, L.W., Ho, S.T., and Tu, C.W., "Improved High-Temperature Performance of 1.3–1.5-μm InNAsP–InGaAsP Quantum-Well Microdisk Lasers," *IEEE Photonics Technol. Lett.* 9: 1072–1074 (1997).

29. Liau, Z.L., Palmateer, S.C., Groves, S.H., Walpole, J.N., and Missaggia, L.J., "Low-Threshold InGaAs Strained-Layer Quantum-Well Lasers ($\lambda = 0.98$ μm) with GaInP Cladding Layers and Mass-Transported Buried Heterostructure," *Appl. Phys. Lett.* 60: 6–8 (1992).

30. Ahmari, D.A., Fresina, M.T., Hartmann, Q.J., Barlage, D.W., Mares, P.J., Feng, M., and Stillman, G.E., "High-Speed InGaP/GaAs HBTs with a Strained $In_xGa_{1-x}As$ Base," *IEEE Electron. Device Lett.* 17: 226–228 (1996).

31. Bertness, K.A., Kurtz, S.R., Friedman, D.J., Kibbler, A.E., Kramer, C., and Olson, J.M., "29.5-Percent-Efficient GaInP/GaAs Tandem Solar Cells," *Appl. Phys. Lett.* 65: 989–991 (1994).

32. Welty, R.J., Mochizuki, K., Lutz, C.R., and Asbeck, P.M., "Tunnel Collector GaInP/GaAs HBTs for Microwave Power Amplifiers," *Proceedings of 2001 IEEE BIPOLAR/BiCMOS Circuits and Technology Meeting* (Washington, DC: IEEE, 2001), 74–77.

33. Hong, Y.G., Andre, R., and Tu, C.W., "Gas-Source Molecular Beam Epitaxy of GaInNP/GaAs and a Study of Its Band Lineup," *J. Vacuum Sci. Technol. B* 19: 1413–1416 (2001).

34. Shur, M., *Physics of Semiconductor Devices* (Englewood Cliffs, NJ: Prentice Hall, 1990), 625.

35. Rao, M.A., Caine, E.J., Kroemer, H., Long, S.I., and Babic, D.I., "Determination of Valence and Conduction-Band Discontinuities at the (Ga,In)P/GaAs Heterojunction by C-V Profiling," *J. Appl. Phys.* 61: 643–649 (1987).

36. Watanabe, M.O., and Ohba, Y., "Interface Properties for GaAs/InGaAlP Heterojunctions by the Capacitance-Voltage Profiling Technique," *Appl. Phys. Lett.* 50: 906–908 (1987).

37. Zunger, A., "Anomalous Behavior of the Nitride Alloys," *Phys. Stat. Solidi B* 216: 117–123 (1999).

38. Bellaiche, L., Wei, S.H., and Zunger, A., "Composition Dependence of Interband Transition Intensities in GaPN, GaAsN, and GaPAs Alloys," *Phys. Rev. B* 56: 10233–10240 (1997).

39. Xin, H.P., Tu, C.W., Zhang, Y., and Mascarenhas, A., "Effects of Nitrogen on the Band Structure of GaN_xP_{1-x} Alloys," *Appl. Phys. Lett.* 76: 1267–1269 (2000).

40. Thomas, D.G., Hopfield, J.J., and Frosch, C.J., "Isoelectronic Traps Due to Nitrogen in Gallium Phosphide," *Phys. Rev. Lett.* 15: 857–859 (1965).

41. Thomas, D.G., and Hopfield, J.J., "Isoelectronic Traps due to Nitrogen in Gallium Phosphide," *Phys. Rev.* 150: 680–686 (1966).

42. Groves, W.O., Herzog, A.H., and Craford, M.G., "The Effect of Nitrogen Doping on $GaAs_{1-x}P_x$ Electroluminescent Diodes," *Appl. Phys. Lett.* 19: 184–186 (1971).

43. Baillargeon, J.N., Cheng, K.Y., Hofler, G.E., Pearah, P.J., and Hsieh, K.C., "Luminescence Quenching and the Formation of the $GaP_{1-x}N_x$ Alloy in GaP with Increasing Nitrogen Content," *Appl. Phys. Lett.* 60: 1540–1542 (1992).

44. Zhang, Y., Mascarenhas, A., Xin, H.P., and Tu, C.W., "Formation of an Impurity Band and Its Quantum Confinement in Heavily Doped GaAs:N," *Phys. Rev. B* 61: 7479–7482 (2000).

45. Shan, W., Walukiewicz, W., Yu, K.M., Wu, J., Ager III, J.W., Xin, H.P., and Tu, C.W., "Nature of the Fundamental Band-gap in GaN_xP_{1-x} Alloys," *Appl. Phys. Lett.* 76: 3251–3253 (2000).

46. Shan, W., Walukiewicz, W., Yu, K.M., Ager III, J.W., Haller, E.E., Geisz, J.F., Friedman, D.J., Olson, J.M., Kurtz, S.R., Xin, H.P., and Tu, C.W., "Band Anticrossing in III-N-V Alloys," *Phys. Stat. Solidi B* 223: 75–85 (2001).

47. Xin, H.P., Welty, R.J., and Tu, C.W., "$GaN_{0.011}P_{0.989}$ Red Light-Emitting Diodes Directly Grown on GaP Substrates," *Appl. Phys. Lett.* 77: 1946–1948 (2000).

48. Xin, H.P., Welty, R.J., and Tu, C.W., "$GaN_{0.011}P_{0.989}$–GaP Double-Heterostructure Red Light-Emitting Diodes Directly Grown on GaP Substrates," *IEEE Photonics Technol. Lett.* 12: 960–962 (2000).

49. Kish, F.A., Steranka, F.M., DeFevere, D.C., Vanderwater, D.A., Park, K.G., Kuo, C.P., Osentowski, T.D., Peanasky, M.J., Yu, J.G., Fletcher, R.M., Steigerwald, D.A., Craford, M.G., and Robbins, V.M., "Very High-Efficiency Semiconductor Wafer-Bonded Transparent-Substrate $(Al_xGa_{1-x})_{0.5}In_{0.5}P/GaP$ Light-Emitting Diodes," *Appl. Phys. Lett.* 64: 2839–2841 (1994).

CHAPTER 11

GaInNAs Long-Wavelength Lasers for 1.3-μm-Range Applications

M. KONDOW

Hitachi Ltd., Central Research Laboratoryy, Tokyo, Japan

GaInNAs was proposed and created in 1995. It can be grown pseudomorphically on a GaAs substrate and is a light-emitting material with a band-gap energy that corresponds to the long-wavelength range. By combining GaInNAs with GaAs, an ideal band lineup for laser-diode application is achieved. This chapter presents the application of GaInNAs in long-wavelength laser diodes, especially in the 1.3-μm-range lasers that have already gone on the market.

1-591-69019-6/04/$0.00+$1.50

11.1 INTRODUCTION

GaInNAs is a member of the family of group V nitride alloys and III-N-V alloys, such as GaNP and GaNAs, novel semiconductor materials that were only developed in the 1990s. Several research groups reported on III-N-V alloy semiconductors before 1994 [1–4], but these reports have been limited to crystal growth and measurements of physical properties. Kondow et al. [5–7] have proposed application to optoelectronic devices because the exceptional physical properties of III-N-V alloy semiconductors, such as their huge degrees of band-gap bowing, facilitate devices with levels of performance greatly superior to those of current devices. The unusual physical properties are consequences of the exceptional chemical characteristics of nitrogen as compared to the other elements in groups III and V. However, these chemical characteristics, in turn, lead to difficulties in the creation of alloys of nitrogen and III-V crystals, i.e., in the growth of III-N-V alloys. A strongly nonequilibrium method of growth and highly reactive nitrogen precursor are essential to overcome the immiscibility of N in III-V alloys. For this reason, no one had succeeded in growing any such material before the early 1990s. Kondow et al. [17] proposed and developed a method of growth for III-N-V alloys in 1994. They adopted molecular-beam epitaxy (MBE) with nitrogen supplied in the form of radicals. Their method has subsequently been widely used as a standard method for the growth of III-N-V alloys with excellent crystallinity. In 1996, Sato et al. [8] succeeded in growing GaInNAs with excellent crystallinity by conventional organometallic vapor-phase epitaxy (OMVPE), despite the previous failure of several groups to do this. In general, the advantages of OMVPE are its applicability to the fabrication of complex layered structures, good reproducibility of growth from run to run, and suitability for use in mass production. The availability of both growth methods has accelerated research into III-N-V alloys and the development of devices based on these materials.

   GaInNAs is a light-emitting material with a band-gap energy suitable for long-wavelength laser diodes (1.3–1.55 μm and longer wavelengths) and can be grown pseudomorphically on a GaAs substrate. By combining GaInNAs with GaAs or any other wide-gap materials that can be grown on a GaAs substrate, it is possible to achieve a Type I band lineup. (In order to apply a material to the quantum-well active layer of a laser diode, a Type I band lineup is essential so that both electrons and holes are confined to the quantum-well layer.) This allows the fabrication of very deep quantum wells, especially in the conduction band. Deep quantum wells bring many advantages in terms of laser performance, including high operating speeds and excellent temperature stability of threshold current and lasing wavelength. GaInNAs is also promising as a material for use in long-wavelength vertical-cavity surface-emitting lasers (VCSELs), because a GaInNAs active layer can be grown on a highly reflective GaAs/AlAs distributed Bragg reflector (DBR) mirror over a GaAs substrate in a single stage of epitaxial growth.

The body of this chapter begins with a description of the band structure of GaInNAs/GaAs deep quantum wells for use as the active layers of laser diodes. We then look at the demonstrated performance of GaInNAs lasers in edge-emitter and VCSEL structures.

11.2 BAND-GAP AND BAND ALIGNMENT

In this section, we start by examining the band-gap structure of GaNAs and GaInNAs. Next, the band lineup of GaNAs and GaInNAs is explained. Finally, quantum levels in a GaInNAs/GaAs well are presented.

Figure 11.1 shows the experimentally obtained relationship between the N content and band-gap energy of GaNAs epilayers grown on a GaAs substrate [7]. The N content was estimated by X-ray diffraction, and the band-gap energy was measured by photoluminescence (PL). As for GaNAs with a low N content, the PL peaks of the GaNAs epilayer and the GaAs substrate overlap with each other in the measurement taken at room temperature. Therefore, PL was also observed at the very low temperature of 77 K. As can be seen in Figure 11.1, the band-gap energy decreases monotonically with increases in the N content. The energy difference between room-temperature and 77-K measurements hardly varies, remaining at approximately 70 meV. The lines denoting room-temperature and 77-K

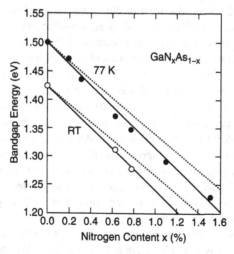

Figure 11.1 The relationship between N content and band-gap energy of GaNAs epilayers grown on a GaAs substrate. The solid lines are experimental results. Broken lines are the result of calibration to remove the effect of strain. (From Ref. [7].)

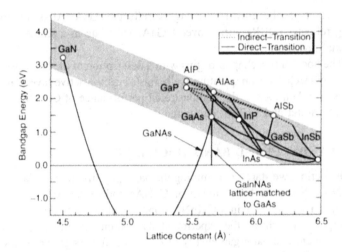

Figure 11.2 The relationship between lattice constant and band-gap energy in III-V alloy semiconductors. (From Ref. [9].)

measurement thus run parallel in Figure 11.1. The slope of each line is 0.184 eV/% N. GaNAs epilayers grown on a GaAs substrate are under tensile strain, and the tensile strain lowers the band-gap energy. To determine the relationship between the N content and the band-gap energy of free-standing, i.e., stress-free GaNAs, the band-gap energy is calibrated by using the deformation potential parameters for GaAs. The results are shown as broken lines in Figure 11.1; these lines have a slope of 0.156 eV/% N.

Figure 11.2 shows the relationship between the lattice constant and band-gap energy in various III-V alloy semiconductors, including GaNAs and GaInNAs [9]. The bowing parameter of GaNAs was derived from the above experimental results for GaNAs with a small content of nitrogen [7], assuming a parabolic curve [10]. With conventional alloy semiconductors, the results for which lie in the shaded area of Figure 11.2, the tendency is for the band-gap energy to increase with decreasing lattice constant. On the other hand, the result for GaNAs departs from the shaded area to which conventional alloy semiconductors are restricted. Until N comes to make up half of the GaNAs, an increase in N content leads to a monotonic decrease in band-gap energy rather than an increase toward the value for cubic GaN (3.2 eV). This curious behavior is supported by both experimental results [4] and theoretical predictions [10]. The responsible factor is the great discrepancy between the electronegativity values for nitrogen and for the other constituent atoms [10]. Phillips [11] states that the electronegativity of N is 3.00 eV, while the values for P, As, and Sb are in the range from 1.31 to 1.64 eV. Since one property of III-N-V alloy semiconductors is that a decrease in the lattice constant leads to a decrease in the band-gap energy, they provide operation over a dramatically broader area than conventional III-V alloy semiconductors, as is shown in Figure 11.2. This leads to significantly greater freedom in the design of semiconductor devices, to the extent that novel devices and dramatic improvements to the performance of current devices become possible. The figure includes results for GaNAs, which has a negative band-gap energy. This might mean that GaNAs can act as a metal or semiconductor, according to the content of N. While such behavior is very interesting, there is no reliable experimental data on this region because of phase-separation and, furthermore, no guarantee that band-gap theory can be expanded to cover this region. Thus, the middle range of composition for III-N-V alloys remains unexplored.

Here, let us focus on the band-gap energy of GaInNAs. Adding In to GaAs, i.e., making a GaInAs-alloy semiconductor, increases the lattice constant, while adding N to GaAs, i.e., making a GaNAs-alloy semiconductor, decreases the lattice constant. GaInNAs can thus be lattice-matched to GaAs by adjusting the contents of In and N. Adding In to GaAs decreases the band gap, and in the same way, adding N to GaAs also decreases the band gap, as shown in Figure 11.2. Since both GaInAs and GaNAs are direct-transition-type semiconductors, GaInNAs is also of this type. Thus, GaInNAs is a light-emitting material that has a band-gap energy suitable for long-wavelength laser diodes (0.8–1.0 eV) and is suitable for formation on a GaAs substrate.

Figure 11.3 shows the nitrogen-content dependence of the band-discontinuity energy in the valence band (ΔE_v) for GaNAs/GaAs as measured by X-ray photoelectron spectroscopy (XPS) [12]. The rather large experimental error was due to the limit on energy resolution in this measurement. The center value of ΔE_v for GaNAs fell as the content of nitrogen increased. This suggests the formation of a Type II band lineup in GaNAs/GaAs. From the slope, ΔE_v for GaNAs/GaAs was estimated as $-(0.019 \pm 0.053)$ eV/%N (the large experimental error means that we still have a possibility that a

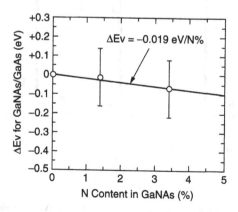

Figure 11.3 Dependence of ΔE_v on the content of N in GaNAs/GaAs. (From Ref. [12].)

Type I band lineup forms in GaNAs/GaAs). Ding et al. [13] reported ΔE_v of -1.84 eV in cubic-GaN/GaAs based on XPS measurement. If the ΔE_v of cubic-GaN/GaAs is linearly interpolated, ΔE_v of GaNAs/GaAs can be calculated as -0.0184 eV/%N. This is in good agreement with the above-measured value of -0.019 eV/%N. Thus, the band-gap bowing seems to have little effect on the ΔE_v of GaNAs, and this result supports the theoretical prediction of Sakai et al. [14], in which bowing of the valence band of GaNAs is negligible.

Next, the band lineup of the GaInNAs materials system is examined from the viewpoint of application to a semiconductor laser, because in order to apply a material to the quantum-well active layer of a laser diode, a Type I band lineup is essential so that both electrons and holes are confined to the quantum-well layer.

A schematic diagram of the band-gap for GaInAs is shown in the right half of Figure 11.4. The band-gap of GaNAs is shown in the left half of the figure. The horizontal axis shows strain, which allows us to draw the diagrams for GaInAs and GaNAs in the same figure. GaAs is therefore located in the center. In Figure 11.4, bowing of the valence band of GaNAs

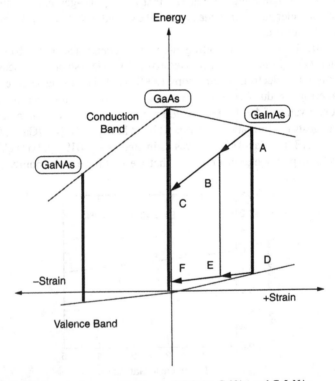

Figure 11.4 Schematic diagram of band-gap of GaInAs, GaNAs, and GaInNAs.

is assumed to be zero on the basis of the experimental result cited above [12]. As a result, both the conduction band and band-gap have the same bowing parameter in the GaNAs system. Thus, the bowing of the conduction band is very large. Increasing the content of In in GaInAs, i.e., increasing the compressive (+) strain, lowers the conduction band and raises the valence band. On the other hand, increasing the N content in GaNAs, i.e., increasing the tensile (−) strain, lowers both the conduction and valence bands. Since the conduction band falls more steeply than the valence band, increasing the content of N decreases the value of the band gap. If a small amount of N is added to GaInAs to form GaInNAs, the conduction and valence bands will be moved from A to B and from D to E in Figure 11.4, respectively. When the composition is such that the GaInNAs is lattice-matched with GaAs, the conduction band will be at C and the valence band will be at F. Note that the valence bands of the GaInNAs and GaAs are at almost the same energy level. Therefore, by combining GaInNAs with wide-gap materials such as AlGaAs, a Type I band lineup is easily achieved. Compressively (+) strained GaInNAs, can also have a Type I band lineup in combination with GaAs, as shown by the bold line between B and E in the figure. In general, a ΔE_v of more than 50 meV is required to confine holes in the quantum-well layer. Since the GaInAs quantum-well layer in the 0.98-μm-range laser diode formed on a GaAs substrate is strained by approximately 1% and has a ΔE_v as high as 50 meV, compressively strained GaInNAs has to be strained by at least 1% to achieve good hole confinement. (In this case, the GaInNAs should be grown as a strained quantum-well layer that is less thick than the critical value where misfit dislocations start to appear.) The conduction band of GaInNAs, whether it is lattice-matched to GaAs or compressively strained, has large discontinuity energy (ΔE_c).

Figure 11.5 shows the band lineups for combinations of materials that produce wavelengths in the 1.3-μm range. As is shown in Figure 11.5a, the ΔE_c for the GaInPAs system on InP is smaller than the ΔE_v. The ΔE_c of 100 meV is too small to suppress electron overflow from the well layer to the barrier layer. The small ΔE_c leads to poor electron confinement, which in turn leads to poor high-temperature performance for a laser diode. In the AlGaInAs system on InP (Figure 11.5b), ΔE_c is sufficiently large for electron confinement. However, this system has a Type II band lineup with the InP cladding layer. The large ΔE_c between AlGaInAs and InP is an obstacle to the injection of electrons from the cladding to the active layer. On the other hand, GaInNAs has an ideal band lineup, as seen in Figure 11.5c. The ΔE_c between the GaAs and GaInNAs, at 350 meV, is sufficient to completely prevent electron overflow, as is later discussed. Therefore, we can expect the GaInNAs material system to produce uncooled laser diodes that operate over a wide range of temperatures.

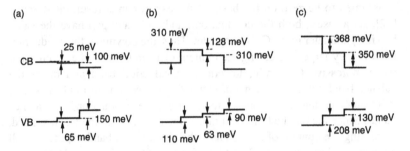

Figure 11.5 Band lineups for materials systems with wavelengths in the 1.3-μm range. (a): $InP/Ga_{0.9}In_{0.1}P_{0.8}As_{0.2}/Ga_{0.2}In_{0.8}P_{0.3}As$, (b): $InP/Al_{0.5}In_{0.5}As/(Al_{0.7}Ga_{0.3})_{0.5}In_{0.5}As/(Al_{0.3}Ga_{0.7})_{0.4}$ $In_{0.6}As$, and (c): $Al_{0.5}Ga_{0.5}As/GaAs/Ga_{0.7}In_{0.3}N_{0.01}As_{0.99}$.

Next, let us discuss the quantum levels in GaInNAs/GaAs wells. Knowledge of the effective masses of individual holes and electrons is necessary if we are to determine these levels. As was mentioned above, alloying Ga(In)As with nitrogen strongly affects the conduction band but has little effect on the valence band. Therefore, we can assume that the effective mass of a hole in GaInNAs is the same as that of a hole in GaInAs with the same content of In. On the other hand, the effective mass of an electron in GaInNAs differs greatly from that of an electron in GaInAs. Figure 11.6 shows this effective mass as a function of the content of nitrogen. This result was obtained by analyzing the optical transitions between the ground levels [15] and high-order quantum levels [16, 17] in GaInNAs/GaAs quantum wells. Note that the transition energies between the high-order quantum levels are sensitive to the mass value. Therefore, we can obtain precise values using this approach The value of effective mass remains in the range $0.08 \pm 0.01 \ m_0$ (where m_0 is the mass of a free electron) almost independently of the nitrogen content. Thi.s behavior is in

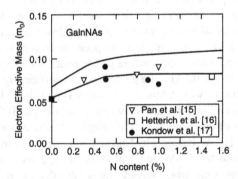

Figure 11.6 The effective mass of an electron in GaInNAs as a function of the content of N. Broken line denotes a theory for GaNAs. (From Ref. [18].)

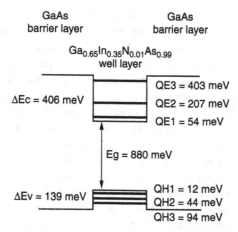

Figure 11.7 The quantum levels in a 7-nm-thick $Ga_{0.65}In_{0.35}N_{0.01}A_{0.99}$/GaAs well with an emission wavelength of 1.31 μm.

good agreement with a theoretical prediction for GaNAs [18]. An electron has a 60% larger effective mass in $Ga_{0.7}In_{0.3}NAs$ than in $Ga_{0.7}In_{0.3}As$.

Figure 11.7 is a sketch of the quantum levels in a $Ga_{0.65}In_{0.35}N_{0.01}As_{0.99}$/GaAs well, a system that produces emission at 1.31 μm. The well is 7 nm thick. The energy difference between electrons at the first and second quantum levels is 153 meV. Thus, most electrons are on the ground level. This leads to a large optical gain for the laser diode, as is later shown.

At the end of this section, it should be noted that many issues on electronic structure of the III-N-V alloys, i.e., band bowing, band lineup, and effective mass values, remain under debate (as noted in other chapters in this book). The values given in this section are mostly derived from the pioneering works by the author's group. They might be modified, more or less, in the future. However, the outline of the material properties of the III-N-V alloys illustrated in this chapter should be close to the real one because the device performances have been improved by using these materials, as might be expected.

11.3 PERFORMANCE OF GaInNAs LASER DIODES

In this section, the applications of GaInNAs in both edge-emitting and vertical-cavity surface-emitting lasers (VCSELs) are demonstrated in the long-wavelength range.

11.3.1 Edge-Emitting Lasers

11.3.1.1 Background

In optical fiber communications today, 1.3- or 1.55-μm-long wavelength semiconductor lasers are used as light sources to minimize the transmission loss at the optical fiber windows. Most of these diodes are GaInPAs-alloy semiconductors formed on InP substrates. However, typical characteristic temperature (T_0) values around 60 K mean that these devices have unsatisfactory lasing properties at higher operating temperatures. This T_0 range is much lower than that for 0.98-μm-range GaInAs laser diodes formed on GaAs substrates; the T_0 range for this system is above 150 K. We thus have to use thermoelectric coolers in most practical applications of GaInPAs laser diodes formed on InP substrates. However, uncooled long-wavelength laser diodes that perform well in high-temperature environments would be very useful. The low T_0 range for the former system is a consequence of poor electron confinement, which in turn is due to the small band-gap discontinuity in the conduction band (ΔE_c). An AlGaInAs laser on an InP substrate [19] and a GaInPAs/GaInP laser on a GaInAs ternary substrate [20, 21] have demonstrated the improved electron confinement of these systems.

Kondow et al. [5, 6] have proposed the use of GaInNAs in the active layer as a way to dramatically improve the high-temperature performance of long-wavelength lasers. As was discussed in the last section, ΔE_c values greater than 300 meV can be achieved in a GaInNAs/GaAs deep quantum well. This should effectively provide perfect electron confinement in the active layer and lead to lasers with high T_0 values, i.e., with values above 150 K. Deep quantum wells for the active layers bring the following benefits to laser diodes:

1. Excellent high-temperature performance
2. Excellent stability of the lasing wavelength despite changes in temperature
3. High-speed operation

Figure 11.8 is a schematic cross-sectional view of a structure for index-guided GaInNAs/GaAs single-qu antum-well (SQW) laser diodes (LDs)

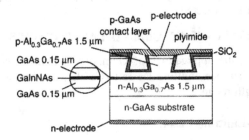

Figure 11.8 Schematic cross-sectional view of index-guided GaInNAs/GaAs SQW-LDs with a ridge-waveguide structure.

with a ridge-waveguide structure, grown on a (1,0,0)-oriented n-GaAs substrate. The SQW active region consisted of a 7- or 10-nm-thick GaInNAs strained well layer and a 150-nm-thick GaAs barrier (or waveguide) layer. Two 1.5-μm-thick $Al_{0.3}Ga_{0.7}As$ layers were used as cladding layers in order to obtain strong confinement of both carriers and light. The carrier density was $7-10 \times 10^{17}$ cm^{-3}. A p-GaAs contact layer ($p = 1 \times 10^{19}$ cm^{-3}) was formed to decrease the contact resistance. The ridge mesa was formed by wet etching. The bottom of the mesa was a few micrometers wide. The distance between the bottom of the mesa and the top of the GaAs barrier layer was 0.15 μm, so the injection current spread, more or less, into a larger area beyond the bottom of the mesa. To suppress mirror loss, highly reflective mirrors were deposited on the front and rear facets (respectively, 70 and 95% reflective) of some devices. Since index-guided-type LDs have a single optical transverse mode, it is easy to couple with a single-mode optical fiber. This structure is thus adopted when the LDs are used for practical application.

Figure 11.9 shows a schematic cross-section of gain-guided-type GaInNAs/GaAs SQW LDs. The stripe for current injection was fabricated by selectively removing the p-GaAs contact layer and inserting a SiO$_2$ layer. The stripe width was varied from 20 to 80 μm. The stripe is broad enough for the spread of injection current to be negligible. Therefore, the threshold current density can be accurately estimated. This LD structure is simple and relatively easy to fabricate, and is thus often used in evaluating device performance.

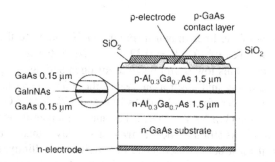

Figure 11.9 Schematic cross-sectional view of gain-guided GaInNAs/GaAs SQW-LDs.

11.3.1.2 Fundamental properties

The novel GaInNAs materials system was investigated through an examination of the optical and physical parameters of GaInNAs LDs [22]. The findings are presented in this subsection. The samples were gain-guided-type GaInNAs-SQW lasers with a stripe width (W) of 60 μm. GaInAs LDs were also examined for reference. To eliminate the effects of heating during lasing operation, the device was in pulsed operation at 5 kHz with a 0.4% duty cycle during the measurements.

To start with, the reciprocal of the external differential quantum efficiency ($1/\eta_D$) versus the cavity length (L) is plotted in Figure 11.10. The solid and broken lines denote data for the GaInNAs and GaInAs LDs, respectively. The external differential quantum efficiency is expressed as

$$\eta_D = \eta_i \left(\frac{\alpha_i + \alpha_m}{\alpha_m} \right), \tag{11.1}$$

where η_i is the internal quantum efficiency, α_i is the internal loss, and α_m is the mirror loss. The mirror loss is described by

$$\alpha_m = \frac{1}{L} \ln \left(\frac{1}{R} \right), \tag{11.2}$$

where R is the power reflectivity at the mirrors of the cavity. The reciprocal of the external differential quantum efficiency ($1/\eta_D$) can thus be written as a function of the cavity length (L) in the following way:

$$\frac{1}{\eta_D} - \frac{1}{\eta_i} \left(1 + \frac{\alpha_i \cdot L}{\ln \frac{1}{R}} \right). \tag{11.3}$$

The slope of the lines drawn in Figure 11.10 indicated that α_i was 6 cm^{-1} for both types of LD; R = 0.32 was used as the reflectivity of an uncoated mirror. This low α_i implies that nitrogen adding brings about no additional scattering and absorption loss in the SQW active layer. The η_i was estimated as 35% for the GaInNAs LDs and 57% for the GaInAs LDs. The low η_i for the GaInNAs devices is not due to a larger amount of strain, because there is less strain in the GaInNAs active layer than in the GaInAs layer. Thus, it must be a consequence of inferior crystallinity. In fact, the greater incidence of crystal defects in the GaInNAs active layer leads to a higher incidence of nonradiative recombination.

Now, let us focus on the gain parameters of the GaInNAs laser diode. Figure 11.11 is a plot of the threshold current density (J_{th}) as a function of the total loss ($\alpha = \alpha_i + \alpha_m$), which is the sum of the internal loss and the mirror loss. The solid and broken lines denote data from GaInNAs and GaInAs LDs, respectively. The optical gain (g) of a quantum-well layer is approximated as a function of the current density (J) in the following way:

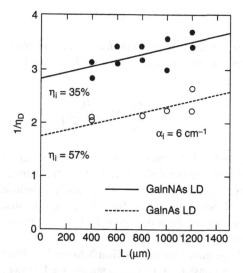

Figure 11.10 The reciprocal of external differential quantum efficiency vs. cavity length for GaInNAs and GaInAs LDs. (From Ref. [22].)

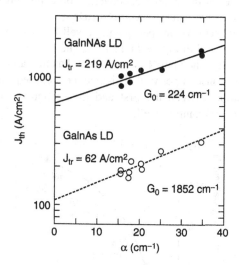

Figure 11.11 Threshold current density vs. total loss for GaInNAs and GaInAs LDs. (From Ref. [22].)

$$g = G_0 \ln\left(\frac{\eta_i \cdot J}{J_{tr}}\right) \qquad (11.4)$$

where G_0 is the gain constant and J_{tr} is the transparency current density. At the lasing threshold, the net gain equals the total loss, i.e.,

$$\Gamma \cdot g = \alpha \tag{11.5}$$

where Γ is the optical confinement factor for the well. Therefore, the threshold current density (J_{th}) can be described as a function of total loss (α):

$$J_{th} = \frac{J_{tr}}{\eta_i} \exp\left(\frac{\alpha}{G_0 \cdot \Gamma}\right) \tag{11.6}$$

From Figure 11.11, J_{tr} was estimated to be 219 A/cm^2 for the GaInNAs LDs and 62 A/cm^2 for the GaInAs LDs. Like the internal quantum efficiency, the large J_{tr} of the GaInNAs LDs is due to insufficient crystallinity. To evaluate G_0, Γ was estimated as 0.0168 through calculation. The slopes of the solid and broken lines in Figure 11.11 then provided estimates for G_0 of 2243 cm^{-1} for the GaInNAs LDs and 1852 cm^{-1} for the GaInAs-LDs. Note that the large gain constant for GaInNAs enables laser operation even if the active layer consists of a single quantum well. However, after a great deal of further evaluation of 1.3-μm-range GaInNAs LDs, it was revealed that the G_0 value for these devices converges on approximately 1800 cm^{-1} and is the same as the value for GaInAs LDs

The gain constant values per quantum well for laser diodes of several kinds are summarized in Figure 11.12 [23]. The figure shows that G_0/well increases with ΔE_c. The GaInNAs LD, which has the largest ΔE_c, also has the highest gain constant per well. This may be a consequence of the large energy separation between the first and the second quantum levels for electrons in a deep quantum well.

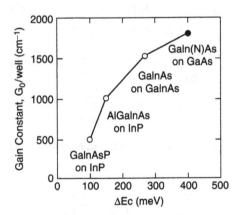

Figure 11.12 The gain constant of several kinds of laser diodes for emission in the 1.3-μm range. (From Ref. [23].)

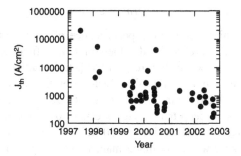

Figure 11.13 Threshold current density of GaInNAs LDs as a function of time in years.

Figure 11.13 is a plot of the threshold current density (J_{th}) of GaInNAs LDs as a function of time in years. Data were collected from worldwide reports on 1.3-μm-range operation. Most of the subject LD structures are gain-guided-type lasers with a broad stripe and long cavity. The crystallinity of GaInNAs has improved with advances in growth techniques. The threshold current is thus getting lower year by year. The lowest value at the time of writing in 2003 is 211 A/cm² for an OMVPE-grown GaInNAs SQW-LD [24]. This value is comparable with those for GaInAs LDs. The crystallinity of GaInNAs has thus reached the same level as for the corresponding conventional III-V semiconductor.

Figure 11.14 shows results for aging tests of index-guided-type GaInNAs LDs grown by MBE. Testing was at high temperature under an automatic power control (APC) condition [25]. The APC test with a chip power level of 3 mW was conducted for six devices without prescreening. After 3200 hours, the ambient temperature was changed from 70°C to 85°C. Even after 8500 hours of total aging time, no measurable increase in operating current, i.e., no degradation, was found. This indicates that GaInNAs LDs will operate for more than 100,000 hours under practical usage conditions.

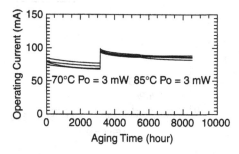

Figure 11.14 Results for aging tests of index-guided-type GaInNAs LDs under an APC condition.

11.3.1.3 High-temperature performance

Results on the high-temperature performance of GaInNAs LDs are presented in this section. Calculated values for T_0 as a function of ΔE_c of the
laser diodes are shown in Figure 11.15 [6]. Calculation is based on the
thermionic emission model [26]. Quasi-Fermi levels for electrons are
assumed to be 50 and 70 meV at 300 and 360 K, respectively. When the
quantum well is so deep that electrons cannot overflow from the well to
the barrier layer, T_0 is assumed to be 180 K, i.e., the intrinsic value for laser
diodes. This value was determined mainly by taking into account Auger
recombination and intervalence-band absorption. The less effect these phenomena have, the greater is the T_0 value, and vice versa. The returning rate
of electrons that have overflowed is estimated by fitting the calculated
results to the experimental T_0 values for GaInPAs and AlGaInAs laser
diodes. Figure 11.15 implies that a ΔE_c greater than 300 meV completely
blocks the overflow of electrons and that T_0 then rises to the intrinsic value
for a laser diode.

Figure 11.16 shows light output power as a function of injected current
for an index-guided-type GaInNAs SQW laser in pulsed operation (at 5
kHz with a 0.4% duty cycle) at various temperatures [27]. The threshold
current increased from 82.5 to 110 mA as the temperature rose from 20°C
to 80°C. Figure 11.17 shows a spectrum produced by laser operation at
room temperature [28]. The lasing wavelength was 1.3045 µm.

The threshold current is plotted as a function of temperature in Figure
11.18, where a line of fit has been applied to the data [28]. The slope of
this line indicates that T_0 is 215 K. This is the best reported value for
long-wavelength lasers operating in the 1.3-µm range and is in agreement
with the theoretical results given in Figure 11.15.

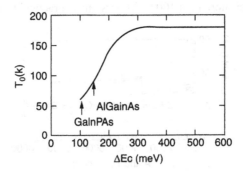

Figure 11.15 Calculated relationship between T_0 and ΔE_c values for a laser diode. (From Ref.
[6].)

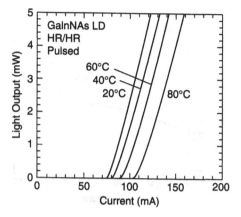

Figure 11.16 Light-output power as a function of injected current for a GaInNAs-SQW laser. (From Ref. [27].)

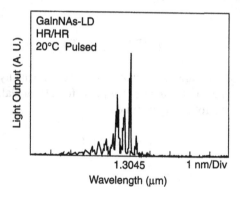

Figure 11.17 Spectrum of emission from a GaInNAs-SQW laser in room temperature operation. (From Ref. [28].)

11.3.1.4 Stability of lasing wavelength

In this subsection, results on the stability of lasing wavelength for GaInNAs LDs with changes in temperature are presented. The International Telecommunications Union (ITU) recommends that the lasing wavelength of a 1.3-µm-range LD be from 1270 to 1340 nm over the entire range of operating temperatures from −40°C to 85°C. Thus, the shift in lasing wavelength with temperature, $d\lambda/dT$, has to be less than 0.56 nm/°C. However, if we assume that fabrication error results in a lasing-wavelength dispersal of ±5 nm, this restriction falls to less than 0.48 nm/°C. Higashi et al. [29] reported a trade-off between wavelength stability with changes in temperature and high-temperature performance in 1.3-µm-range operation for conventional quan-

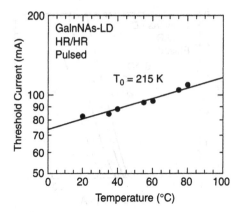

Figure 11.18 Temperature dependence of the threshold current for a GaInNAs-SQW laser. (From Ref. [28].)

tum-well GaInPAs LDs. They found the following relationship between $d\lambda/dT$ and T_0:

$$d\lambda/dT = a_1 - (a_2/T_0) \qquad (11.7)$$

where a_1 and a_2 are parameters that can be experimentally evaluated and do not depend on the LD structure. They also found that GaInPAs quantum-well LDs have the following parameters:

$$a_1 = 0.8 \text{ nm/°C} \qquad (11.8)$$

and

$$a_2 = 24 \text{ nm} \qquad (11.9)$$

Therefore, we have to find out whether GaInNAs quantum-well LDs with a very high T_0 have a low enough $d\lambda/dT$.

Figure 11.19 shows the dependence on temperature of the lasing wavelength of a GaInNAs LD [30]. To eliminate the effects of heating during lasing operation, the measurements were of a device in pulsed operation at 5 kHz with a 0.4% duty cycle. The T_0 of this LD was 214 K. From the slope in the figure, $d\lambda/dT$ was measured as 0.39 nm/°C.

Figure 11.20 summarizes the dependence of measured $d\lambda/dT$ values on T_0 for 1.2-μm- and 1.3-μm-range GaInNAs LDs (filled symbols) [30]. This figure also shows experimental data on 1.3-μm-range conventional GaInPAs LDs (open circles) and a curve calculated by using Equations 11.7–11.9 (dashed line). The average $d\lambda/dT$ for GaInPAs LDs with T_0 values around 70 K is approximately 0.5 nm/°C, which is almost within the limit

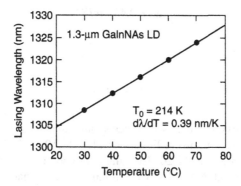

Figure 11.19 Temperature dependence of lasing wavelength for a GaInNAs LD. (From Ref. [30].)

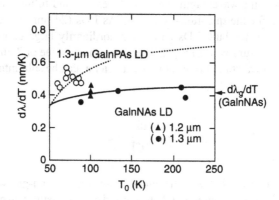

Figure 11.20 The dependence of $d\lambda/dT$ on T_0 for 1.2-μm- and 1.3-μm-range GaInNAs LDs (filled symbols) and 1.3-μm-range conventional GaInPAs LDs (open circles). (From Ref. [30].)

described above. In contrast, $d\lambda/dT$ for GaInNAs LDs is almost independent of T_0. Data for both types of LDs are scattered to roughly the same degree. So, $d\lambda/dT$ of GaInNAs LDs does not depend on the LD structure, i.e., whether the structure is gain-guided or index-guided, in the same way as for GaInPAs LDs. The average $d\lambda/dT$ for GaInNAs LDs is close to the temperature shift of the band-gap wavelength of GaInNAs $d\lambda_g/dT$ (0.42 nm/°C). This is lower than both the value for a GaInPAs LD and the required value. We can thus conclude that the lasing wavelength of a GaInNAs LD with high T_0 has sufficiently small dependence on temperature.

To investigate the reason behind the low $d\lambda/dT$ value for GaInNAs LDs, parameters a_1 and a_2 of the GaInNAs LDs were experimentally estimated in the same way as was used by Higashi et al. [29]. This produced the following parameters of GaInNAs LDs [30]:

$$a_1 = 0.48 \text{ nm/}°\text{C} \tag{11.10}$$

and

$$a_2 = 7 \text{ nm} \tag{11.11}$$

Both of these parameters are much smaller than those of GaInPAs LDs. A curve calculated by using Equations 11.7, 11.10, and 11.11 is given as the solid line in Figure 11.20. This line is in good agreement with experimental data on GaInNAs LDs. We can thus conclude that Equation 11.7 is also applicable to GaInNAs LDs.

The reason for the small a_1 and a_2 parameters of GaInNAs LDs has been determined and is explained below. Parameter a_2 corresponds to the blueshift in lasing wavelength that is produced by band-filling. Furthermore, the a_2 value for the quantum-well GaInPAs LDs (24 nm) agrees with the value expected for bulk LDs and is extraordinarily large compared with values for quantum-well LDs [9]. Therefore, the a_2 value of 7 nm GaInNAs LDs is not unrealistic. Furthermore, parameter a_1 can be split into two terms in the following way:

$$a_1 = \frac{d\lambda_g}{dT} + a_3 \tag{11.12}$$

The first term, $d\lambda_g/dT$, is the temperature shift of band-gap wavelength. The second term, a_3, corresponds to the redshift of lasing wavelength due to reduced band-filling. Parameter a_1 is estimated by measuring the temperature dependence of the gain-peak wavelength under a constant injected current. Since the current for nonradiative recombination through carrier overflow, Auger recombination, etc. increases with temperature, the constant injected current implies that the current that contributes to lasing falls as the temperature increases. This reduces band filling. The lasing wavelength is thus redshifted by more than $d\lambda_g/dT$ when the temperature is increased, even if a constant current is being injected into the LD. Since the small a_2 parameter for the GaInNAs LDs indicates that a band filling has little effect on the on the lasing wavelength of a GaInNAs LD, we can expect a_3, and thus a_1, to also be small. The value of a_3 is estimated as 0.06 nm/°C by subtracting $d\lambda_g/dT$ (0.42 nm/°C) from a_1 (0.48 nm/°C) according to Equation 11.12. Hence, the low $d\lambda/dT$ of the GaInNAs LDs is a result of small a_3 (0.06 nm/°C) and a_2/T_0 (0.035–0.07 nm/°C in the 100–200-K T_0 range) as well as of mutual cancellation. As a result of the cancellation, $d\lambda/dT$ of the GaInNAs LDs is in almost complete agreement with $d\lambda_g/dT$

and is independent of T_0. The reason for the small a_1, a_2, and a_3 parameters of GaInNAs LDs is that band filling has little effect on the lasing wavelength. This is a further consequence of the deep quantum well, which is one characteristic of the GaInNAs materials system.

11.3.1.5 High-speed operation

This subsection covers the high-speed operation of GaInNAs LDs. The explosive growth of Internet/intranet traffic has created a strong demand for cost-effective, high-speed light sources to be used in metropolitan-area and local access networks and other types of high-speed data links. The frequency of relaxation oscillation (f_r) is a major factor that restricts the high-speed operation of laser diodes. It is represented as

$$f_r = \frac{1}{2\pi} \sqrt{\frac{S \cdot V \cdot \Gamma \cdot \frac{dg}{dn}}{\tau_p}} \qquad (11.13)$$

where S is the photon density, V is the group velocity (the velocity of light in the material), Γ is the optical confinement factor, dg/dn is the differential gain (where g is the optical gain and n is the carrier density), and τ_p is the photon lifetime. To achieve a high f_r the material should have a large differential gain. As was noted above, GaInNAs has a large optical gain. We would thus expect GaInNAs to also have a large differential gain. Shimizu et al. [31] have experimentally examined the differential gain of GaInNAs SQW-LDs. They estimated it as 1.06×10^{15} cm^{-2}. As could be expected, this value is greater than those for other 1.3-μm-range LDs that have active layers of materials other than GaInNAs.

An uncooled GaInNAs SQW-LD was used in 10-Gb/s direct modulation. In Figure 11.21, (a) and (b) show the 10-Gb/s eye diagrams measured at 25°C and 80°C, with a modulation-current swing of 36 mA (*p-p*) and the dc bias set at 30% above the threshold [25]. Reflecting its high differential gain, this *single QW* device achieved an eye opening even at 80°C. This could never be expected of a conventional GaInPAs/InP or AlGaInAs/InP QW laser. Increasing the number of GaInNAs quantum wells and decreasing the cavity volume will further improve the high-speed performance.

11.3.2 Vertical-Cavity Surface-Emitting Laser

11.3.2.1 Background

Long-wavelength vertical-cavity surface-emitting lasers (VCSEL) emitting at 1.3 or 1.55 μm are very attractive as light sources for low-cost optical

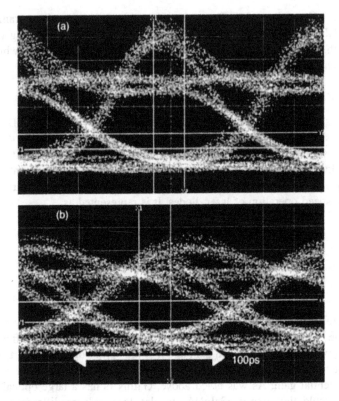

Figure 11.21 Eye diagrams (10 Gb/s) produced by a GaInNAs LD, as measured at (a) 25°C and (b) 80°C. (From Ref. [25].)

access network and high-speed data links. While these wavelengths are accessible to GaInPAs/InP-based semiconductor active layers, the limited contrast between the refractive indices of these materials prevents the formation of highly reflective distributed Bragg reflector (DBR) mirrors. Two techniques have been used to circumvent this obstacle: dielectric DBR combined with backside etching and processing [32–34], and wafer fusion to form the active layer, with GaAs/AlAs DBRs grown on separate GaAs substrates [35–37]. Devices produced by the former technique suffer from nonuniform current injection around the dielectric (insulating) mirrors and require a thick cavity layer for mechanical stability, thereby introducing excess optical loss. The latter technique requires multiple stages of epitaxial growth as well as complicated bonding and substrate etching, and the long-term stability of bonded interfaces so near to the active layer is yet to be determined. In either case, continuous wave (CW) output power and high-temperature performance have not been sufficient for practical application, particularly at 1.3 μm.

One novel alternative is the use of a GaInNAs quantum-well active layer [38, 39], which is grown pseudomorphically on a highly reflective GaAs/AlAs DBR mirror over a GaAs substrate. This allows integration of a long-wavelength active layer with high-gain and high-contrast DBRs in a single stage of epitaxial growth. Note that the deep quantum wells seen in the GaInNAs active layers of edge emitters, as described in section 11.3.1.1, are also seen in the VCSEL structure, which thus gains the same advantages.

11.3.2.2 Device performance

Laser operation of a GaInNAs VCSEL diode was achieved for the first time in 1997 by Hitachi, Ltd. Lasing was at room temperature and pulsed, with a wavelength of 1.2 μm [40]. In 2000, 1.3-μm-range, room-temperature, continuous-wave operation was realized by a joint group formed by Sandia National Laboratories and Cielo Communications, Inc. [41]. The first demonstration of 10-Gb/s, 10-km communication was by this group. In 2001, GaInNAs VCSELs went on the market. At the time of writing in 2003, they were supplied by several vendors.

Figure 11.22 is a schematic view of the first GaInNAs VCSEL diodes to be demonstrated. These devices have an etched-pillar structure [40]. The bottom mirror was a 25.5-period, Si-doped GaAs/AlAs n-type DBR with a center wavelength near 1.2 μm. A single 7-nm-thick $Ga_{0.7}In_{0.3}N_{0.004}As_{0.996}$ QW was located at the center of the λ (one wavelength) GaAs cavity, which was doped to form a pin diode for current injection. Above the cavity was a 21-period, Be-doped GaAs/AlAs p-DBR with parabolically graded interfaces to reduce series resistance. A Ti/Au contact electrode, combined with a $p+$ phase-matching layer, increased the reflectance of the p-DBR. Circular electrodes with diameters of approximately 25 and 45 μm were used to define the pillar's dimensions, and device isolation was done by formation of the pillar. Electrical contact with the n-substrate was made by using indium solder, and light was emitted through the bottom surface of the substrate, since the top surface was covered by a metal contact.

The VCSELs were operated at room temperature with a pulse width of 800 ns and a repetition rate of 5 kHz. Light was coupled with a broad-area GaInAs calibrated photodetector or brought, through a multimode fiber, to an optical spectrum analyzer. Figure 11.23 shows the output power vs. injection current for two devices of two different sizes in room-temperature operation without heat sinks. Both devices were emitting at approximately 1.18 μm. The device with a 45-μm electrode diameter had a threshold current I_{th} of 61 mA, corresponding to a current density of 3.8 kA/cm². The maximum output power was 5 mW. The voltage–current characteristics are also shown; the high threshold voltage of 12 V, due primarily to the series resistance of the p-DBR, probably precluded continuous-wave oper-

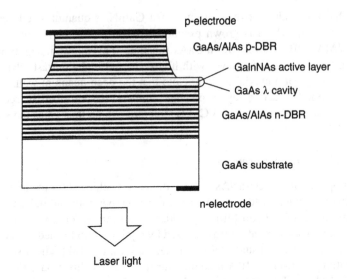

p-electrode

GaAs/AlAs p-DBR

GaInNAs active layer

GaAs λ cavity

GaAs/AlAs n-DBR

GaAs substrate

n-electrode

Laser light

Figure 11.22 Schematic diagram of the first GaInNAs VCSEL diodes to be demonstrated.

ation. By comparison, a device with a diameter of approximately 25 μm exhibited a threshold current of 22.3 mA and a pulsed output-power level above 1.1 mW.

Figure 11.24 shows the dependence on temperature of I_{th} for one of the VCSELs described above. Threshold current varied by only 20% between 20°C and 70°C. This is a consequence of the detuning of gain and cavity-resonant wavelengths, and of an active layer with excellent high-temperature performance. This could never be expected of conventional GaInPAs/InP active layers, with their poor high-temperature performance.

The gain bandwidth of GaInNAs active layers was investigated by forming planar VCSELs that consisted of stacked layers on a wafer and then applying photo-pumping to these structures. Such structures were used because of the simplicity of their design and fabrication, from which the need for doping, forming of contacts, and device isolation are eliminated. The need for optimized current conduction through the DBR hetero-interfaces is also eliminated, since electron-hole pairs are generated directly in the cavity region and then diffuse to the active layer to radiatively recombine in the process of stimulated emission. However, these structures introduce additional constraints by requiring that (1) the top mirror be transparent at the pump wavelength and (2) the cavity or active layer have efficient absorption at the pump wavelength. Thus, transparency to a Ti:sapphire pump (wavelength ≈ 750 nm) was obtained by using a top mirror in the form of a ten-period SiO_2/TiO_2 DBR; the bottom mirror was an (Al)GaAs/AlAs DBR. Spectra and output power were measured at several

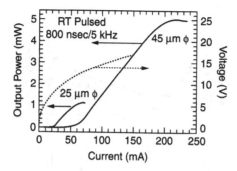

Figure 11.23 Light-output power and voltage against injection current for GaInNAs VCSELs of two different sizes. (From Ref. [40].)

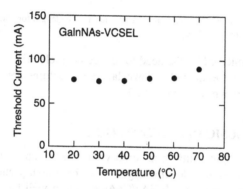

Figure 11.24 The dependence of threshold current on temperature for a GaInNAs VCSEL.

points across the wafer. A radial layer-thickness gradient across the grown wafer introduces a position-dependent shift in the cavity wavelength and gain spectrum. Disregarding secondary effects, the position of minimum threshold pump power should correspond with the point where the gain spectrum is optimally aligned with the cavity mode. Figure 11.25 is a plot of the threshold pump power against lasing wavelength for the wafer [42]. Laser oscillation was seen over an extremely wide range, from 1.146 μm to 1.256 μm, which may reflect the large amount of available gain and the wide gain bandwidth of the GaInNAs active layer. This 110-nm span may be the widest yet achieved by a VCSEL array. This result means that GaInNAs has a large gain bandwidth as well as a large gain.

In the same way as 0.98-μm GaInAs VCSELs grown on a GaAs substrate, most practical GaInNAs VCSELs include a current-confinement aperture of AlO_x above the GaInNAs active layer [43]. The AlO_x is formed from an epitaxially grown Al(Ga)As layer by selective oxidation. The aperture leads to a low-threshold current and single optical transverse mode, qualities that are useful in practical application. While these device struc-

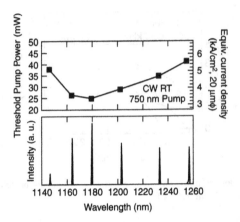

Figure 11.25 The threshold pump power against lasing wavelength for photo-pumped GaIn-NAs VCSELs. (From Ref. [42].)

tures are complicated by the need to reduce heat, electrical resistance, and capacitance, their epitaxially grown layered structures are essentially the same as was shown in Figure 11.22.

11.4 CONCLUSIONS AND OUTLOOK

GaInNAs — a light-emitting material with a band-gap energy that corresponds to the long-wavelength range — can be grown pseudomorphically on a GaAs substrate. A GaInNAs/GaAs quantum well has an ideal band lineup for laser-diode application. This facilitates the formation of very deep quantum wells, especially in the conduction band. The deep quantum wells bring about many advantages in terms of laser performance, including high-speed operation and excellent temperature stability of threshold current and lasing wavelength. GaInNAs is also applicable to the fabrication of long-wavelength VCSELs, because a GaInNAs active layer can be grown on a highly reflective GaAs/AlAs DBR mirror over a GaAs substrate in a single stage of epitaxial growth. These characteristics of GaInNAs lasers fulfill market demands for cost-effective, high-performance light-sources to be used in optical access network and high-speed data links. In 2001, GaInNAs VCSELs went on the market. At the time of writing in 2003, they were supplied by several vendors.

As has been presented, GaInNAs and other III-N-V alloys are very promising and their future looks bright. However, maturations in crystal growth and device fabrication are necessary before the tree of III-N-V materials can bear full fruit. Thus, further effort is required of the research community.

To develop the III-N-V materials in the long term, new applications and markets are necessary. Growth on a Si wafer is promising. Integration with Si integrated circuits will bring a huge market to III-V alloys. Even in discrete devices, the use of large and cheap wafers is effective in reducing the cost. In the future, environmental issues might play a key role, because the GaAs wafer is toxic. Development of the photonics crystal is an interesting and hopeful theme in III-V research. The III-N-V materials can also be applied to this field.

References

1. Baillargeon, J.N., Cheng, K.Y., Hofler, G.E., Pearch, P.J., and Heigh, C., "Luminescence Quenching and the Formation of the GaPN Alloy in GaP with Increasing Nitrogen Content," *Appl. Phys. Lett.* 60: 2540–2542 (1992).
2. Igarashi, O., "Heteroepitaxial Growth of GaNP on Sapphire Substrates," *Jpn. J. Appl. Phys.* 31: 3791–3793 (1992).
3. Miyoshi, S., Yaguchi, H., Onabe, K., Ito, R., and Shiraki, Y., "Metalorganic Vapor Phase Epitaxy of GaPN Alloys on GaP," *Appl. Phys. Lett.* 63: 3506–3508 (1993).
4. Sato, M., and Weyers, M., "GaAsN Alloys: Growth and Optical Properties," in *19th Int. Symp. GaAs and Related Compound Semiconductors*, Karuizawa, Japan, 1992; in *Inst. Phys. Conf. Ser.*, vol. 129 (Philadelphia: Institute of Physics, 1993) 555–560.
5. Kondow, M., Uomi, K., Niwa, A., Kitatani, T., Watahiki, S., and Yazawa, Y., "A Novel Material of GaInNAs for Long-Wavelength-Range Laser Diodes with Excellent High-Temperature Performance," in *1995 Solid State Devices and Materials*, Osaka, 1995, p. D-8-1.
6. Kondow, M., Uomi, K., Niwa, A., Kitatani, T., Watahiki, S., and Yazawa, Y., "GaInNAs: A Novel Material for Long-Wavelength-Range Laser Diodes with Excellent High-Temperature Performance," *Jpn. J. Appl. Phys.* 35: 1273–1275 (1996).
7. Kondow, M., Uomin, K., Hosomi, K., and Mozume, T., "Gas-Source Molecular Beam Epitaxy of GaNAs Using an n-Radical as the N Source," *Jpn. J. Appl. Phys.* 33: L1056–L1058 (1994).
8. Sato, S., "Room Temperature Operation of InGaNAs/InGaP DH Lasers Grown by MOCVD," in *57th Autumn Meeting Jpn. Soc. Appl. Phys.*, Fukuoka, Japan, 1996, p. 951.
9. Kondow, M., Kitatani, T., Larson, M.C., Nakahara, K., Uomi, K., and Inoue, H., "Gas-Source MBE of GaInNAs for Long-Wavelength Laser Diodes," *J. Crystal Growth* 188: 255–259 (1998).
10. Sakai, S., Ueta, Y., and Terauchi, Y., "Band-gap Energy and Band Lineup of III-V Alloy Semiconductors Incorporating Nitrogen and Boron," *Jpn. J. Appl. Phys.* 32: 4413–4417 (1993).
11. Phillips, J.C., *Bonds and Bands in Semiconductors* (New York: Academic Press, 1973), 54.
12. Kitatani, T., Kondow, M., Kikawa, T., Yazawa, Y., Okai, M., and Uomi, K., "Analysis of Band Offset in GaNAs/GaAs by X-ray Photoelectron Spectroscopy," *Jpn. J. Appl. Phys.* 38: 5003–5006 (1999).
13. Ding, S.A., Barman, S.R., Horn, K., Yang, H., Yang, B., Brandt, O., and Ploog, K., "Valence Band Discontinuity at a Cubic GaN/GaAs Heterojunction Measured by Synchrotron-Radiation Photoemission Spectroscopy," *Appl. Phys. Lett.* 70: 2407–2409 (1997).
14. Sakai, S., and Abe, T., "Band Lineup of Nitride-Alloy Heterostructures," in *41st Spring Meeting Jpn. Soc. Appl. Phys.*, Tokyo, 1994, p. 186 (in Japanese).

15. Pan, Z., Li, L.H., Lin, Y.W., Sun, B.Q., Jiang, D.S., and Ge, W.K., "Conduction Band Offset and Electron Effective Mass in GaInNAs/GaAs Quantum-Well Structures with Low Nitrogen Concentration," *Appl. Phys. Lett.* 78: 2217–2219 (2001).
16. Hetterich, M., Dawson, M.D., Egorov, A.Y., Bernklau, D., and Riechert, H., "Electronic States and Band Alignment in GaInNAs/GaAs Quantum-Well Structures with Low Nitrogen Content," *Appl. Phys. Lett.* 76: 1030–1032 (2000).
17. Kondow, M., Fujisaki, S., Shirakata, S., Ikari, T., and Kitatani, T., "Electron Effective Mass of $Ga_{0.7}In_{0.3}N_xAs_{1-x}$," in *30th International Symposium on Compound Semiconductors*, San Diego, 2003, p. MB 3.8.
18. Skierbiszewski, C., Perlin, P., Wisniewski, P., Suski, T., Walukiewicz, W., Shan, W., Ager, J.W., Haller, E.E., Geisz, J.F., Friedman, D.J., Olson, J.M., and Kurtz, S.R., "Effect of Nitrogen-Induced Modification of the Conduction Band Structure on Electron Transport in GaAsN Alloys," *Phys. Stat. Solidi B* 216: 135–140 (1999).
19. Zah, C.E., Bhat, R., Pathak, B.N., Favier, F., Lin, W., Wang, M.C., Andreadakis, N.C., Hwang, D.M., Koza, M.A., Lee, T.P., Wang, Z., Darby, D., Flanders, D., and Hsieh, J.J., "High Performance Uncooled 1.3-µm $Al_xGa_yIn_{1-x-y}As$/InP Strained-Layer Quantum-Well Lasers for Subscriber Loop Applications," *IEEE J. Quantum Electron.* 30: 511–523 (1994).
20. Shoji, H., Uchida, T., Kusunoki, T., Matsuda, M., Kurakake, H., Yamazaki, S., Nakajima, K., and Ishikawa, H., "Fabrication of $In_{0.25}Ga_{0.75}As$/GaInPAs Strained SQW Lasers on $In_{0.05}Ga_{0.95}As$ Ternary Substrate," *IEEE Photonics Technol. Lett.* 6: 1170–1172 (1994).
21. Kurakake, H., Uchida, T., Kubota, K., Ogita, S., Soda, H., and Yamazaki, S., "High T_0 1.3-µm InGaAs Strained Single Quantum Well Laser with InGaP Wide Band-Gap Clad Layers," in *14th IEEE Int. Semiconductor Laser Conf.* (New York: IEEE,1994), 24–25.
22. Kondow, M., Nakatsuka, S., Kitatani, T., Yazawa, Y., and Okai, M., "Room-Temperature Pulsed Operation of GaInNAs Laser Diodes with Excellent High-Temperature Performance," *Jpn. J. Appl. Phys.* 35: 5711–5713 (1996).
23. Kondow, M., Aoki, M., Nakatsuka, S., Kitatani, T., and Tsuji, S., "High-Speed Operation of GaInNAs Edge Emitting Lasers," in *IEEE/LEOS 2002 Annual Meeting Proceedings*, Glasgow, 2002, p. TuC1.
24. Tansu, N., and Mawst, L.J., "High Performance 1300-nm dilute nitride Quantum Well Lasers by MOCVD," in *18th IEEE Int. Semiconductor Laser Conference Proceedings*, Garmisch-Partenkirchen, Germany, 2002, p. TuA1.
25. Aoki, M., Kondow, M., Kitatani, T., Nakatsuka, S., Kudo, M., and Tsuji, S., "Uncooled Reliable 10-Gbit/s Operation of GaInNAs-SQW Laser," in *18th IEEE Int. Semiconductor Laser Conference Proceedings*, Garmisch-Partenkirchen, Germany, 2002, p. TuC4.
26. Suemune, I., "Theoretical Estimation of Leakage Current in II-VI Heterostructure Lasers," *Jpn. J. Appl. Phys.* 31: L95–98 (1992).
27. Kitatani, T., Kondow, M., Nakahara, K., and Tanaka, T., "Recent Progress in GaInNAs Laser," *IEICE Trans. Electron.* E83-C: 830–837 (2000).
28. Kitatani, T., Nakahara, K., Kondow, M., Uomi, K., and Tanaka, T., "High Characteristic Temperature (over 200 K) of a 1.3-µm GaInNAs/GaAs Single-Quantum-Well Laser Diode," *Jpn. J. Appl. Phys.* 39: L86–87 (2000).
29. Higashi, T., Yamamoto, T., Ogita, S., and Kobayashi, M., "Experimental Analysis of Temperature Dependence of Oscillation Wavelength in Quantum-Well FP Semiconductor Lasers," *IEEE J. Quantum Electron.* 34: 1680–1689 (1998).
30. Kondow, M., Kitatani, T., Nakahara, K., and Tanaka, T., "Temperature Dependence of Lasing Wavelength in a GaInNAs Laser Diode," *IEEE Photonics Technol. Lett.* 12: 777–779 (2000).
31. Shimizu, H., Kumada, K., Uchiyama, S., and Kasukawa, A., "High-Performance CW 1.26-µm GaInNAsSb-SQW Ridge Lasers," *IEEE J. Selected Top. Quantum Electron.* 7: 355–363 (2001).

32. Baba, T., Yogo, Y., Suzuki, K., Koyama, F., and Iga, K., "Near Room Temperature Continuous Wave Lasing Characteristics of GaInAsP/InP Surface Emitting Laser," *Electron. Lett.* 29: 913–914 (1993).

33. Uomi, K., Yoo, S.J.B., Scherer, A., Bhat, R., Andreadakis, N.C., Zah, C.E., Koza, M.A., and Lee, T.P., "Low Threshold Room Temperature Pulsed Operation of 1.5-µm Vertical-Cavity Surface-Emitting Lasers with an Optimized Multi-Quantum Well Active Layer," *IEEE Photonics Technol. Lett.* 6: 317–319 (1994).

34. Uchiyama, S., Yokouchi, N., and Ninomiya, T., "Continuous-Wave Operation up to 36°C of 1.3-µm GaInAsP-InP Vertical-Cavity Surface-Emitting Lasers," *IEEE Photonics Technol. Lett.* 9: 141–142 (1997).

35. Dudley, J.J., Babic, D.I., Mirin, R., Yang, L., Miller, B.I., Ram, R.J., Reynolds, T., Hu, E.L., and Bowers, J.E., "Low Threshold Wafer Fused Long Wavelength Vertical Cavity Lasers," *Appl. Phys. Lett.* 64: 1463–1465 (1994).

36. Babic, B.I., Streubel, K., Mirin, R.P., Margalit, N.M., Bowers, J.E., Hu, E.L., Mars, D.E., Yang, L., and Carey, K., "Room-Temperature Continuous-Wave Operation of 1.54-µm Vertical-Cavity Lasers," *IEEE Photonics Technol. Lett.* 7: 1225–1227 (1995).

37. Qian, Y., Zhu, Z.H., Lo, Y.H., Hou, H.Q., Wang, M.C., and Lin, W., "1.3-µm Vertical-Cavity Surface-Emitting Lasers with Double-Bonded GaAs-AlAs Bragg Mirrors," *IEEE Photonics Technol. Lett.* 9: 8–10 (1997).

38. Kondow, M., Nishinura, S., Shinoda, K., and Uomi, K., "Vertical Cavity Surface Emitting Laser, Optical Transmitter-Receiver Module Using the Laser, and Parallel Processing System Using the Laser," Japanese patent AP07-340520 (27 Dec. 1995), U.S. patent 5 912 913 (15 June 1999).

39. Iga, K., "Vertical Cavity Surface Emitting Lasers Based on InP and Related Compounds: Bottleneck and Corkscrew," in *Proceedings of 8th Int. Conf. InP and Related Materials*, Schwabisch Gmund, Germany, 1996, ThA1-1.

40. Larson, M.C., Kondow, M., Kitatani, T., Nakahara, K., Tamura, K., Inoue, H., and Uomi, K., "GaInNAs/GaAs Long-Wavelength Vertical-Cavity Surface-Emitting Laser Diodes," *IEEE Photonics Technol. Lett.* 10: 188–190 (1998).

41. Choquette, K.D., Klem, J.F., Fischer, A.J., Blum, O., Allerman, A.A., Fritz, I.J., Kurtz, S.R., Breiland, W.G., Sieg, R., Geib, K.M., Scott, J.W., and Naone, R.L., "Room Temperature Continuous Wave InGaAsN Quantum Well Vertical-Cavity Lasers Emitting at 1.3 µm," *Electron. Lett.* 36: 1388–1389 (2000).

42. Larson, M.C., Kondow, M., Kitatani, T., Tamura, K., Yazawa, Y., and Okai, M., "Photopumped Lasing at 1.25 µm of GaInNAs/GaAs Multiple Quantum Well Vertical-Cavity Surface-Emitting Lasers," *IEEE Photonics Technol. Lett.* 9: 1549–1551 (1997).

43. Steinle, G., Mederer, F., Kicherer, M., Michalzik, R., Kristen, G., Egorov, A.Y., Riechert, H., Wolf, H.D., and Ebeling, K.J., "Data Transmission up to 10 Gbit/s with 1.3-µm Wavelength InGaAsN VCSELs," *Electron. Lett.* 37: 632–634 (2001).

CHAPTER 12

Recombination Processes in GaInNAs Long-Wavelength Lasers

R. FEHSE and A.R. ADAMS

University of Surrey, United Kingdom

12.1 INTRODUCTION

A major commercial force driving the development of GaInNAs alloys is the need to develop semiconductor lasers on GaAs substrates that can operate in the 1.3–1.55-μm wavelength range, which is used for optical-fiber communications. For example, monolithic vertical-cavity surface-emitting lasers (VCSELs), with their cheaper batch processing and simple optics, can be grown on GaAs. Of crucial importance to the operation of diode lasers is the value of the current density, J_{th}, that must be passed through the device in order to reach the required optical gain. Also, since the lasers must be able to operate over quite a wide temperature range, the variation of J_{th} with temperature is another important parameter. It is therefore extremely important to be able to quantify the electron-hole recombination mechanisms that lead to J_{th} and their variation with temperature. It should be emphasized that this does not just involve identifying which recombination mechanisms may be present, but also determining their relative magnitudes at the injected carrier density required by the lasing process.

This chapter is primarily concerned with presenting results obtained by the authors using an experimental method, which enables the determination of the monomolecular-, radiative-, and Auger-related contributions to the threshold current. This was achieved by simultaneously monitoring the stimulated emission from the end facets and the spontaneous emission (SE) from a window in the laser substrate. These were studied as a function of injection current, temperature, and hydrostatic pressure. The devices studied were MBE (molecular-beam epitaxy)- and MOVPE (metalorganic vapor-phase epitaxy)-grown 1.3-μm GaInNAs/GaAs-based edge-emitting lasers and MBE-grown VCSELs produced by Infineon Technologies AG, Munich, Germany. We will first concentrate on the results for edge-emitting lasers, and finally, in Section 12.6, we show how the data obtained can be applied to studies of the characteristics of VCSELs with similar active regions.

12.2 DETAILS OF THE EDGE-EMITTING DEVICES

The MBE-grown GaInNAs-based laser structures used in this study were grown on n^+-GaAs substrates. Conventional solid sources were used, with the only modification being a radio frequency- (RF) coupled plasma source to generate reactive nitrogen from N_2. The active GaInNAs quantum wells (QW) were grown at low temperature, as described elsewhere [1], and were sandwiched within a 300-nm-thick, undoped GaAs waveguide layer. The p- and n-type cladding layers are composed of 1.5-μm-thick $Al_{0.3}Ga_{0.7}As$, doped with Be and Si to a concentration of 4×10^{17} and 5×10^{17} cm^{-3}, respectively. The 0.6-μm-thick p-type GaAs contact layer is doped to about 1×10^{19} cm^{-3} in the top 200 nm.

The GaInNAs QWs have an In content of about 36% and a N content of about 1.7%. The composition of the wells can be conveniently set after separate growth runs for the In and N fluxes, since it has been found that in MBE, the In and N are incorporated independently of each other. The necessary thermal annealing of the active region was effectively performed during the growth of the p-AlGaAs layer, which takes place at 680°C.

Three different laser structures were grown. The first was a 3-QW structure, where the wells had a thickness of 6.2 nm. These wells were separated by 25-nm barrier layers consisting of InGaAsN with 1.8% N and ≈5% In content to make them approximately lattice-matched to GaAs. The second was a single-quantum-well (SQW) structure, where the QW was 6.4-nm thick and had the same composition as the 3QW structure, but within GaAs. The SQW structures were fabricated into 50-μm-wide and 700-μm-long broad-area devices, and the 3QW structures were processed into 4-μm ridge-waveguide devices (350-μm width and 700-μm length). The ridges were processed using an argon ion dry-etching technique and passivated with RF-sputtered SiN_x [2].

The third laser structure considered in this chapter was grown by low-pressure MOVPE on n^+-GaAs substrates. Triethylgallium (TEG), trimethylaluminum (TMA), trimethylindium (TMI), tertiary butyl arsine (TBA), and 1,1-dimethyl hydrazine (uDMHy) were used as precursors, and the samples were grown at temperatures between 520°C (QW) and 680°C (cladding). The structure consists of two GaInNAs quantum wells between two 150-nm-thick undoped GaAs waveguide layers. The p- and n-type cladding layers are composed of 1.5-μm-thick $Al_{0.38}Ga_{0.62}As$, doped with C and Te to 5×10^{17} cm^{-3}. The 0.150-μm-thick p-type GaAs contact layer is heavily doped with Zn to about 1×10^{19} cm^{-3}. The active region consists of two 8-nm-thick GaInNAs QWs separated by a 20-nm-thick GaAs barrier, where the In content in the wells is 34% and N content lies between 1 and 1.7%. The composition values of the quaternary QW could not be determined more precisely because, in MOVPE growth, the incorporation of N decreases with increasing In content [3], and determination of the In and N compositions in such quantum-well structures is difficult [4]. The structures were fabricated into 50-μm-wide and 700-μm-long broad-area devices.

12.3 RADIATIVE CURRENT

The integrated SE rate of a semiconductor laser is directly proportional to the radiative current flowing through the device. Measuring the integrated SE rate of a laser as a function of injected current, temperature, and pressure can therefore reveal useful information about the radiative processes in the device. In this section, the methods used for measuring and analyzing the SE are introduced, and the temperature dependence of the radiative current for various GaInNAs-based edge-emitting lasers are compared. It will be shown that the characteristic temperature of the radiative current is significantly larger than the characteristic temperature of the total current at threshold, indicating that temperature-sensitive nonradiative recombination must be present in the devices.

12.3.1 Collecting the Spontaneous Emission

The SE was collected from a window etched into the substrate contact of the laser chips and detected via an optical fiber and an optical spectrum analyzer (OSA). This was achieved by placing the laser chips in a specially designed laser clip consisting of a copper base plate, which provides both heat sinking and the connection to the n-side contact. In the front part of the base plate, a 150-μm hole is drilled, which holds an optical fiber mounted flush to surface of the base. In addition, a 900-μm-diameter counterbored hole is drilled, which holds the fiber sheathing to ensure high mechanical stability. The fiber is held in place using a high-strength epoxy-based glue. The laser chip is now placed, n-side down, on top of the base plate with its substrate window carefully aligned with the optical fiber core. The window in the substrate of the device has a typical dimension of approximately 50–70-μm width and 150–200-μm length and is processed using an ion-beam milling technique that has been described elsewhere [5]. The laser chip is now held in place by means of a metal spring clip, which connects to the p-contact of the device and presses the chip tightly to the base plate. The latter is very important, as it is absolutely vital for the experiments that the relative position of the contact window and the optical fiber (and hence the light-collection efficiency) stay constant throughout the experiment. Extensive tests were performed to ensure that the setup described here reliably met this requirement.

In the configuration described, the SE will not have undergone any significant amplification or absorption, and hence the collected pure SE will be directly proportional to the radiative current. The SE spectra, measured with an optical spectrum analyzer, are numerically integrated to obtain the integrated spontaneous emission rate, L_{SE}, according to Equation 1:

$$L_{SE} = \frac{1}{hc\Delta\lambda} \int_{\lambda_1}^{\lambda_2} P_{SE}(\lambda)\lambda \, d\lambda \qquad (12.1)$$

where h is the Planck constant, c is the velocity of light, $\Delta\lambda$ is the spectral resolution of the optical spectrum analyzer, λ_1 and λ_2 are the integration limits determined by the sweep limits of the OSA, and P_{SE} is the measured optical power of the spontaneous emission at wavelength λ.

By measuring the spontaneous recombination from a window in the substrate, the spontaneous recombination of electrons with holes having x- or y-like transverse electric (TE) character is detected, assuming negligible photon scattering. Radiative recombination via z-like transverse magnetic (TM) valence-band states will not be detected because the polarization vector is in the direction of light collection. The resulting integrated spontaneous emission rate would therefore underestimate the total radiative current. However, in the GaInNAs lasers studied here, this is not expected to be a problem. Given the large heavy-hole ($^1/_2$x, $^1/_2$y) to light-hole ($^1/_6$x, $^1/_6$y, $^2/_3$z) subband splitting (=114 meV) in these compressively strained devices, the hole density in the light-hole band is expected to be negligible. From gain calculations (as discussed later in section 12.4.2), it was calculated that recombination via light-hole states, resulting in spontaneous emission of TM-like character, was less than 0.05% of the TE-like spontaneous emission occurring via the heavy-hole states. Hence carrier recombination effectively occurs via the heavy-hole band and is consequently of only x- and y-like character. It is therefore assumed that the measured SE spectra are representative of the total spontaneous recombination in the devices.

12.3.2 Temperature Variation of the Radiative Current

In Figure 12.1, we plot the measured total spontaneous emission rate L as a function of current for a triple-quantum-well GaInNAs ridge-waveguide device with a ridge width of 4 μm at 288 K and 335 K. It can be clearly observed that L increases with increasing injection current. At threshold (≈17.5 mA at 288 K and ≈29 mA at 335 K), the stimulated emission process pins the carrier concentration and hence the spontaneous emission. Because L is directly proportional to the radiative current flowing through the device, this analysis delivers the temperature dependence of the radiative current. The corresponding characteristic temperature for this process is $T_0(I_{rad}) \approx$ 313 K. In contrast, the characteristic temperature for the total current at threshold is much lower, with a value of $T_0(I_{th}) \approx$ 90 K. Performing the same analysis for the broad-area devices, we obtain $T_0(I_{rad}) \approx$ 230 K and $T_0(I_{th}) \approx$ 85 K for the MBE-grown SQW laser, and we obtain $T_0(I_{rad}) \approx$ 220 K and $T_0(I_{th}) \approx$ 80 K for the MOVPE-grown DQW device. These results clearly indicate that the devices are not dominated by radiative recombination in this temperature range but, rather, by a more-temperature-sensitive, nonradiative recombination path. The origin and magnitude of this nonradiative recombination process will be analyzed in detail later in this chapter.

Figure 12.2 shows plots of the temperature variation of the radiative current in arbitrary units as determined from the integrated SE rate at threshold for the ridge-waveguide (gray triangles), the MBE-grown broad-area laser (black circles), and the MOVPE-grown broad-area laser (open

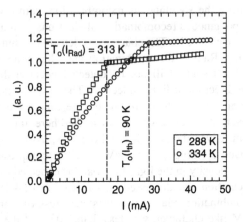

Figure 12.1 Total spontaneous emission rate as a function of injection current at 288 K and 334 K. From the pinning level of the spontaneous emission, it is clear that $T_0(I_{Rad}) = 313$ K while $T_0(I_{th}) = 90$ K. This shows that the strong temperature dependence of the threshold current is due to a nonradiative process over this temperature range.

squares). For comparison, the plots were normalized at T = 300 K. It can be observed that I_{rad} increases in a similar way for all three devices for temperatures T > 250 K. For temperatures between 160 and 240 K, the increase is much reduced, and in the case of the MBE broad-area device, a decrease in I_{rad} is observed, leading to a bump in the I_{rad} vs. T plot, which is a very unusual observation for a QW laser. Qualitatively, this behavior of the radiative current can be observed in this temperature range in most of the GaInNAs lasers that were studied. It is assumed that this effect has its origin in strong nitrogen-induced spatial fluctuations of the band-gap

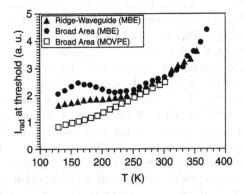

Figure 12.2 Temperature variation of the radiative current in arbitrary units as determined from the integrated spontaneous emission rate at threshold for the ridge waveguide (gray circles), the MBE-grown broad-area laser (black circles), and the MOVPE-grown broad-area laser (open squares). Plots are normalized at T = 300 K.

energy, leading to carrier localization effects, as discussed in Chapter 9. At low temperatures, carriers localize in potential minima higher than the lasing energy and cannot take part in the lasing process, as there is no effective transport mechanism for carriers from one potential minimum to another. Since a quasi-equilibrium cannot be achieved, this leads to a higher carrier density at threshold and, hence, a larger radiative current. With increasing temperature, thermal excitation out of the potential minima can provide the necessary transport mechanism to more nearly achieve quasi-equilibrium. Carriers that contributed to a "wasted" parallel current path at lower temperatures can now take part in the lasing process, which effectively reduces the number of carriers that have to be additionally injected to reach threshold, leading to a weaker increase or even a decrease in the radiative current, as observed in the experiment. A similar effect has been observed for quantum-dot lasers, which can be regarded as extreme examples of strongly inhomogeneous devices [6].

12.4 NONRADIATIVE CURRENT PATHS

In Section 12.3, the temperature dependence of the radiative current was determined (in arbitrary units) by measuring the pure spontaneous emission from a window in the substrate contact. The results showed that the lasers are dominated by a temperature-sensitive nonradiative recombination processes at room temperature (RT). In order to gather more information about the absolute values of the radiative current and the nonradiative processes at threshold, the carrier-density dependence of the threshold current as a function of temperature was investigated.

Assuming that $n = p$ in the active region, the total current injected into a semiconductor quantum-well laser can be written as

$$I = eV \ (An + Bn^2 + Cn^3) \qquad (12.2)$$

where V is the pumped volume of the active region, which is calculated by multiplying the surface area of the contact stripe by the total quantum-well thickness, and e is the electronic charge. Coefficient A is the monomolecular recombination coefficient, describing recombination through defects. Assuming the defect density N_D is small compared with n and that the capture cross-section of one type of carrier is larger than for the other, the recombination rate is directly proportional to N_D and the carrier density n. Coefficient B is the coefficient describing the direct radiative recombination of an electron in the conduction band with a hole in the valence band, which is therefore a bimolecular process and hence proportional to n^2 [7]. Auger recombination involves three carriers (hence $I_{Auger} \propto n^3$) and is described by the Auger coefficient C. Here we ignore carrier leakage due to the very large offset of GaInNAs–QW-to-AlGaAs–cladding-layer in these devices.

Also, at room temperature and at laser threshold, we calculate the energy difference between the electron quasi-Fermi level and the GaAs-barrier conduction band edge to be 198 meV, reducing to 165 meV at 360 K. This large offset should ensure negligible electron occupation probability in the barrier layers, even up to high temperature. Thus, despite the large volume of GaAs material in the barrier and separate confinement heter-structure layers (SCH) layers, the number of carriers in these layers is negligible. This is in contrast to InGaAsP/InP lasers, where — due to the lower conduction-band offset — the presence of electrons in the barrier layers can degrade their high-temperature characteristics [8].

Over a limited current range, we can now rewrite Equation 12.2 as

$$I \propto n^Z \qquad (12.3)$$

where $1 \leq Z \leq 3$.

The value of Z depends on the relative importance of monomolecular, radiative, and Auger recombination processes. As mentioned earlier, the integrated SE rate L_{SE}, as determined experimentally, is directly proportional to the radiative recombination rate, hence $L \propto Bn^2$. Thus $n \propto L^{1/2}$ and $I \propto (L^{1/2})^Z$, and consequently

$$\ln(I) = Z \ln(L^{1/2}) \qquad (12.4)$$

Hence, Z can be determined by evaluating the slope of a plot $\ln(I)$ vs. $\ln(L^{1/2})$. A more-detailed description of the underlying basic theory and assumptions can be found in the literature [9].

12.4.1 The Current-Spreading Problem

An important assumption for the analysis of the carrier-density dependence of the threshold current is that Equation 12.2 can be rewritten over a limited range of n as $I \propto n^Z$. This is only meaningful if the pumped volume, V, of the active region stays constant with varying current and temperature. However, as the carriers have to travel a distance of about 1–2 μm from the current-confining ridge (ridge-waveguide device) or contact stripe (broad-area device) to the active region, a certain amount of current spreading will occur. Because the extent of current spreading depends on the carrier mobility and density, this effect will be sensitive to temperature and current density, possibly resulting in a change of V with varying temperature and current. Accurate determination of the degree of current spreading is a complex task. However, in order to estimate an upper limit for the magnitude of this effect, the maximum current spreading is assumed to be about 1 μm on each side of the ridge or contact stripe. In the case of the broad-area devices with their 50-μm-wide contact stripes, this leads to a maximum variation in the pumped volume of 4%. As will be seen later, this is well within the experimental uncertainty of the analysis presented here and is therefore considered to be acceptable. However, the ridge width of the ridge-waveguide laser is only 4 µm, and assuming the same amount of current spreading as in the previous case, this could result in a maximum error in the pumped volume of up to 50%. This is the reason why all the key analysis presented in the following sections will be based on the broad-area devices.

12.4.2 Testing whether $J_{rad} \propto n^2$

In the analysis of the carrier-density dependence of the threshold current, it is assumed, that $I_{mono} \propto n$, $I_{rad} \propto n^2$, and $I_{Aug} \propto n^3$. These expressions are based upon the Boltzmann approximation, and while it is safe to assume that this is valid for low carrier densities, this might not be the case for the high carrier densities required to reach threshold in a semiconductor laser. To estimate how well the Boltzmann approximation holds, it was calculated how the spontaneous emission rate varies as a function of carrier density for the SQW GaInNAs/GaAs (MBE) broad-area laser. The gain model used in this analysis has been developed (and the calculations were performed) by Tomic and O'Reilly and is discussed in detail in Chapter 3 and elsewhere [10, 11]. The model is based on a realistic ten-band $\mathbf{k \cdot p}$ Hamiltonian. The basis states (which are each doubly spin-degenerate) include the highest valence bands (i.e., heavy hole, light hole, and spin-split-off) and the lowest CB of the $In_yGa_{1-y}As$ host material, and two additional spin-degenerated basis states representing the nitrogen resonant level. The model has been applied successfully to predict the gain characteristics of dilute nitride QW lasers [12, 13].

To determine the nitrogen, conduction, and valence-band edge energies at the Γ-point, model solid theory [10, 14] was first used to give the relative band alignment and valence-band offset of the unstrained GaAs barrier and "host" $In_yGa_{1-y}As$ well materials. The effect of replacing a fraction x of As by N was then calculated, assuming that the unstrained alloy conduction band edge and valence-band edge in the well shift linearly with x, where the parameters necessary were constrained using tight-binding supercell calculations [15]. The built-in hydrostatic and axial strain in the QW was included using interpolated alloy-deformation potentials, and the calculations were further refined by fitting the $\mathbf{k \cdot p}$ results to the interband transitions observed in microphotomodulated reflectance measurements performed on the processed laser devices. The theory neglects direct N-VB coupling and any changes in the spin-orbit splitting with N content. The gain and SE spectra were then calculated using the density matrix formulation [16], including Lorentzian broadening.

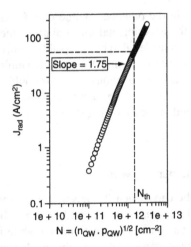

Figure 12.3 Calculated radiative J_{rad} vs. carrier density for a SQW broad-area device.

In Figure 12.3, the calculated radiative current density, J_{rad}, vs. carrier density n is plotted on a logarithmic scale, where n is the square root of the product of the density of electrons, n_{qw}, and holes, p_{qw}, in the quantum well. In the Boltzmann regime, the slope of this plot should be equal to 2. The calculated slope for the SQW GaInNAs device was determined to be 1.75, where the slope was measured over the range $1/2\ n_{th}$ to n_{th} in order to be consistent with the experimentally determined Z in Section 12.4.3. This means that in our case, $J_{rad} \propto n^{1.75}$ over this range. As there are inherent uncertainties involved in band-structure and gain calculations, we have continued to assume that $J_{rad} \propto n^2$ and accept that the possible 12.5% error in n^2 is a measure of our error in determining Z at high current densities.

12.4.3 Power Factor Z at Threshold

In order to obtain information about the nonradiative processes present, the power factor Z at threshold (Z_{th}) was determined as described in Section 12.4. To give an example how Z_{th} is determined, in Figure 12.4 ln(I) vs. $\ln(L_{SE}^{1/2})$ at four different temperatures is plotted as measured for a MBE broad-area device. The spectra were artificially offset for better clarity. Here Z_{th} was determined by measuring the slope of a linear regression of the data over a current range of approximately $1/2\ I_{th}$ up to I_{th}.

In this case, Z_{th} was observed to increase from $Z_{th} \approx 1.58$ at 240 K to $Z_{th} \approx 1.94$ at 300 K. In Figure 12.5, the results from T = 130 K to T = 370 K for the MBE-grown (squares) and the MOVPE-grown (triangles) GaIn-NAs devices are plotted. For comparison, the measured Z_{th} for a 1.5-μm InGaAs-based laser (circles) are included as measured by Sweeney [17]. The InGaAs lasers measured by Sweeney showed no evidence of defect-related, monomolecular recombination and are dominated by Auger recom-

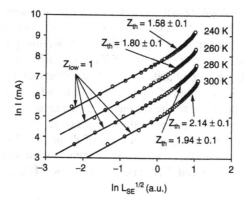

Figure 12.4 Determination of the power factor Z at threshold (Z_{th}) by measuring the slope of a linear regression of the data over a current range of approximately 1/2 I_{th} up to I_{th}. For currents smaller than about $I_{th}/5$, one finds $Z_{low} = 1$, indicating that the device is dominated by monomolecular recombination through defects at low currents.

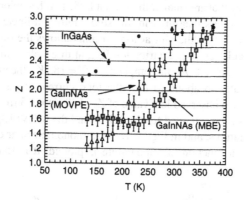

Figure 12.5 Power factor close to threshold Z_{th} as a function of temperature for InGaAs (circles) and GaInNAs MBE (squares) and GaInNAs MOVPE (triangles). The InGaAs devices show a transition from a radiatively dominated I_{th} ($\propto n^2$) at low temperature to an Auger-dominated I_{th} ($\propto n^3$) at high temperature. In contrast, I_{th} for the GaInNAs devices are strongly influenced by monomolecular recombination at low temperature ($\propto n$). With increasing temperature, Z_{th} increases due to the presence of radiative ($\propto n^2$) and Auger ($\propto n^3$) recombination processes.

bination around room temperature. These properties are reflected in the measured Z_{th} of these lasers. For the InGaAs/InP laser, the power factor Z_{th} increases from a value of $Z_{th} \approx 2.1$ at low temperatures, where the device is dominated by radiative recombination, to a value of $Z_{th} \approx 2.9$ at high temperatures, indicating that Auger recombination is the dominant process at high temperature, if leakage currents are negligible. For the MBE-grown GaInNAs/GaAs device, we observe a value of $Z_{th} \approx 1.6$ at low temperatures, which decreases slightly from 130 K to 200 K. This low value of Z_{th} signifies the presence of a significant monomolecular current contribution at threshold. For temperatures higher then 200 K, the power factor Z_{th} increases strongly, reaching a value of $Z_{th} \approx 2.8$ at 370 K, suggesting the presence of Auger recombination. The MOVPE-grown GaInNAs device shows an even lower value of $Z_{th} \approx 1.3$ at 130 K, which increases slightly to $Z_{th} \approx 1.4$ at 180 K, again indicating the presence of a significant mono-molecular current contribution at threshold. For temperatures higher then 180 K, Z_{th} increases quickly up to a value of about $Z_{th} \approx 2.8$ at 300 K. This result suggests that at RT, Auger recombination is even more important in these MOVPE-grown GaInNAs lasers than in the MBE-grown GaInNAs devices.

12.4.4 Monomolecular Current

When analyzing the power factor Z for low injection currents (Z_{low}), it is observed that the power factor $Z_{low} = 1$ (Figure 12.4), indicating that the device is dominated by monomolecular recombination through defects for injection currents that are smaller than about $I_{th}/5$. This behavior was observed in all GaInNAs lasers measured in this study. Fitting a linear function with the slope $Z_{low} = 1$ onto the plot at low currents allows the quantification of the monomolecular current contribution I_{mono} at threshold. This procedure is illustrated in Figure 12.6. At this temperature (T = 180 K), the monomolecular current contributes to about 70% of the total current at threshold. This is consistent with the observed low value of Z_{th} (180 K) ≈ 1.5 from Figure 12.5. This analysis is now extended for the MBE- and the MOVPE-grown devices over all temperatures, and in Figure 12.7 the monomolecular current density at threshold as a function of temperature is plotted. It is found that at the

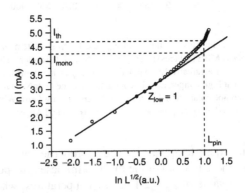

Figure 12.6 Plot of ln(I) vs. ln($L^{1/2}$) at T = 180 K for a GaInNAs (MBE) laser. At low currents, I μ n due to the strong influence of monomolecular recombination. By extrapolating this behavior, it is possible to determine the magnitude of the monomolecular recombination current, I_{mono}, at laser threshold.

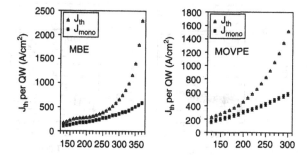

Figure 12.7 Total threshold current density per QW (triangles) and monomolecular recombination current density per QW (circles) at threshold for (a) MBE-grown and (b) MOVPE-grown GaInNAs lasers.

lowest temperature of T = 130 K, the monomolecular current contributes about 72% to the total current threshold in the MOVPE-grown device, compared with about 60% in the MBE-grown laser. This finding is consistent with the lower value of $Z_{th} \approx 1.3$ (Figure 12.5) of the MOVPE device at this temperature compared with $Z_{th} \approx 1.6$ for the MBE-grown laser. As will be discussed later, it is interesting to note that the defect-related current, which in principle might vary over many orders of magnitude, is in fact not very different despite the very different growth conditions for MOVPE and MBE. We take this to imply that the defect density is quite closely linked to the nitrogen density. More information about defects in N-containing materials can be found in Chapters 8 and 9.

12.4.5 Auger Current

The results from Figure 12.7 enable the removal of the monomolecular current contribution from the total current at threshold. Equation 12.2 can now be modified to become

$$I_{rest} = I_{total} - I_{mono} = eV\left(Bn^2 + Cn^3\right) \tag{12.5}$$

where I_{rest} equals the remaining current and is assumed to consist only of a radiative and an Auger-related current contribution. The current I_{rest} as a function of temperature is then plotted for the MBE- and the MOVPE-grown device in Figure 12.8 (open squares). Assuming that the Auger process is negligible at low temperatures [18], I_{rest} will be due entirely to radiative recombination in this temperature range. As the radiative current and its variation with temperature has been determined before in arbitrary units from the SE spectra (Figure 12.2), these results (closed circles) now can be normalized onto the plots in Figure 12.8 at low temperature.

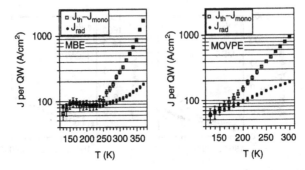

Figure 12.8 Rest-current density $J_{rest} = J_{th} - J_{mono}$ (open squares) and radiative-current density J_{rad} (closed circles) vs. temperature for (a) MBE-grown and (b) MOVPE-grown GaInNAs lasers. The radiative current density J_{rad} was determined by normalizing the integrated spontaneous emission at threshold to J_{rest} at low temperature.

It is observed that for temperatures $T \leq 230$ K, the radiative current determined from the spontaneous-emission spectra of the MBE-grown device (Figure 12.8a) shows the same temperature dependence as the current component J_{rest} within experimental error, including the anomalous bump. This supports the assumption that this current component is entirely radiative at low temperatures. For temperatures $T > 230$ K, J_{rest} increases much more strongly than J_{rad}, indicating the presence of an additional, strongly temperature-dependent, nonradiative current contribution. In the same temperature range, a strong increase in the power factor Z_{th} (Figure 12.5) up to $Z_{th} \approx 2.8$ at 370 K is observed, identifying a process that is proportional to n^3. This is strong evidence that the identified, additional nonradiative process for temperatures $T > 230$ K is due to Auger recombination. For the MOVPE-grown device (Figure 12.8b) a similar behavior can be observed. However, in these lasers, the rest current, J_{rest}, starts to deviate from the radiative current line at a much lower temperature of $T \approx 180$ K. This observation is again consistent with the strongly increasing value of Z_{th} (Figure 12.5) in the same temperature range, again indicating a process that is proportional to n^3 and hence identifying Auger recombination processes.

Subtracting the radiative contribution J_{rad} from J_{rest} leaves the Auger current J_{Auger}. In Figure 12.9 the current densities per QW at threshold of the three recombination processes J_{mono} (circles), J_{rad} (triangles), and J_{Auger} (squares) are plotted for the MBE-grown (closed symbols) and the MOVPE-grown (open symbols) GaInNAs lasers. These results reveal that at around room temperature, which is the temperature range of interest for applications, the Auger current density per well at threshold in the MOVPE-grown devices ($J_{Auger} \approx 760$ A/cm²) is about four times larger than in the MBE-grown lasers ($J_{Auger} \approx 180$ A/cm²). The radiative current density per well in the MOVPE lasers ($J_{rad} \approx 195$ A/cm²) is about two times greater than in the

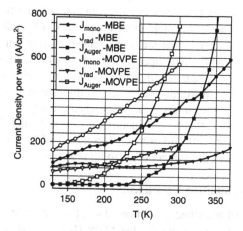

Figure 12.9 Monomolecular (circles), radiative (triangles), and Auger (squares) current density at threshold for the GaInNAs/GaAs MBE-grown (closed symbols) and MOVPE-grown (open symbols) lasers. J_{Auger} was calculated from $J_{th} = J_{mono} - J_{rad}$.

MBE devices ($J_{rad} \approx 110$ A/cm²), and the monomolecular contribution accounts for $J_{mono} \approx 565$ A/cm² (MOVPE) in comparison with $J_{mono} \approx 360$ A/cm² (MBE).

Assuming that the recombination coefficients A, B, and C are the same in the MOVPE- and the MBE-grown lasers, the ratio of the threshold carrier densities in the two devices $n_{th}^{(MOVPE)}/n_{th}^{(MBE)}$ can be determined from the measured current densities.

$$\frac{A n_{th}^{MOVPE}}{A n_{th}^{MBE}} = \frac{565}{360} = \frac{n_{th}^{MOVPE}}{n_{th}^{MBE}} = 1.56 \qquad (12.6)$$

$$\sqrt{\frac{B\left(n_{th}^{MOVPE}\right)^2}{B\left(n_{th}^{MBE}\right)^2}} = \sqrt{\frac{195}{110}} = \frac{n_{th}^{MOVPE}}{n_{th}^{MBE}} = 1.33 \qquad (12.7)$$

$$\sqrt[3]{\frac{C\left(n_{th}^{MOVPE}\right)^3}{C\left(n_{th}^{MBE}\right)^3}} = \sqrt[3]{\frac{760}{180}} = \frac{n_{th}^{MOVPE}}{n_{th}^{MBE}} = 1.61 \qquad (12.8)$$

The fact that the ratio $n_{th}^{(MOVPE)}/n_{th}^{(MBE)}$ is not very different for the three experimentally determined recombination processes indicates that a larger threshold carrier density in the MOVPE-grown device (in comparison with the MBE laser) could alone explain the larger current densities of all three recombination processes at room temperature. From this it is concluded that the larger threshold current densities in the MOVPE material is not due to a hugely different defect-related current but is more likely associated with additional temperature-dependent optical-loss processes in the MOVPE-grown material. Because of the large spin-orbit splitting in GaAs, longer wavelength devices based on this material system are likely to be very sensitive to intervalence band absorption [19] and its variation brought about by different distributions in the p-doping. Therefore, this alone may account for the different characteristics of the MOVPE and MBE devices, and this is the subject of ongoing research.

12.4.6 Recombination Coefficients

Using the model described in Section 12.4.2, the carrier density at threshold as a function of temperature for the SQW (MBE) broad-area device was calculated by Tomic [11]. The results are plotted in Figure 12.10 (solid line). For calculations of the temperature dependence of n_{th}, the internal losses were assumed to be invariant with temperature and equal to 4 cm^{-1}. The confinement factor was calculated to be 1.5%. To get better agreement with the experiment, a temperature-dependent Lorentzian line-broadening parameter δ was varied from $\delta = 17.5$ meV at T = 300 K up to $\delta = 33$ meV at T = 370 K. This is consistent with the work of Hofmann et al. [13], who obtained a value of 18 meV at room temperature. This variation was also justified by comparing the measured spontaneous emission spectra at threshold with the spectra calculated in the framework of the ten-band $\mathbf{k \cdot p}$ model, assuming δ as a fitting parameter.

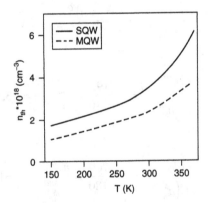

Figure 12.10 Calculated carrier density at threshold for SQW (solid line) and 3QW (dashed line) devices. The n_{th} for the SQW devices increases more strongly at high temperature due to the lower optical-confinement factor and lower differential gain.

Using the experimentally determined I_{mono}, I_{rad}, and I_{Auger}, the recombination coefficients of the three processes for the SQW device could be estimated by dividing the measured current component by the appropriate power 1, 2, or 3 of the calculated threshold carrier density. In Figure 12.11a, the monomolecular recombination coefficient, A (closed circles), as a function of temperature is plotted. At threshold, assuming a simple model of free carriers with a thermal velocity, v_{th}, that can be captured by defects with a capture cross section, s (which is assumed to be invariant with temperature), we can write the monomolecular recombination rate I_{mono} as

$$I_{mono} = n_{th}/\tau_{mono} = s\, v_{th}\, N_D\, n_{th} = eVA n_{th} \qquad (12.9)$$

where τ_{mono} is the monomolecular recombination lifetime and N_D is the number of defect centers. For $mv_{th}^2/2 \propto kT$, it follows that $v_{th} \propto T^{1/2}$ and hence $A \propto T^{1/2}$. For comparison, a normalized $A \propto T^{1/2}$ plot (solid line) is included in Figure 12.11a, and it is observed that, up to room temperature, this simple model is in good agreement with the determined temperature variation of A. In Figure 12.11b, the radiative recombination coefficient, B, as a function of temperature (closed squares) is plotted. In addition, in Figure 12.11b the theoretical values for B (open squares) derived employing the ten-band model as described earlier are included. An excellent agreement is observed, except for the lowest temperatures. This is due to the anomalous bump in the radiative current in that temperature range, which cannot be described by the theoretical model because the model did not include inhomogeneities in the device.

Figure 12.11c shows the Auger coefficient C (closed triangles). Interestingly, these values are of similar magnitude to reported values of the Auger coefficient for 1.3-μm InGaAsP-based devices, for which C is typically in the range $1-8 \times 10^{29}$ cm^{6}s^{-1} [20, 21]. This indicates that Auger recombination is also an important intrinsic recombination mechanism in GaInNAs. The results show that for the SQW device at around room temperature, which is the temperature range of main interest for applications, about 55% of the total current at threshold is due to monomolecular recombination through defects. This implies that removing the monomolecular component by improving the growth process could halve the room temperature threshold current of the devices. In these lasers, the radiative component contributes about 20%, and Auger recombination accounts for about 25% of the total current at threshold at RT. This may seem to be a disappointing result, as the Auger contribution is not reduced in comparison with conventional multiple-quantum-well (MQW) InGaAsP/InP-based lasers. However, the lasers analyzed here are single-quantum-well (SQW) devices, which will have a significantly higher threshold carrier density

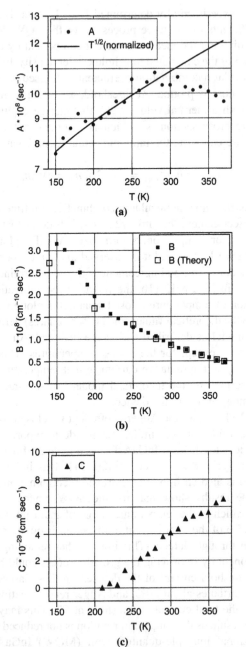

Figure 12.11 Recombination coefficients — monomolecular, A (closed circles); radiative, B (closed squares); and Auger, C (closed triangles) — for the SQW device. These coefficients were calculated using the experimentally measured temperature variations of J_{mono}, J_{rad}, and J_{Auger} and the calculated temperature variation of n_{th}.

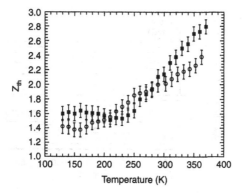

Figure 12.12 Power factor at threshold vs. temperature for SQW (squares) and 3QW (circles) devices. At high temperature, the SQW devices exhibit higher Z_{th} due to the higher n_{th} and correspondingly higher Auger-current contribution.

compared with MQW devices, due to the lower optical-confinement factor and lower differential gain at threshold (since the gain vs. carrier-density relationship is approximately logarithmic in a QW). With a reduced n, the Auger process, which is proportional to n^3, will be reduced much more strongly than the radiative ($\propto n^2$) and the monomolecular contributions ($\propto n$). Consequently, the relative contribution of the Auger current at threshold in an optimized MQW device should be lower than observed in the SQW devices.

In Figure 12.12 the power factor Z_{th} vs. temperature for the GaIn-NAs/GaAs SQW laser (closed squares) and the 3QW ridge-waveguide device (open circles) is plotted. At low temperature, for the 3QW device a lower value of Z_{th} than for the SQW device is observed, which is consistent with an expected larger relative contribution of the monomolecular recombination at lower n. For the 3QW laser, Z_{th} increases less rapidly with temperature compared with the SQW laser, leading to a maximum value of $Z_{th} \approx 2.4$ (3QW) in contrast to $Z_{th} \approx 2.8$ (SQW). This lower value of Z_{th} indicates that increasing the number of quantum wells has reduced the relative contribution of the Auger recombination current. This effect can also be seen when comparing the temperature dependence of the threshold current of the two devices.

In Figure 12.13, the threshold current density per QW vs. temperature for the 3QW (open circles) and SQW (closed circles) devices is plotted. It is clearly visible that J_{th} increases much more strongly with temperature for the SQW laser than for the 3QW device. This again indicates that the Auger recombination contribution to the total current at threshold is reduced in the 3QW lasers. It is interesting to note here that in the GaInNAs/GaAs materials system, the SQW device still exhibits reasonably low values of J_{th}. This is in contrast to InGaAsP-based devices where, due to the effects

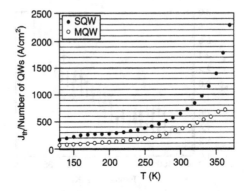

Figure 12.13 Threshold current density per well vs. temperature for GaInNAs/GaAs 3QW (open circles) and SQW (closed circles) devices.

of both low optical confinement and carrier spill-over, it is usually necessary to include at least four quantum wells for optimized room temperature operation [22]. Using the ten-band model, as described earlier, the carrier density at threshold as a function of temperature for the 3QW ridge-waveguide laser is calculated. For the calculations, the internal losses were assumed to be equal to 10 cm^{-1} and invariant with temperature [23]. The confinement factor was calculated to be 5.3%. The same broadening parameters were used as for the SQW laser. In Figure 12.10, the calculated n_{th} vs. temperature for the 3QW laser (dashed line) is plotted.

Assuming negligible leakage currents and that the recombination coefficients determined are the same in the SQW and the 3QW device, the calculated threshold carrier density, n_{th}, of the 3QW laser and Equation 12.2 can now be used to plot (Figure 12.14) the expected variation of J_{th}

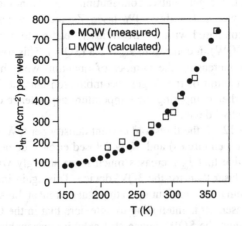

Figure 12.14 Measured and calculated J_{th} for the 3QW device.

with temperature (open squares) compared with the measured results (closed circles). Good agreement is observed, indicating that the different threshold carrier density of the SQW and the 3QW devices can reasonably explain the different temperature variation of the threshold current.

12.5 PRESSURE DEPENDENCE OF THE RECOMBINATION PROCESSES

In this section, measurements of the threshold current, I_{th}, and the integrated SE rate, L_{SE}, of GaInNAs-based semiconductor diode lasers are discussed as a function of hydrostatic pressure. The direct band-gap of III-V semiconductor alloys increases with pressure, providing a powerful tool to investigate band-gap-dependent recombination mechanisms. A piston-in-cylinder system was used to apply hydrostatic pressure up to 1.5 GPa. The light from the facet was collected via an optical fiber and was analyzed with an optical-spectrum analyzer to determine the lasing wavelength or with a fiber-coupled optical power meter to measure the LI characteristics. The devices have a cavity length of 350 μm and emit at 1.29 μm at room temperature and atmospheric pressure (RTP). For all of our measurements, the lasers were operated under pulsed conditions and in bare-chip form.

12.5.1 Hydrostatic Pressure Dependence of the Threshold Current

Previous studies have shown that the threshold current of 1.3-μm and 1.5-μm InGaAsP-based devices always decreases with increasing hydrostatic pressure due to a reduction of the Auger recombination rate with increasing band gap. At 1.3 μm, Auger recombination forms ≈50% of I_{th} in InGaAsP devices at RTP. The first reversible increase in I_{th} with pressure in a 1.3-μm semiconductor laser was observed with an AlGaInAs-based device [24], where it was found that nonradiative recombination forms only ≈20% of I_{th} at RTP.

In Figure 12.15, the normalized results for the threshold current vs. pressure for InGaAsP (triangles) and AlGaInAs (circles) devices (taken from [24]) are plotted, where the solid lines are a guide for the eye. Ideally, the radiative current (I_{Rad}) varies as $I_{Rad} \propto E_g^2$, as indicated for the GaInNAs device by the dashed line. The measured threshold current as a function of pressure for the GaInNAs-based device is represented by the square symbols. Although it was shown earlier that Auger recombination at RTP contributes approximately 20% to the total current at threshold, it can be observed that I_{th} of the GaInNAs device increases even more quickly with pressure than expected for a purely radiative device, which is a very unusual finding. While the different pressure dependencies of the InGaAsP and AlGaInAs devices can be readily explained by using a combination of

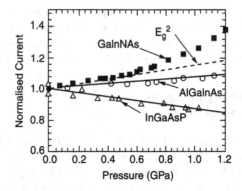

Figure 12.15 Normalized variation of I_{th} as a function of pressure for the GaInNAs-based devices (squares). For comparison, typical data for InGaAsP-based (triangles) and AlGaInAs-based (circles) devices are also shown. The solid lines are provided as a guide for the eye. The dashed line represents the ideal variation of the radiative current, $I_{Rad} \propto E_g^2$, for the GaInNAs devices. Initially, $I_{th}(P) \approx I_{Rad}(P)$, but it increases more rapidly at the higher pressures due to the strong increase in n_{th}. This is in contrast to the behavior of InGaAsP lasers, for which I_{th} decreases with pressure, and for AlGaInAs lasers, where I_{th} increases only slightly. In all three devices, J_{th} is influenced by the presence of Auger recombination.

radiative and Auger recombination and standard models, this is not possible for the GaInNAs lasers. A possible explanation for the strong increase of I_{th} in the GaInNAs devices is the increasing interaction of the nitrogen level and the CB edge with increasing pressure, which will increase the electron effective mass, m_c^*, resulting in an increased density of states and hence an increased threshold carrier density, as was discussed in Chapters 2–4.

To verify these assumptions, the pressure variation of n_{th} for a 1.3-μm GaInNAs QW laser was calculated by Tomic [25], employing the model based on the ten-band **k·p** Hamiltonian described earlier. The results predict that n_{th} increases by approximately 12% over a pressure range of 10 kbar. It is therefore possible that the increase in n_{th} — and the resulting even stronger influence on the n²- (radiative) and n³- (Auger) dependent processes — will more than compensate for the decrease in the Auger coefficient, C, with increasing pressure.

That the strong increase of I_{th} with hydrostatic pressure in 1.3-μm GaInNAs laser diodes is indeed due to an increasing CB effective mass has recently been confirmed experimentally by Jin et al. [25]. An analysis similar to that introduced in the previous sections was performed by measuring the spontaneous emission from a window in the substrate of the laser as a function of hydrostatic pressure, which enabled the determination of the absolute monomolecular-, radiative-, and Auger-related current contributions at threshold as a function of pressure, as well as the corresponding recombination coefficients. Here the lasers investigated were the same

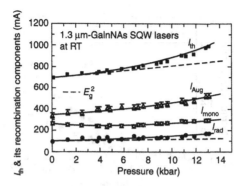

Figure 12.16 Threshold current (squares) and the corresponding monomolecular- (diamonds), radiative- (circles), and Auger- (triangles) related current components vs. pressure for 1.3-μm GaInNAs SQW broad-area laser.

MBE-grown SQW broad-area devices used in the previous sections. The results, taken from Jin et al. [25], are shown in Figure 12.16.

It can be observed that the monomolecular contribution (diamonds) increases only slightly with pressure, while the radiative current (circles) increases even more strongly than expected for an ideal quantum-well laser ($\propto E_g^2$). The most surprising result is that the Auger recombination process (triangles) shows a strong increase with pressure as well, in contrast to the InGaAsP and AlGaInAs devices [24], where a decreasing Auger current was observed. Jin et al. [25] show that the Auger coefficient, C, in the GaInNAs devices is decreasing by ≈12% (this compares with ≈30% in InGaAsP) over the pressure range studied, but the strong increase of n_{th} of ≈12% leads to a rise in the Auger-related current, which is proportional to n^3. These findings are also consistent with the observation that the radiative current ($\propto n^2$) is increasing even more quickly than E_g^2. The monomolecular recombination coefficient A was found to be approximately constant with increasing hydrostatic pressure.

From these results, one would expect to observe an even stronger increase of I_{th} with pressure in GaInNAs devices, which initially have a higher carrier density than the device shown in Figure 12.15, as the Auger-related current ($\propto n^3$) would be larger in this case. To verify this, in Figure 12.17 the normalized I_{th} vs. pressure is plotted for an MOVPE-grown MQW broad-area laser (circles) that has a three-times-larger threshold current density at RTP. This is associated with an increased n_{th} due to additional optical-loss processes. For comparison, we include the data for the SQW MBE-grown device from Figure 12.15. It can be clearly observed that, as expected, there is a much stronger increase in I_{th} with pressure for the MOVPE lasers than for the MBE-grown devices, supporting our model.

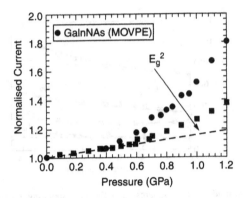

Figure 12.17 Normalized variation of I_{th} as a function of pressure for the GaInNAs-based MBE-grown (squares) and MOVPE-grown (circles) devices.

12.6 1.3-μm GaInNAs/GaAs VCSELs

As was mentioned in the introduction, monolithically grown GaIn-NAs/GaAs 1.3-μm vertical-cavity surface-emitting lasers (VCSELs) are particularly promising for communications applications. This is for two key reasons. First, appropriate mixtures of N and In can be used to produce VCSELs, which have already been demonstrated, with an output power of 500 μW at 1.28 μm under continuous wave (CW) operation at room temperature [26]. Second, this type of application requires stable operation over a wide range of temperatures. This presents a particular problem for the VCSEL, because the lasing energy of the cavity mode has a different temperature dependence than that of the peak of the optical gain spectrum, leading to a tuning/detuning effect. However, experimental evidence [27] suggests that GaInNAs has an intrinsically broad gain spectrum, potentially leading to a wider operating temperature range.

The VCSELs studied here have an intracavity contacted design [26]. The cavity is formed by a top DBR (distributed Bragg reflector) consisting of 28 pairs of undoped $Al_{0.8}Ga_{0.2}As$/GaAs mirrors and a bottom DBR consisting of 34 pairs of undoped AlAs/GaAs mirrors. The active region consists of two GaInNAs/GaAs quantum wells with a nominal nitrogen and indium content of 1.8% and 35%, respectively. Finally, a layer of AlAs adjacent to the cavity [28] was incorporated and oxidized to form a current aperture of approximately 11-μm diameter.

Because the VCSELs studied had active regions of MBE-grown GaIn-NAs similar to the edge-emitting lasers (EELs), we use the same recombination coefficients A, B, and C and their variation with temperature to analyze their characteristics. As when comparing with MOVPE-grown lasers, the main difference between the two types of device will be the value of the carrier density at threshold. To determine this, we have used the same ten-band $\mathbf{k \cdot p}$ Hamiltonian and calculated the expected temperature and pressure dependence of the threshold current density (J_{th}) of the VCSELs.

In an edge-emitting device, the lasing wavelength is determined by the energy of the peak of the gain spectrum, E_p. In contrast, the lasing wavelength of a VCSEL is determined by the photon energy determined by the cavity mode, E_c. The lowest threshold current at a given temperature is achieved when $E_p = E_c$. However, E_p decreases more quickly than E_c with increasing temperature, and E_p and E_c detune with changing temperature. If the energy spread of the gain spectrum is narrow, the gain at E_c changes quickly with temperature, and the devices are able to operate only over a narrow temperature window. Therefore, for a wide temperature range of operation, it is desirable to have a broad, flat gain spectrum. Additionally, E_p should be designed to be larger than E_c in the lower part of the desired operating temperature range, e.g., at room temperature.

Since E_p varies quickly and much faster than E_c with hydrostatic pressure, this provides a very convenient adjustable parameter to study the effects of gain–cavity alignment and the width of the gain spectrum. Application of hydrostatic pressure affects the GaInNAs active region of the VCSELs and EELs in exactly the same manner. This is shown experimentally for the EELs in Figure 12.18. The energy of the peak of the gain spectrum increases at 73 meV(GPa)$^{-1}$. In contrast, the variation of the lasing energy of the VCSEL is determined by the variation of the cavity mode. This variation is determined by the compressibility of the cavity material and the change with pressure of the refractive index, both of which are slower. The variation of the VCSEL wavelength is also shown in Figure 12.18 and corresponds to an increase of 17 meV(GPa)$^{-1}$. Figure 12.18 also shows that the VCSEL has the cavity mode on the low-energy side of the gain peak at room temperature and so detunes at a net rate of 56 meV(GPa)$^{-1}$.

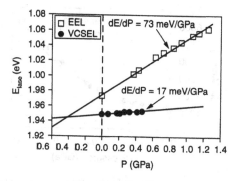

Figure 12.18 Lasing energy vs. hydrostatic pressure for GaInNAs-based VCSEL (closed circles) and edge-emitter (open squares) with nominally identical active regions.

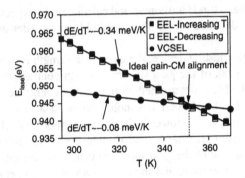

Figure 12.19 Lasing energy, E_{lase}, vs. temperature, T, for GaInNAs-based edge-emitting laser (squares) and VCSEL (circles) with nominally identical active regions. E_{lase} for the edge-emitting laser provides the energy for the gain peak.

The temperature variation of the peak of the gain determined from the lasing wavelength of the EELs is shown in Figure 12.19. Also shown is the measured emission energy of the VCSELs as a function of temperature. Gain–cavity alignment improves from RT at 0.26 meVK^{-1}, which has the advantage that it compensates the deteriorating intrinsic characteristics of the active region up to 350 K.

Figure 12.20 shows the observed increase in threshold current density (circles), J_{th}, as E_p and E_c detune with increasing pressure as shown in Figure 12.18. Figure 12.21 shows how J_{th} at first decreases as the temperature is increased from low temperature and E_p and E_c tune as shown in Figure 12.19.

The theoretical modeling was carried out using (a) the energies for the peak gain and the cavity mode and their variation with pressure and temperature shown in Figures 12.18 and 12.19, (b) the gain spectrum calculated using the ten-band k·p Hamiltonian, and (c) the recombination coefficients

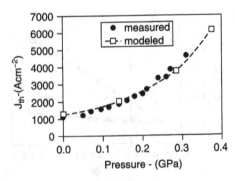

Figure 12.20 Measured (closed circles) and calculated (open squares) J_{th} vs. pressure for GaInNAs-based VCSEL.

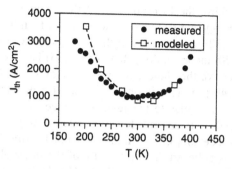

Figure 12.21 Measured (closed circles) and calculated (open squares) J_{th} vs. temperature for GaInNAs-based VCSEL.

A, B, and C for the monomolecular, radiative, and Auger processes, respectively, determined from the edge-emitting lasers as described above.

Using these parameters, the expected variation of the threshold current with pressure and temperature was modeled and compared with the measured results in Figures 12.20 and 12.21, respectively. As can be seen, good agreement is obtained in both cases. Thus, the gain model and recombination-rate parameters appear to give a good description of GaInNAs VCSELs and shows that they do indeed operate over a wide temperature range, making them promising devices for communications applications.

12.7 LASERS OPERATING AT WAVELENGTHS LONGER THAN 1.3-μm

In more standard materials systems such as GaInAsP/InP, as we move from 1.3-μm to 1.55-μm wavelength and beyond, the threshold current density, J_{th}, is always observed to increase despite the decrease in electron effective mass and n_{th}. This is because the Auger recombination coefficient increases swiftly as the band-gap decreases. In the GaInNAs/GaAs system, if the move to longer wavelengths is accompanied by an increase in the N content, then J_{th} can be expected to increase even more quickly. This is because there will be little or no decrease in electron effective mass and n_{th}, while both A and C will increase. The first GaInNAs lasers operating at 1.5 μm were achieved by Fischer et al. [29], primarily by increasing the N content, but at room temperature the threshold current density of ≈60 kA/cm² (3QW structure) was indeed very large.

One exciting approach to overcome these problems is by the addition of Sb in the growth process. This may have three beneficial effects. First, the Sb acts as a surfactant that reduces strain-induced defect formation and hence allows quantum wells with higher strain (and thus reduced N content) to be grown [30]. Second, the Sb is incorporated into the well, also causing a lower band-gap and longer wavelength for a given N content [31]. Third, the incorporation of the Sb may increase the hole confinement energy and so reduce hole escape, which has been proposed by Tansu et al. [32] to be a cause of the low T_0 observed in their devices. By the use of Sb incorporation, Bank et al. [33] have recently achieved the very impressive threshold current density of 1.1 kA/cm^2 for a laser operating at 1.49 μm. This is only a factor ≈1.5 increase from 1.3-μm devices, which is similar to the relative change in J_{th} with wavelength observed in GaInAsP lasers. However, interestingly, the T_0 of 62 K, observed by Bank et al. [33], is similar to that of 1.5-μm GaInAsP devices [34], where it is believed that up to 80% of the threshold current is via Auger recombination, resulting in a $T_0 < T/3$ [34]. For a fuller discussion of long-wavelength dilute nitride lasers, see Chapter 14.

12.8 SUMMARY AND CONCLUSIONS

It has been possible to determine the magnitudes of the recombination currents at lasing threshold due to monomolecular (An), radiative (Bn^2), and Auger (Cn^3) processes in 1.3-μm GaInNAs/GaAs-based semiconductor lasers. Hence, by calculating the magnitude of the carrier density at threshold, n_{th}, and its temperature variation, it has been possible to determine the values of the coefficients A, B, and C and their temperature variation.

Surprisingly, it has been found that approximately the same value of A, which depends on the defect density, can explain the behavior of edge-emitting lasers grown by both MBE and MOVPE and in vertical-cavity surface-emitting lasers. This can be taken to imply that the density of defects giving rise to the monomolecular recombination is associated with the density of nitrogen atoms. Less surprisingly, the same values of B and C can be used to describe the behavior of all the devices investigated. Clearly, the relative values of the monomolecular, radiative, and Auger currents will depend on the value of n_{th} for the particular structure and temperature considered. In the case of the single-quantum-well lasers grown by MBE [1, 2], monomolecular recombination accounted for about 50% of the total current at room temperature, and the radiative and Auger currents each amounted to about 25%. Because n_{th} increases with increasing temperature, Auger recombination is more important at higher temperatures, while monomolecular recombination becomes more important as the temperature is lowered. By using several quantum wells, the optical-confinement factor can be increased and n_{th} decreased, resulting in a decrease of Auger recombination.

Another approach is to reduce the N content and increase the In content to maintain the required wavelength. Tomic and O'Reilly [35] showed that, theoretically, the ideal nitrogen content is zero! Even a small increase in the N content decreases the optical matrix element for radiative transitions and also increases the electron effective mass and density of states so that

n_{th} is increased. For the radiative current, the reduction in B is offset almost exactly by the increase in n_{th}^2. However, as N is increased, both A (if the defects are related to the N concentration) and n_{th} increase, so that An_{th} increases with respect to Bn_{th}^2. Since the Auger recombination is most likely the CHSH process (a conduction band to heavy hole band transition excites a heavy hole into the spin split-off band), the Auger coefficient C will not change with N content because this has little effect on the valence band. However, n_{th}^3 clearly increases strongly. We therefore conclude that both nonradiative processes decrease as the N content is decreased. However, this leads to increasing compressive strain and eventually to further growth defects.

For each device application, careful design is clearly required for optimum performance, but this can now be achieved using the parameters presented in this chapter.

Acknowledgments

The authors are particularly grateful to Dr. Henning Riechert and his group at Infineon Technologies for providing the laser samples on which the authors' experimental results are based. They would also like to thank Drs. Stanko Tomi, Shirong Jin, Gareth Knowles and Stephen J. Sweeney and Prof. Eoin O'Reilly for their collaboration. Thanks also to Prof. J.S. Harris for providing data prior to publication.

References

1. Egorov, A.Y., Bernklau, D., Borchert, B., Illek, S., Livshits, D., Rucki, A., Schuster, M., Kaschner, A., Hoffmann, A., Dumitras, G., Amann, M.C., and Riechert, H., "Growth of High Quality InGaAsN Heterostructures and Their Laser Application," *J. Crystal Growth* 227–228: 545–552 (2001).
2. Riechert, H., Egorov, A.Y., Livshits, D., Borchert, B., and Illek, S., "InGaAsN/GaAs Heterostructures for Long-Wavelength Light-Emitting Devices," *Nanotechnology* 11: 201–205 (2000).
3. Ramakrishnan, A., Ebbinghaus, G., and Stolz, W., "MOVPE-Growth and Characterization of Metastable (GaIn)(NAs)/GaAs Heterostructures for 1.3-μm Lasers," in *Proc. Int. Symp. Compound Semicond. (ISCS 2001)*, Tokyo, 2001.
4. Hoehnsdorf, F., Koch, J., Gerd, C.A., and Stolz, W., "Investigation of GaInNAs Bulk Layers and GaInNAs Multiple Quantum Well Structures Grown Using Tertiary Butyl Arsine (TBAs) and 1,1-Dimethyl Hydrazine (uDMHy)," *J. Crystal Growth* 195: 391 (1998).
5. Phillips, A.F., "Temperature Dependence of the Radiative and Non-Radiative Currents in Visible and Near Infra-Red Semiconductor Lasers," Ph.D. thesis (Surrey, U.K.: University of Surrey, 1996).
6. Zhukov, A.E., Ustinov, V.M., Egorov, A.Y., Kovsh, A.R., Tsatsulnikov, A.F., Ledentsov, N.N., Zaitsev, S.V., Gordeev, N.Y., Kopev, P.S., and Alferov, Z.I., "Negative Characteristic Temperature of InGaAs Quantum Dot Injection Lasers," *Jpn. J. Appl. Phys.* 36: 4216–4218 (1997).

7. Haug, A., "Relations between the T_0 Values of Bulk and Quantum Well GaAs," *Appl. Phys. B, Photophys. Laser Chem.* 44: 151–153 (1987).

8. Chen, T.R., Margalit, S., Koren, U., Yu, K.L., Chiu, L.C., Hasson, A., and Yariv, A., "Direct Measurement of the Carrier Leakage in an InGaAsP/InP Laser," *Appl. Phys. Lett.* 42: 1000–1002 (1983).

9. Higashi, T., Sweeney, S.J., Phillips, A.F., Adams, A.R., O'Reilly, E.P., Uchida, T., and Fujii, T., "Experimental Analysis of Temperature Dependence in 1.3-µm AlGaInAs-InP Strained MQW Lasers." *IEEE J. Selected Top. Quantum Electron.* 5: 413–419 (1999).

10. Tomi, S., and O'Reilly, E.P., "Gain Characteristics of Ideal Dilute Nitride Quantum Well Lasers," *Physica E* 13: 1102–1105 (2002).

11. Fehse, R., Tomic, S., Adams, A.R., Sweeney, S.J., O'Reilley, E.P., Andreev, A., and Riechert, H., "A Quantitative Study of Radiative, Auger and Defect Related Recombination Processes in 1.3-µm GaInNAs-Based Quantum-Well Lasers," *IEEE J. Selected Top. Quantum Electron.* 8: Vol. 8, No. 4, 801–810 (2002).

12. Hader, J., Koch, S.W., Moloney, J.V., and O'Reilly, E.P., "Gain in 1.3-µm Materials: InGaNAs and InGaPAs Semiconductor Quantum-Well Lasers," *Appl. Phys. Lett.* 77: 630–632 (2000).

13. Hofmann, M., Wagner, A., Ellmers, C., Schlichenmeier, C., Schäfer, S., Höhnsdorf, F., Koch, J., Stolz, W., Koch, S.W., Rühle, W.W., Hader, J., Moloney, J.V., O'Reilly, E.P., Borchert, B., Egorov, A.Y., and Riechert, H., "Gain Spectra of (GaIn)(NAs) Laser Diodes for the 1.3-µm-Wavelength Regime," *Appl. Phys. Lett.* 78: 3009–3012 (2001).

14. Van de Walle, C.G., "Band Lineups and Deformation Potentials in the Model-Solid Theory," *Phys. Rev. B* 39: 1871–1883 (1989).

15. Klar, P.J., Gruning, H., Koch, J., Schafer, S., Volz, K., Stolz, W., Heimbrodt, W., Kamal Saadi, A.M., Lindsay, A., and O'Reilly, E.P., "(Ga, In)(N, As)-Fine Structure of the Band-gap due to Nearest-Neighbour Configurations of the Isovalent Nitrogen," *Phys. Rev. B* 64: 121203/1-4 (2001).

16. Ahn, D., and Chuang, S.L., "Optical Gain and Gain Suppression of Quantum-Well Lasers with Valence Band Mixing," *IEEE J. Quantum Electron.* 26: 13–24 (1990).

17. Sweeney, S.J., "Radiative and Non-Radiative Recombination in 1.3-µm and 1.5-µm Semiconductor Diode Lasers," Ph.D. thesis (Surrey, U.K.: University of Surrey, 1999).

18. Phillips, A.F., Sweeney, S.J., Adams, A.R., and Thijs, P.J.A., "The Temperature Dependence of 1.3- and 1.5-µm Compressively Strained InGaAs(P) MQW Semiconductor Lasers," *IEEE J. Selected Top. Quantum Electron.* 5: 401–412 (1999).

19. Adams, A.R., Asada, M., Suematsu, Y., and Arai, S., "The Temperature Dependence of the Efficiency and Threshold Current of InGaAsP Lasers Related to Intervalence Band Absorption," *Jpn. J. Appl. Phys.* 19: 621–627 (1980).

20. Agrawal, G.P., and Dutta, N.K., *Long Wavelength Semiconductor Lasers* (New York: Van Nostrand Reinhold, 1986).

21. Pikal, J.M., Menoni, C.S., Temkin, H., Thiagarajan, P., and Robinson, G.Y., "Carrier Lifetime and Recombination in Long-Wavelength Quantum-Well Lasers," *IEEE J. Selected Top. Quantum Electron.* 5: 613–619 (1999).

22. Silver, M., and O'Reilly, E.P., "Optimisation of Long Wavelength InGaAsP Strained Quantum-Well Lasers," *IEEE J. Selected Top. Quantum Electron.* 31: 1193 (1995).

23. Riechert, H., Egorov, A.Y., Livshits, D., Borchert, B., and Illek, S., "InGaAsN/GaAs Heterostructures for Long-Wavelength Light-Emitting Devices," *IOP Nanotechnology* 11: 201–205 (2000).

24. Sweeney, S.J., Higashi, T., Adams, A.R., Uchida, T., and Fujii, T., "Improved Temperature Dependence of 1.3-µm AlGaInAs-Based MQW Semiconductor Diode Lasers Revealed by Hydrostatic Pressure," *Electron. Lett.* 34: 2130–2131 (1998).

25. Jin, S.R., Sweeney, S.J., Tomic, S., Adams, A.R., and Riechert, H., "Quantifying Pressure Dependent Recombination Currents in GaInNAs Lasers Using Spontaneous Emission Measurements," *Phys. Stat. Solidi B* 235 (2): 486–490 (2003).

26. Steinle, G., Riechert, H., and Egorov, A.Y., "Monolithic VCSEL with InGaAsN Active Region Emitting at 1.28-μm and CW Output Power Exceeding 500 μW at Room Temperature," *Electron. Lett.* 37: 93–95 (2001).

27. Choulis, S.A., Hosea, T.J.C., Klar, P.J., Hofmann, M., and Stolz, W., "Influence of Varying N-Environments on the Properties of (GaIn)(NAs) Vertical-Cavity Surface-Emitting Lasers," *Appl. Phys. Lett.* 79: 4277–4279 (2001).

28. Choquette, K.D., Schneider, R.P., Hagerott-Crawford, M.., Geib, K.M., and Figiel, J.J., "Continuous Wave Operation of 640–660-nm Selectively Oxidized AlGaInP Vertical Cavity Lasers," *Electron. Lett.* 31: 1145–1146 (1995).

29. Fischer, M., Reinhardt, M., and Forchel, A., "GaInAsN/GaAs Laser Diodes Operating at 1.52 μm," *IEEE Electron. Lett.* 36: 1208–1209 (2000).

30. Yang, X., Heroux, J.B., Jurkovic, M.J., and Wang, W.I., "Low-Threshold 1.3-μm InGaAsN:Sb-GaAs Single-Quantum-Well Lasers Grown by Molecular Beam Epitaxy," *IEEE Photonics Tech. Lett.* 12: 128–130 (2000).

31. Harris, J.S., "GaInNAs Long-Wavelength Lasers: Progress and Challenges," *Semicond. Sci. Technol.* 17: 880–891 (2002).

32. Tansu, N., Kirsch, N.J., and Mawst, L.J., "Low-Threshold-Current-Density 1300-nm dilute nitride Quantum Well Lasers," *Appl. Phys. Lett.* 81: 2523–2525 (2002).

33. Bank, S.R., Wistey, M.A., Yuen, H.B., Goddard, L.L., Ha, W., and Harris, J.S., "A Low Threshold CW GaInNAsSb/GaAs Laser at 1.49 μm," submitted to *IEEE* Electron. Lett., 39 (20), 1445–1446 (2003).

34. Phillips, A.F., Sweeney, S.J., Adams, A.R., and Thijs, P.J.A., "The Temperature Dependence of 1.3- and 1.5-μm Compressively Strained InGaAs(P) MQW Semiconductor Lasers," *IEEE J. Selected Top. Quantum Electron.* 5: 401–412 (1999).

35. Tomic, S., and O'Reilly, E.P., "Optimization of Material Parameters in 1.3-μm InGaAsN-GaAs Lasers," *IEEE Photonics Tech. Lett.* 15: 6–8 (2003).

CHAPTER 13

Photovoltaic Applications of Dilute Nitrides

D.J. FRIEDMAN, J.F. GEISZ, and A.J. PTAK

National Renewable Energy Laboratory, Golden, Colorado

1-591-69019-6/04/$0.00+$1.50
© 2004 by CRC Press LLC

13.1 MOTIVATION FOR DILUTE NITRIDES IN PHOTOVOLTAICS

13.1.1 Introduction

With the discovery [1] of the giant band-gap bowing in the GaN-GaAs system, and the subsequent realization [2] that the material can be lattice-matched to GaAs with the addition of In in the proportion $Ga_{1-3x}In_{3x}N_xAs_{1-x}$, a whole new range of low-band-gap devices lattice-matched to GaAs became possible. This development is potentially of great significance for high-efficiency multijunction solar cells, where these low-band-gap materials can, in principle, be used to efficiently collect the low-photon-energy portion of the solar spectrum [3].

However, implementing GaInNAs into solar cells has proven much more problematic than originally anticipated [4–8]. The incorporation of N into GaAs or GaInAs dramatically reduces the minority-carrier lifetime and mobility (see Chapters 2, 9, and 10). Because collection of minority carriers is central to the function of solar cells, the adverse effects of N on the minority-carrier properties has thus far prevented the realization of efficient cells based on GaInNAs. This chapter will provide a detailed introduction to the operation and materials requirements of high-efficiency III-V solar cells in order to motivate further research into both the opportunities and the challenges provided by the dilute III-V-N system. The main body of this chapter will then address the question of the origin of the poor minority-carrier properties, and then discuss the potential for next-generation high-efficiency cells should these problems prove possible to mitigate. Finally, we will discuss another dilute nitride system, GaPN (see also

Chapter 10), which opens up the possibility of high-efficiency multijunction cells lattice-matched to silicon.

13.1.2 Collection in Solar-Cell Junctions

13.1.2.1 Constraints on solar-cell length scales

A solar-cell junction works by absorbing photons, converting them into electron-hole pairs, and then collecting the minority carriers as current at the junction. Photons that are not absorbed, or minority carriers that recombine at traps rather than being converted to current at the junction, represent losses in cell efficiency. There are three important length scales of the junction:

1. The thickness d of the absorbing region
2. The absorption length $1/\alpha(\lambda)$ at the photon wavelength λ of interest
3. The collection length l from the junction in which minority carriers are collected without significant recombination loss

The need for efficient photovoltaic conversion sets constraints on these length scales as design rules for an efficient solar cell. For most of the incident photons to be absorbed, it is necessary that $d > 1/\alpha(\lambda)$. Furthermore, for complete collection of the minority carriers generated by the absorbed photons, it is necessary that $l > d$. For GaAs cells, these constraints call for a thickness $d \approx 3$ μm and a collection length $l > d$. Indeed, solar cells, with their critical dependence on minority-carrier collection, can be used as test structures for the minority-carrier properties of their component semiconductors.

13.1.2.2 Diffusion and drift collection

Minority-carrier collection in the solar cell, as characterized by the collection length l, can be accomplished either by diffusion or by field-aided collection (i.e., by drift in the junction depletion region). For conventional III-V alloys such as GaAs and GaInP, the former is readily achievable with modern growth technology. GaAs can readily be grown with diffusion lengths several times the ≈3-μm cell thickness. In contrast, for a material having high trap densities, and therefore a diffusion length too small for adequate carrier collection, it would be necessary to collect carriers in the depletion region. This approach is more difficult than collecting carriers by diffusion, because a background carrier concentration of less than ≈10^{14}/cm^3 is needed to obtain a sufficiently wide depletion region. Also, a wider depletion region will increase Shockley-Read-Hall recombination, lowering the open-circuit voltage of the cell. We emphasize these points because among the most distinctive properties of the dilute nitride III-Vs are high trap densities, high recombination/low diffusion lengths, and high background-carrier densities. These characteristics will be a primary focus of this chapter. A detailed discussion of the theory of defects in these materials is given in Chapter 8.

13.1.3 Conversion Losses in Single-Junction Solar Cells

The great challenge in efficiently converting sunlight to electricity is the broadness of the solar spectrum, which contains photons ranging in energy from 0 to 4 eV. A conventional single-junction cell is characterized by a single band gap, E_g. For photons with $h\nu > E_g$, the excess energy $h\nu - E_g$ will be converted to heat in the cell, whereas photons with $h\nu < E_g$ will not be absorbed by the cell, and their energy is completely lost. Thus, a conventional single-junction cell will most efficiently convert those photons of energy $h\nu = E_g$, putting a fundamental limit on the efficiency of such a cell.

13.1.4 Multijunction Concept

The way around this limit is, in principle, straightforward: divide the solar spectrum into several ranges, and convert each range with a band-gap tuned to that range. Henry [9] has calculated the theoretical efficiencies for ideal multijunction cells as a function of the number of junctions at 1000-suns concentration. The resulting efficiencies for 1, 2, 3, and 36 junctions are 37%, 50%, 56%, and 72%, respectively. In other words, there is much to be gained by the first few additional junctions added, but the returns diminish rapidly after that. Therefore, for practical purposes, one is led to consider a configuration with only a few junctions.

Conceptually, it is probably easiest to imagine accomplishing this spectrum splitting by using a prism to spatially distribute the solar spectrum, and placing an array of cells with appropriate band gaps at the output of the prism. This approach is illustrated in the left-hand panel in Figure 13.1. However, although this approach is simple in concept, in practice it is mechanically, optically, and electrically complex to the point of impracticality. Fortunately, there is a far simpler and better approach based on the fact that, for an ideal semiconductor layer with band-gap E_g, essentially all photons with $h\nu < E_g$ are transmitted through the layer. Thus, the desired spectrum splitting can be accomplished simply by stacking the junctions in order of decreasing band gap, as illustrated in the right-hand panel in Figure 13.1. An incident photon travels through each junction in turn until being absorbed by the one with the appropriate band gap. For example, in Figure 13.1b, an incident photon with energy $E_{g1} < h\nu < E_{g2}$ will be transmitted through the top two cells and be absorbed by the bottom cell.

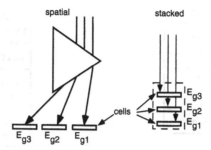

Figure 13.1 Spatial and stacked approaches for distributing the solar spectrum to junctions of different band gaps.

13.1.5 Heteroepitaxy and Lattice Matching

13.1.5.1 Monolithic multijunction structure

A critical advantage of the configuration of Figure 13.1b is that it allows for the junctions to be grown monolithically, one on top of the next, with monolithic series interconnects. The result is a *monolithic two-terminal* package, desirable as a plug-in replacement for a single-junction solar cell. The structure of such a monolithic multijunction cell consists of layers of semiconductors of different band gaps and compositions. Growth of such a structure involves the complexities of heteroepitaxy.

13.1.5.2 Lattice-matched device structures

Unlike light-emitting devices with active layers that can be mismatched but whose thickness is less than the critical thickness for formation of extended structural defects, the thickness of the absorbing layer of a III-V solar cell needs to be on the order of a micron, as determined by α. The need for long diffusion lengths means that structural defects such as misfit dislocations are very harmful to the operation of high-efficiency solar cells. For this reason, it is important that the entire multilayer structure be lattice-matched to its substrate. (Progress has been made toward achieving high-quality mismatched multijunction structures, but the lattice-mismatched approach is beyond the scope of this chapter [10–12].) The best example of a lattice-matched structure is the $Ga_{0.5}In_{0.5}P/GaAs/Ge$ (E_g = 1.8/1.4/0.7 eV) three-junction solar cell, illustrated schematically in Figure 13.2a, which has been adopted for commercial use and has achieved highest cell efficiencies in excess of 35% under concentration [13–15]. All the layers in this structure are lattice-matched to the Ge substrate, and all can be grown with high quality in conventional metalorganic vapor-phase epitaxy (MOVPE) reactors.

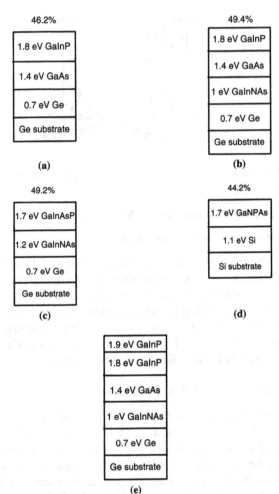

Figure 13.2 Present (a) and proposed (b–e) multijunction solar-cell structures, lattice matched to the substrate. Structures a–d are labeled with their idealized efficiencies at 500 suns, air mass (AM)1.5 Global spectrum. The tunnel-junction interconnects are not shown. The schematics are not to scale.

13.1.5.3 Current limiting in series-connected junctions

In the series-connected configuration of solar-cell junctions, the photocurrent generated by the full multijunction structure will be limited to the smallest of the photocurrents of the individual junctions [15]. Thus it is useful to add an additional junction to an existing design only if the performance of the added junction is sufficiently high. This again illustrates how critical the minority-carrier properties of a semiconductor are to its usefulness in a multijunction solar cell.

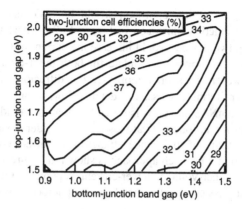

Figure 13.3 Efficiency vs. top- and bottom-junction band gaps for a series-connected idealized two-junction solar cell under the AM1.5 Global spectrum at 300 K and one sun.

In the $Ga_{0.5}In_{0.5}P/GaAs/Ge$ cell, the number of photons available to the Ge junction is almost twice the number of photons available to each of the other two junctions. Therefore, the current through the cell is limited by the top two junctions, and the extra light absorbed by the Ge junction is essentially wasted. The Ge junction could therefore be replaced by a higher-band-gap junction that would absorb fewer photons but produce more energy from each one than does the Ge junction.

13.1.5.4 Choice of band gaps

Figure 13.3 illustrates the tradeoffs involved in choosing semiconductor band-gap combinations for a practical lattice-matched multijunction cell. The figure shows the efficiency as a function of junction band gaps for a two-junction cell [16] (only two junctions are considered in this figure for simplicity). Clearly, the GaInP/GaAs band-gap combination of 1.8/1.4 eV is not ideal, but the ability to actually grow these materials with excellent minority-carrier properties made this materials combination the practical choice for a two-junction cell. The next generation of this device, with a third Ge junction, was a straightforward step.

13.1.6 Role of Dilute Nitrides in Next-Generation Cells

13.1.6.1 Devices lattice-matched to GaAs or Ge

The existence of a semiconductor lattice-matched to GaAs (or Ge, which has essentially the same lattice constant) but with a lower band-gap presents a variety of options for next-generation multijunction solar cells [3, 6, 7, 17, 18]. One option is the insertion of a 1-eV junction into the GaInP/GaAs/ Ge device to give a GaInP/GaAs/1-eV/Ge four-junction structure, as illustrated in Figure 13.2b. Another approach would be a GaInAsP(1.7 eV)/1.2-eV/Ge(0.7 eV) junction illustrated in Figure 13.2c, where the lower band gaps of the top two devices would increase the current of the device [7].

GaInNAs is an obvious candidate for both the 1-eV and the 1.2-eV junctions. However, the short minority-carrier diffusion lengths observed to date in this material, discussed in detail below, mean that photocurrent collection in GaInNAs junctions is much less than for an ideal material. It is therefore desirable, for practical reasons, to consider device designs that lower the sensitivity of the device efficiency to current collection in the GaInNAs junction. One such approach is shown in Figure 13.2e [19]. This structure is a higher-voltage, lower-current version of the Figure 13.2b design that puts less stringent demands on the photocurrent from the GaIn-NAs junction.

13.1.6.2 Devices lattice-matched to Si: GaNPAs/Si

Dilute nitride GaPN provides, as an alternative to GaAs lattice-matched devices, the attractive possibility for devices lattice-matched to Si [20–22]. An ideal two-junction 1.7/1.1-eV GaNPAs/Si cell (Figure 13.2d) would be nearly as efficient as the present GaInP/GaAs/Ge device and would have the major cost and ruggedness advantages of the Si substrate, whereas a 1.8/1.4/1.1-eV GaNPAs/GaNPAs/Si cell would be even more efficient. Such approaches, which introduce the additional complication of the epitaxy of III-Vs on Si, are discussed near the end of this chapter. The dilute nitride phosphide materials are discussed further in Chapter 10.

13.2 RECOMBINATION AND TRANSPORT

13.2.1 Diffusion Length

The addition of N to GaAs to form the dilute nitride alloy has a dramatic effect not only on the band gap, but also on the minority-carrier diffusion lengths L_e and L_h for electrons and holes, respectively. For GaAs, L_e for electrons can easily be 10 μm or more, whereas for Ga(In)NAs, L falls into the 0.1–1-μm range, at best. L is typically measured by measuring the internal spectral quantum efficiency QE(λ) of a p-n or Schottky-barrier junction, and then fitting to the appropriate model for QE(λ) [23]. QE(λ) depends on the absorption coefficient $\alpha(\lambda)$, which must be determined by some complementary means such as transmission measurements. For an infinitely thick absorber with no surface recombination, QE(λ) = $\alpha L/(1+\alpha L)*\exp(-\alpha W_d) + [1 - \exp(-\alpha W_d)]$, where the first term represents collection in the flat-band region due to diffusion and the second term represents collection in the depletion width W_d due to drift. For junctions made of dilute nitride materials, L is typically so small that it is comparable with or smaller than W_d [24]. The functional form for $QE(\lambda)$ does not permit unambiguous determination of both L and W_d, so W_d must be determined by an alternative method such as capacitance-voltage profiling.

The first indication of small diffusion lengths in GaInNAs was the observation [24] that, for an extensive set of GaInNAs epilayers grown by MOVPE over a range of growth conditions, the magnitude of the QE was correlated with the depletion width, implying that $L < W_d$. This set an upper bound of L 0.02–0.5 µm for this set of epilayers. Subsequent studies of L and its dependence on postgrowth annealing led to estimates of hole diffusion length L_h 0.2–0.3 µm in material grown by MOVPE, increasing to 0.6–0.8 µm after postgrowth annealing [8]. Intriguingly, for material grown by molecular-beam epitaxy (MBE), L_h was reported to be only 0.03 µm, with $L_e \approx 0.25$ µm/0.5 µm before/after *ex situ* anneal [25]. These results would suggest that diffusion lengths in these materials are limited to <1 µm for both MOVPE and MBE materials by a mechanism or defect that could be common to both materials, and then further limited to <0.1 µm by other defects that are different for MOVPE and MBE material.

Accurate determination of L when $L < W_d$ is difficult. Further difficulties in determining L arise from the optical and doping properties of GaIn-NAs. The determination requires a precise knowledge of absorption coefficient $\alpha(\lambda)$ [26–28], which is a function of nitrogen content [29, 30]. Another difficulty in determining L from $QE(\lambda)$ stems from the observed tendency (discussed below) of GaInNAs to change conduction type, e.g., from *p*-type to *n*-type, during growth or subsequent annealing (see also Chapters 5 and 10). In a *p-n* junction, such type conversion can change the thicknesses of the *p*- and *n*-layers, as illustrated in Figure 13.4, and can adversely affect their uniformity, adding further ambiguity to the determination of L from the QE. These and other considerations, discussed in detail elsewhere [30], suggest that any reported determination of a diffusion length in these dilute nitride materials from QE should be examined to see that these issues have been accounted for.

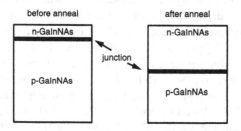

Figure 13.4 Schematic illustration of how type conversion during anneal can change the location of the junction in a GaInNAs *p-n* structure.

The diffusion length L, mobility μ, and lifetime τ of minority carriers are related by $L = (\mu \tau k_B T/e)^{1/2}$. Thus, studies of the mobility and lifetime, discussed next, can shed light on the origin of the short diffusion length.

13.2.2 Lifetime

As discussed in Chapter 9, studies of recombination in GaInNAs have tended to focus on the low-temperature regime and on quantum wells, which are largely dominated by well and interface effects. For solar cells, however, we are interested in the room-temperature regime and bulk layers, where less work has been done to elucidate the trends and mechanisms of recombination in this material. Keyes et al. [31] used time-resolved photoluminescence (TRPL) at room temperature to study the minority-carrier lifetimes of MOVPE-grown GaInP-clad GaInNAs double heterostructures with [N] < 1.3%, and found the lifetime to decrease with increasing doping density. Kaschner et al. [32] performed similar measurements on the complementary composition range of 1.5% < [N] < 2.6%. Both works found PL lifetimes in the range of ≈ 0.2 to 1 ns, a factor of 10 to 100 lower than GaAs lifetimes for comparably doped material. For a set of epilayers with very low [N] = 0.2%, Kurtz et al. found photoluminescence (PL) lifetimes [33] to increase with decreasing background carbon concentration, to as high as 9 ns for [C] 10^{17}/cm^3. Both the Kurtz [33] and Keyes [31] results suggest recombination through a center involving a complex of nitrogen and a doping impurity, e.g., a C-N complex. It is not clear to what extent these results depend on the specifics of the growth conditions. However, Ptak et al. [34] compared MBE- and MOVPE-grown materials and found very comparable subnanosecond lifetimes for both, even though the concentrations of contaminants such as carbon are very different for the two growth processes.

13.2.3 Mobility

Majority-carrier mobilities are typically measured by the Hall effect. It is expected that the minority-carrier mobilities on which the minority-carrier diffusion length depend are closely related to the majority-carrier mobilities. Both n-type and p-type mobilities in GaInNAs are found to be decreasing functions of [N] [34, 35]. However, the electron mobility falls off with the addition of nitrogen far faster than does the hole mobility [24, 35, 36]. For annealed material with [N] $\approx 2\%$ and carrier concentration in the 10^{16}–10^{18} cm^3 range, typical room-temperature hole mobilities are on the order of ≈ 100 cm^2/Vs and electron mobilities ≈ 300 cm^2/Vs [24, 35, 36]. For holes, values are a factor of ≈ 2 less than for GaAs, but for electrons, the decrease compared with the GaAs case is closer to a factor of ten. The reduction in electron mobility is presumed to be due to a combination of alloy scattering and enhanced electron effective mass compared with GaAs

[37]; large-scale inhomogeneities in the material have also been proposed to be important [36]. Mobility is discussed further in Chapters 2 and 10.

13.2.4　$\mu\tau$ Product

It is useful to get an idea of how close the lifetimes and mobilities described above come to giving the desired several-μm diffusion lengths. For 1-eV GaInNAs, a lifetime of $\tau = 0.3$ ns and mobility of $\mu = 300$ cm^2/Vs would be typical of "good" material. Such parameters would give a diffusion length $L = (\mu\tau k_B T/e)^{1/2} = 0.5$ μm (assuming that μ and τ as measured by Hall and TRPL are indeed the quantities that determine L, an assumption that may not be completely justified). To improve this to a more desirable value of 3 μm, the $\mu\tau$ product would have to be improved by more than an order of magnitude. There is no known fundamental reason that such an improvement (most likely through an improvement in τ) cannot be achieved, but achieving such an improvement constitutes a significant challenge.

13.3　IMPURITIES AND DOPING

Background carrier concentrations are typically 10^{17} cm^{-3} in MOVPE-grown GaInNAs. This background is often associated with high concentrations of carbon originating from the source materials [24, 38]. Additionally, concentrations of hydrogen can easily surpass 10^{19} cm^{-3} for some growth conditions [24, 38, 39]. Although these problems are not insurmountable for MOVPE growth, carrier concentrations below 10^{15} cm^{-3} can be obtained fairly easily by MBE due to the much lower levels of background impurities. This enables the growth of p-i-n structures with depletion widths of greater than 2 μm, improving the field-aided collection of photocarriers.

Typically, concentrations of both hydrogen and carbon for GaInNAs grown by solid-source MBE are less than for MOVPE-grown material [34]. MBE growth is discussed in detail in Chapter 10. Figure 13.5 compares typical background carbon and hydrogen concentrations for similar 1-eV GaInNAs samples grown by MBE and MOVPE. However, the growth of fully depleted p-i-n structures is still not easily accomplished, despite the lower impurity concentrations. During the growth of the n-type emitter layer on the nominally undoped base, zinc from the p-type GaAs substrate can diffuse into the base through a Ga interstitial/Zn kick-out process [40]. Some form of diffusion barrier is necessary at the back surface of the device to block the Zn diffusion.

13.3.1　Gallium Vacancies

MOVPE growth has a further complication in terms of background carrier concentration. The addition of hydrogen and nitrogen to GaAs is expected to lower the formation energy of the gallium vacancy (V_{Ga}) by more than

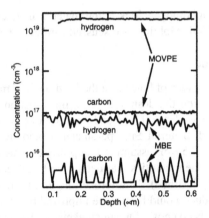

Figure 13.5 Comparison of the carbon and hydrogen concentrations for MBE- and MOVPE-grown GaInNAs, as measured by secondary-ion mass spectrometry. The impurities in the MBE-grown sample are background limited.

2 eV compared with H in GaAs alone [41]. The V_{Ga}-N-H complex acts as a double acceptor and can play a significant role in the background p-type conductivity. Chapter 8 discusses the Fermi-level dependence of the formation energy for various vacancy-related defects in GaNAs and GaAs. V_{Ga} are formed in significant concentrations only when both H and N are present. In the hydrogen-rich MOVPE growth environment, this increases the concentration of V_{Ga} in the V_{Ga}-N-H complex by several orders of magnitude over that expected in GaAs.

Positron annihilation spectroscopy (PAS) shows experimental evidence for this effect [42]. Figure 13.6a compares the normalized S parameter determined from PAS beam measurements for three samples grown by MBE: GaAs, GaInNAs with a band-gap of ≈1.1 eV, and GaInNAs with nominally the same band-gap grown under a flux of atomic hydrogen (a-H). The S parameter is related to the concentration and type of open-volume defects in a material, here assumed to be V_{Ga}. Only the GaInNAs sample grown under a-H shows a signal above background, implying that the hydrogen is facilitating the formation of V_{Ga}. Figure 13.6b shows a similar plot of MOVPE-grown samples, including a GaAs reference, an as-grown GaInNAs sample, and a similar GaInNAs sample that has been annealed. The MBE-grown GaInNAs sample (without a-H) is included for reference. Clearly, the MOVPE-grown GaInNAs sample contains the highest concentration of V_{Ga}, and annealing reduces, but does not completely remove, them. Due to the lack of hydrogen present in the growth atmosphere, MBE has a much lower concentration of V_{Ga} and has the potential to grow low-carrier-concentration material. Theoretical discussions of cation vacancies and of hydrogen-related defects are given in Chapter 8.

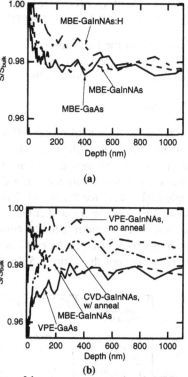

Figure 13.6 (a) Comparison of the vacancy concentrations in MBE-grown GaInNAs samples as measured by positron-annihilation spectroscopy. Only the GaInNAs sample grown under an atomic hydrogen flux shows vacancies. (b) As in (a), but for MOVPE-grown samples. The as-grown GaInNAs sample has the highest concentration of vacancies, with a visible reduction due to annealing. The difference in defects measured by positrons in the first 200 nm of the measurement is due to different passivating cap layers.

13.3.2 Type Conversion

Another effect involving the background carriers is type conversion [30, 43]. Annealing can convert GaInNAs from p-type to n-type. Kurtz et al. [30] showed that a GaInP$_2$/GaInNAs heterojunction turned into a GaInNAs homojunction by annealing the structure. It is likely that this transformation takes place due to the reorganization or injection of hydrogen during the annealing process. Theoretically, hydrogen can act as a donor in p-type semiconductors [44], and specifically, H has been shown to be a donor in GaInNAs [45]. Xin et al. [43] also observed an increase in n-type conductivity that was correlated with increased hydrogen concentrations. Because annealing is commonly used to improve the quality of as-grown GaInNAs layers, this effect must be taken into account when designing devices.

13.4 EFFECT OF RECOMBINATION/TRANSPORT PROPERTIES ON SOLAR-CELL PERFORMANCE

Thus far, we have focused on the basic materials properties of the dilute nitrides. In this section, we illustrate how these properties manifest themselves in the device characteristics of solar cells.

13.4.1 Quantum Efficiency

Figure 13.7a shows the QE for a GaInNAs solar cell with a depletion width $W_d = 0.1$ μm. The small magnitude of the QE is due to a small (<0.1 μm) diffusion length for this particular device, combined with the small W_d. Also shown is another GaInNAs cell with a much wider $W_d = 0.5$ μm. The QE is correspondingly higher than for the $W_d = 0.1$-μm cell due to collection in the depletion region. However, in neither GaInNAs cell does the QE approach the ideal limit of 1, in contrast to the conventional GaAs cell, which is shown for comparison. This GaAs cell has a thin $W_d = 0.2$ μm, yet it collects carriers throughout the entire 3.5-μm-thick absorber layer,

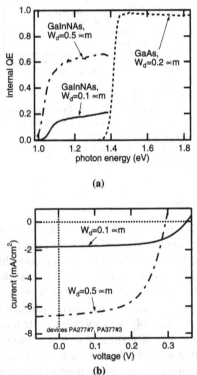

(a)

(b)

Figure 13.7 (a) QE and (b) IV characteristics for GaInNAs junctions with depletion widths $W_d = 0.1$ μm and $W_d = 0.5$ μm.

due to a diffusion length L > 5 μm. In principle, near-unity QE could be obtained in the GaInNAs cells, even with a small diffusion length, if the depletion width could be increased to ≈2–3 μm. However, the background doping in GaInNAs, and its sensitivity to growth conditions, make achievement of such depletion widths difficult in practice.

13.4.2 Current-Voltage Relation

13.4.2.1 Illuminated IV and field-aided collection

For a solar cell, the single most important measure of performance is the current-voltage (IV) relation under illumination. For the two GaInNAs cells whose QEs were compared above, their illuminated and dark IV curves are shown in Figure 13.7b. The $W_d = 0.5$-μm cell has a much greater short-circuit photocurrent J_{SC} than does the $W_d = 0.1$-μm cell, reflecting the greater extent of field-aided collection for the $W_d = 0.5$-μm cell. (The $W_d = 0.5$-μm cell also has a slightly lower band-gap than the $W_d = 0.1$-μm cell, which contributes somewhat to the increased J_{SC}.) For both cells, the illuminated IV curves have a distinct nonzero slope at zero bias, which is not seen in the corresponding dark IV curve. This slope is a signature of field-aided collection, because the depletion width decreases with increasing forward bias.

13.4.2.2 Dark current and depletion-region recombination

Recombination affects not only collection of photocarriers under illumination, but also the dark current. Ideality factors n for GaInNAs p-n junction IV curves are typically observed to fall into the $n = 1.2$–1.5 range [5, 8, 46], in contrast to the textbook $n = 1$ for a junction dominated by recombination of diffusing carriers. Furthermore, the magnitude of the GaInNAs junction dark current is orders of magnitude greater than would be expected for the diffusion recombination, as illustrated in Figure 13.8 [46]. Thus, diffusion recombination is a very minor contribution to the dark current.

One expects trap states to play an important role in recombination in GaInNAs (see Chapter 9). The dark current in these junctions can be very well explained using the classic Sah-Noyce-Shockley [47] model of recombination through midgap trap states in the depletion region, using the trap energy and carrier-capture lifetimes as parameters. Typical agreement with the measured dark current is shown in Figure 13.8 [46]. The deduced trap parameters for this junction and a variety of others grown under similar MOVPE growth conditions are summarized in Figure 13.9 [46]. The recombination is sensitive to the energy difference between the trap and the intrinsic level, but not to the sign of that difference. There are therefore two possibilities for the trap energy, at roughly 0.4–0.5 eV

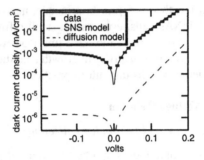

Figure 13.8 Dark IV data for a typical GaInNAs junction (dots) and its fit to a Sah/Noyce/Schockley calculated IV curve. Shown for comparison is a calculation of the contribution of diffusion recombination to the dark IV for this junction.

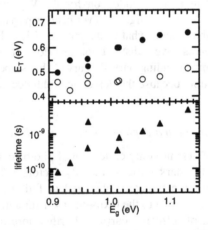

Figure 13.9 Summary of trap parameters: energy E_T (top panel) and lifetime τ (bottom panel) for a set of MOVPE-grown, nonannealed GaInNAs junctions. The two possibilities for E_T are shown as filled and open circles.

from the conduction-band maximum or valence-band minimum, as shown in the figure. Using an analysis of the reverse-saturation current, Kaplar et al. [48] obtained a comparable trap energy of 0.35 eV. It should be emphasized that although there may be several traps in the material, this technique identifies the most important one that dominates the dark current. The trap energies can be used in combination with theoretical studies of trap states (discussed in Chapter 8) to help identify the physical origin of the trap.

The magnitude of the deduced carrier-capture lifetimes is comparable with the PL lifetimes discussed above. For this sample set, the carrier-capture lifetime decreases very rapidly with decreasing band gap/increasing

[N]. If this trend holds in general, it suggests that the lower the band-gap of the GaInNAs, the more problematic will be its application in devices such as solar cells, for which a large dark current is harmful.

For solar cells, a large dark current translates into a low open-circuit voltage V_{OC} when the device is illuminated, adversely affecting the cell's power output. Because the trap-recombination dark current increases with the width of the depletion region, the trick of improving the quantum efficiency (and thus, the cell's current) by increasing the depletion region comes at the price of decreased cell voltage. The IV and QE curves in Figure 13.7 show the results of this tradeoff. The open-circuit voltage V_{OC} of the $W_d = 0.5$-μm junction is 60 mV less than that of the $W_d = 0.1$-μm junction. Half of this difference is due to the 30-meV-lower band-gap of the $W_d = 0.5$-μm junction, but the other half is due to increased recombination in the depletion region. For comparison, a "GaAs-like" 1-eV junction with only band-to-band recombination (as assumed in the idealized device-efficiency projections of Figure 13.2) would be expected to have a significantly higher $V_{OC} = 0.6$ V.

13.5 MATERIALS AND DEVICES LATTICE-MATCHED TO SI

Our discussion thus far has focused on 1-eV GaInNAs lattice-matched to GaAs. However, another dilute nitride system, GaNP, is also of great interest due to its potential for lattice-matching to silicon. Multijunction III-V solar cells fabricated on silicon substrates would be very exciting because they would reduce the substrate cost, increase mechanical robustness, and allow for integration with existing Si technology. dilute nitride phosphides are also discussed in Chapter 10.

13.5.1 Heteroepitaxy on Silicon

The main difficulties in the growth of high-quality III-V semiconductors on Si are related to the difference in the number of valence electrons, lattice constant, and thermal expansion [21]. Lattice-mismatched III-V solar cells on Si substrates have been studied extensively, but the reduction of defect densities resulting from lattice mismatch remains a significant challenge that typically requires complex graded buffer layers [49]. Recently, Yonezu and coworkers [20, 21] have demonstrated dislocation-free III-V-N alloys grown lattice-matched to Si substrates by MBE. Figure 13.10a shows transmission-electron micrographs (TEM) of such high-quality material. MOVPE-grown GaNP(As) remains challenging due to higher growth temperatures necessary to reduce carbon incorporation. These higher growth temperatures may result in more serious problems due to thermal-expansion mismatch.

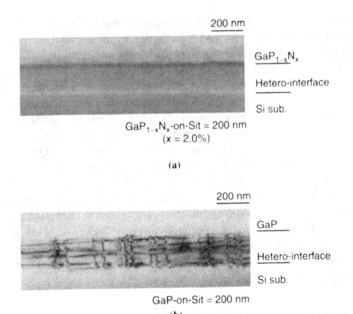

Figure 13.10 TEM images of (a) nearly lattice-matched $GaN_{0.02}P_{0.98}$ on Si and (b) GaP on Si. (Reproduced from Ref. [21] with permission.)

13.5.2 Optical Properties of GaNP

An important property of nitrogen-containing GaP materials is that the indirect GaP quickly converts to a direct (or direct-like) material as nitrogen is added [50, 51]. Thus, GaNPAs or GaInNP alloys with compositions close to GaP can be used in thin absorbing layers in a multilayer solar-cell structure lattice-matched to a silicon substrate (see absorption coefficient in Figure 13.11). These III-V-N alloys are also being studied for light-emitting applications on silicon substrates [20, 52, 53].

13.5.3 Solar-Cell Applications

A proposed solar-cell structure based on lattice-matched GaNPAs alloys grown on silicon [54] is shown in Figure 13.2d. The structure, composed of a lattice-matched III-V cell grown on a Si cell, could potentially rival the efficiencies of high-efficiency cells on GaAs or Ge, with significant cost savings and improvements in mechanical stability, if sufficiently good minority-carrier transport can be achieved in the III-N-V alloy. In fact, for two-junction devices, the band gaps of 1.1 eV and 1.7 eV are nearly optimal for maximum efficiency (see Figure 13.3) [16].

GaNPAs solar cells with band gaps ranging from 1.5 to 2.0 eV have been fabricated on GaP substrates [54] as an initial step toward GaNPAs/Si cells. Figure 13.12 shows the performance of one such cell. Relatively low quantum

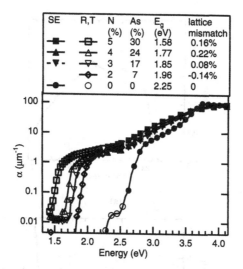

SE	R,T	N (%)	As (%)	E_g (eV)	lattice mismatch
■	▢	5	30	1.58	0.16%
▲	△	4	24	1.77	0.22%
▼	▽	3	17	1.85	0.08%
◆	◇	2	7	1.96	-0.14%
●	○	0	0	2.25	0

Figure 13.11 Absorption coefficient of GaNPAs layers grown on GaP substrates. The GaP substrate is shown as circles.

efficiencies (QE) and open-circuit voltages (relative to the band gap) indicate problems with the material quality. The low internal QE likely indicates short diffusion lengths similar to the GaInNAs on GaAs materials. The QE shown in Figure 13.12 has been improved by using a *p-i-n* (enhanced depletion width) structure similar to the strategy used in GaInNAs solar cells.

13.5.4 Carbon and Hydrogen in MOVPE-Grown GaNP

Growth temperature and source materials (e.g., triethylgallium vs. trimethylgallium) dominate the behavior of carbon and hydrogen incorporation into GaNP [55]. Carbon and hydrogen incorporation from gallium and nitrogen-sources in MOVPE-grown GaNP can reach levels of 10^{20} cm^{-3}. The carbon contamination appears to adversely affect the PL lifetimes [55] and solar-cell performance. Figure 13.13 shows trends of PL lifetime in $GaN_{0.02}P_{0.98}$ with carbon and hydrogen contamination. This incorporation can be minimized by optimizing growth conditions [55]. MBE-grown GaNP(As) may not suffer from problems of unintentional carbon and hydrogen incorporation.

13.6 CONCLUSIONS

The dilute nitride III-V-N materials open up new possibilities for future generations of high-efficiency multijunction solar cells. Realization of these potential benefits awaits further improvement in the minority-carrier properties of these materials.

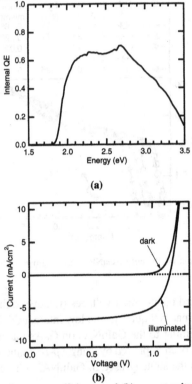

(a)

(b)

Figure 13.12 (a) Internal quantum efficiency and (b) current-voltage curves of a 1.90-eV GaNPAs solar cell grown nearly lattice-matched to the GaP substrate. Dashed curve: measured in the dark; solid curve: measured in the light under AM1.5G conditions.

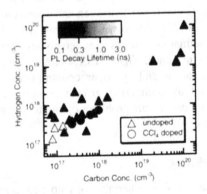

Figure 13.13 Room temperature PL lifetime of $GaN_{0.02}P_{0.98}$ layers grown on GaP at various growth conditions as a function of carbon and hydrogen concentrations measured by secondary-ion mass spectrometry. White-filled data points indicate PL lifetimes as long as 3 ns; darker data points indicate shorter lifetimes.

References

1. Weyers, M., Sato, M., and Ando, H., "Red Shift of Photoluminescence and Absorption in Dilute GaAsN Alloy Layers," *Jpn. J. Appl. Phys.* 31: 853–855 (1992).
2. Kondow, M., Uomi, K., Niwa, A., Kitatani, T., Watahiki, S., and Yazawa, Y., "GaInNAs: A Novel Material for Long-Wavelength-Range Laser Diodes with Excellent High-Temperature Performance," *Jpn. J. Appl. Phys.* 35: 1273–1275 (1996).
3. Kurtz, S.R., Myers, D., and Olson, J.M., "Projected Performance of Three- and Four-Junction Devices Using GaAs and GaInP," in *26th IEEE Photovoltaic Specialists Conference* (New York: IEEE, 1997).
4. Friedman, D.J., Geisz, J.F., Kurtz, S.R., and Olson, J.M., "1-eV Solar Cells with GaIn-NAs Active Layer," *J. Crystal Growth* 195: 409–415 (1998).
5. Friedman, D.J., Geisz, J.F., Kurtz, S.R., and Olson, J.M., "1-eV GaInNAs Solar Cells for Ultrahigh-Efficiency Multijunction Devices," in *2nd World Conf. on Photovoltaic Energy Conversion* (New York: IEEE, 1998).
6. Hou, H.Q., Reinhardt, K.C., Kurtz, S.R., Gee, J.M., Allerman, A.A., Hammons, B.E., Chang, P.C., and Jones, E.D., "Novel InGaAsN pn Junction for High-Efficiency Multiple-Junction Solar Cells," in *2nd World Conf. on Photovoltaic Energy Conversion* (New York: IEEE, 1998).
7. Li, N.Y., Sharps, P.R., Stan, M., Newman, F., Hills, J.S., Hou, H.Q., Gee, J.M., and Aiken, D.J., "Development of 1.25 eV InGaAsN for Triple Junction Solar Cells," in *28th IEEE Photovoltaic Specialists Conference* (New York: IEEE, 2000).
8. Kurtz, S.R., Allerman, A.A., Jones, E.D., Gee, J.M., Banas, J.J., and Hammons, B.E., "InGaAsN Solar Cells with 1.0 eV Band Gap, Lattice Matched to GaAs," *Appl. Phys. Lett.* 74 (5): 729–731 (1999).
9. Henry, C.H., "Limiting Efficiencies of Ideal Single and Multiple Energy Gap Terrestrial Solar Cells," *J. Appl. Phys.* 51: 4494–4500 (1980).
10. Dimroth, F., Beckert, R., Meusel, M., Schubert, U., and Bett, A.W., "Metamorphic $Ga_yIn_{1-y}P/Ga_{1-x}In_xAs$ Tandem Solar Cells for Space and for Terrestrial Concentrator Applications at C >1000 Suns," *Prog. Photovolt.* 9: 165–178 (2001).
11. Dimroth, F., Lanyi, P., Schubert, U., and Bett, A.W., "MOVPE Grown $Ga_{1-x}In_xAs$ Solar Cells for GaInP/GaInAs Tandem Applications," *J. Electron. Mater.* 29: 42–46 (2000).
12. King, R.R., Haddad, M., Isshiki, T., Colter, P., Ermer, J., Yoon, H., Joslin, D.E., and Karam, N.H., "Metamorphic GaInP/GaInAs/Ge Solar Cells," in *28th IEEE Photovoltaic Specialists Conference* (New York: IEEE, 2000).
13. Bertness, K.A., Kurtz, S.R., Friedman, D.J., Kibbler, A.E., Kramer, C., and Olson, J.M., "29.5%-Efficient GaInP/GaAs Tandem Solar Cells," *Appl. Phys. Lett.* 65: 989–991 (1994).
14. King, R.R. et al., "High-Efficiency Space and Terrestrial Multijunction Solar Cells through Bandgap Control in Cell Structures," in *29th IEEE Photovoltaic Specialists Conference* (New York: IEEE, 2002).
15. Olson, J.M., Kurtz, S.R., Kibbler, A.E., and Faine, P., "A 27.3% efficient $Ga_{0.5}In_{0.5}P$/GaAs Tandem Solar Cell," *Appl. Phys. Lett.* 56: 623–625 (1990).
16. Kurtz, S.R., Faine, P., and Olson, J.M., "Modeling of Two-Junction, Series-Connected Tandem Solar Cells Using Top-Cell Thickness as an Adjustable Parameter," *J. Appl. Phys.* 68: 1890–1896 (1990).
17. Friedman, D.J., Kurtz, S.R., and Geisz, J.F., "Analysis of the GaInP/GaAs/1-eV/Ge Cell and Related Structures for Terrestrial Concentrator Application," in *29th Photovoltaic Specialists Conference* (New York: IEEE, 2002).
18. Geisz, J.F., and Friedman, D.J., "III-N-V Semiconductors for Solar Photovoltaic Applications," *Semicond. Sci. Technol.* 17: 769–777 (2002).

19. King, R.R., Colter, P.C., Joslin, D.E., Edmondson, K.M., Krut, D.D., Karam, N.H., and Kurtz, S., "High-Voltage, Low-Current GaInP/GaInP/GaAs/GaInNAs/Ge Solar Cells," in *29th IEEE Photovoltaic Specialists Conference* (New York: IEEE, 2002).

20. Fujimoto, Y., Yonezu, H., Utsumi, A., Momose, K., and Furukawa, Y., "Dislocation-Free GaAsyP$_{1-x-y}$N$_x$/GaP$_{0.98}$N$_{0.02}$ Quantum-Well Structure Lattice-Matched to a Si Substrate," *Appl. Phys. Lett.*, 79: 1306–1308 (2001).

21. Yonezu, H., "Control of Structural Defects in III-V-N Alloys Grown on Si," *Semicond. Sci. Technol.* 17: 762–768 (2002).

22. Tu, C.W., "III-N-V Low-Bandgap Nitrides and Their Device Applications," *J. Phys. Condensed Matter* 13: 7169–7182 (2001).

23. Kurtz, S.R., and Olson, J.M., "Rapid Characterization of Photovoltaic Materials Using Photoelectrochemical Techniques," in *19th IEEE Photovoltaic Specialists Conference* (New York: IEEE 1987).

24. Geisz, J.F., Friedman, D.J., Olson, J.M., Kurtz, S.R., and Keyes, B.M., "Photocurrent of 1-eV GaInNAs Lattice-Matched to GaAs," *J. Crystal Growth* 195: 401–408 (1998).

25. Kurtz, S.R., Klem, J.F., Allerman, A.A., Sieg, R.M., Seager, C.H., and Jones, E.D., "Minority Carrier Diffusion and Defects in InGaAsN Grown by Molecular Beam Epitaxy," *Appl. Phys. Lett.* 80: 1379–1381 (2002).

26. Perlin, P., Wisniewski, P., Skierbiszewski, C., Suski, T., Kaminska, E., Subramanya, S.G., Weber, E.R., Mars, D.E., and Walukiewicz, W., "Interband Optical Absorption in Free Standing Layer of Ga$_{0.96}$In$_{0.04}$As$_{0.99}$N$_{0.01}$," *Appl. Phys. Lett.* 76: 1279–1281 (2000).

27. Wagner, J., Kohler, K., Ganser, P., and Herres, N., "GaAsN Interband Transitions Involving Localized and Extended States Probed by Resonant Raman Scattering and Spectroscopic Ellipsometry," *Appl. Phys. Lett.* 77: 3592–3594 (2000).

28. Matsumoto, S., Yaguchi, H., Kashiwase, S., Hashimoto, T., Yoshida, S., Aoki, D., and Onabe, K., "Optical Characterization of Metalorganic Vapor-Phase Epitaxy-Grown GaAs$_{1-x}$N$_x$ Alloys Using Spectroscopic Ellipsometry," *J. Crystal Growth* 221: 481–484 (2000).

29. Takeuchi, K., Miyamoto, T., Kageyama, T., Koyama, F., and Iga, K., "Chemical Beam Epitaxy Growth and Characterization of GaNAs/GaAs," *Jpn. J. Appl. Phys. Pt. 1* 37: 1603–1607 (1998).

30. Kurtz, S., Geisz, J.F., Friedman, D.J., Olson, J.M., Duda, A., Karam, N.H., King, R.R., Ermer, J.H., and Joslin, D.E., "Modeling of Electron Diffusion Length in GaInAsN Solar Cells," in *28th IEEE Photovoltaic Specialists Conference* (New York: IEEE, 2000).

31. Keyes, B.M., Geisz, J.F., Dippo, P.C., Reedy, R., Kramer, C., Friedman, D.J., Kurtz, S.R., and Olson, J.M., "Optical Investigation of GaNAs," *AIP Conf. Proc.* 462: 511–516 (1999).

32. Kaschner, A., Luttgert, T., Born, H., Hoffmann, A., Egorov, A.Y., and Riechert, H., "Recombination Mechanisms in GaInNAs/GaAs Multiple Quantum Wells," *Appl. Phys. Lett.* 78: 1391–1393 (2001).

33. Kurtz, S., Geisz, J.F., Keyes, B.M., Metzger, W.K., Friedman, D.J., Olson, J.M., Ptak, A.J., King, R.R., and Karam, N.H., "Effect of Growth Rate and Gallium Source on GaAsN," *Appl. Phys. Lett.* 82: 2634–2636 (2003).

34. Ptak, A.J., Johnston, S.W., Kurtz, S., Friedman, D.J., and Metzger, W.K., "A Comparison of MBE- and MOCVD-Grown GaInNAs," *J. Crystal Growth* 251: 392–398 (2003).

35. Li, W., Pessa, M., Toivonen, J., and Lipsanen, H., "Doping and Carrier Transport in Ga$_{1-3x}$In$_{3x}$N$_x$As$_{1-x}$ Alloys," *Phys. Rev. B* 64: 113308/1-3 (2001).

36. Kurtz, S.R., Allerman, A.A., Seager, C.H., Sieg, R.M., and Jones, E.D., "Minority Carrier Diffusion, Defects, and Localization in InGaAsN, with 2% Nitrogen," *Appl. Phys. Lett.* 77: 400–402 (2000).

37. Skierbiszewski, C., Perlin, P., Wisniewski, P., Suski, T., Walukiewicz, W., Shan, W., Ager III, J.W., Haller, E.E., Geisz, J.F., Friedman, D.J., Olson, J.M., and Kurtz, S.R., "Effect of Nitrogen-Induced Modification of the Conduction Band Structure on Electron Transport in GaAsN Alloys," *Phys. Stat. Solidi B* 216: 135–139 (1999).
38. Kurtz, S.R., Allerman, A.A., Jones, E.D., Gee, J.M., Banas, J.J., and Hammons, B.E., "InGaAsN Solar Cells with 1.0 eV Band Gap, Lattice-Matched to GaAs," *Appl. Phys. Lett.* 74: 729–731 (1999).
39. Li, N.Y., Hains, C.P., Yang, K., Lu, J., Cheng, J., and Li, P.W., "Organometallic Vapor Phase Epitaxy Growth and Optical Characteristics of Almost 1.2 µm GaInNAs Three-Quantum-Well Laser Diodes," *Appl. Phys. Lett.* 75: 1051–1053 (1999).
40. Deppe, D.G., "Thermodynamic Explanation to the Enhanced Diffusion of Base Dopant in AlGaAs-GaAs npn Bipolar Transistors," *Appl. Phys. Lett.* 56: 370–372 (1990).
41. Janotti, A., Wei, S.-H., Zhang, S.B., Kurtz, S., and Walle, C.G.V., "Interactions between Nitrogen, Hydrogen, and Gallium Vacancies in GaAsN Alloys," *Phys. Rev. B* 67: 161201/1-4 (2003).
42. Ptak, A.J., Kurtz, S., Lynn, K.G., and Weber, M.H., "The Effect of Hydrogen on Gallium Vacancies in GaInNAs," *J. Vac. Sci. Technol.* B22 in press (2004).
43. Xin, H.P., Tu, C.W., and Geva, M., "Effects of Hydrogen on Doping of GaInNAs Grown by Gas-Source Molecular Beam Epitaxy," *J. Vac. Sci. Technol. B* 18: 1476–1479 (2000).
44. Van de Walle, C.G., Denteneer, P.J.H., Bar-Yam, Y., and Pantelides, S.T., "Theory of Hydrogen Diffusion and Reactions in Crystalline Silicon," *Phys. Rev. B* 39: 10791–10808 (1989).
45. Janotti, A., Zhang, S.B., Wei, S.H., and VandeWalle, C.G., "Effects of Hydrogen on the Electronic Properties of Dilute GaAsN Alloys," *Phys. Rev. Lett.* 89: 6403–6406 (2002).
46. Friedman, D.J., Geisz, J.F., Metzger, W.K., and Johnston, S.W., "Trap-Dominated Minority-Carrier Recombination in GaInNAs pn Junctions," *Appl. Phys. Lett.* 83, p. 698, in press (2003).
47. Sah, C.T., Noyce, R.N., and Shockley, W., "Carrier Generation and Recombination in p-n Junctions and p-n Junction Characteristics," *Proc. IRE* 45: 228–1243 (1957).
48. Kaplar, R.J., Arehart, A.R., Ringel, S.A., Allerman, A.A., Sieg, R.M., and Kurtz, S.R., "Deep Levels and Their Impact on Generation Current in Sn-Doped InGaAsN," *J. Appl. Phys.* 90: 3405–3408 (2001).
49. Soga, T., Kato, T., Yang, M., Umeno, M., and Jimbo, T., "High Efficiency AlGaAs/Si Monolithic Tandem Solar Cell Grown by Metalorganic Chemical Vapor Deposition," *J. Appl. Phys.* 78: 4196–4199 (1995).
50. Xin, H.P., Tu, C.W., Zhang, Y., and Mascarenhas, A., "Effects of Nitrogen on the Band Structure of GaN$_x$P$_{1-x}$ alloys," *Appl. Phys. Lett.* 76: 1267–1269 (2000).
51. Shan, W., Walukiewicz, W., Yu, K.M., Wu, J., Ager, J.W., Haller, E.E., Xin, H.P., and Tu, C.W., "Nature of the Fundamental Band-gap in GaN$_x$P$_{1-x}$ Alloys," *Appl. Phys. Lett.* 76: 3251–3253 (2000).
52. Xin, H.P., Welty, R.J., and Tu, C.W., "GaN$_{0.011}$P$_{0.989}$ Red Light-Emitting Diodes Directly Grown on GaP Substrates," *Appl. Phys. Lett.* 77: 1946–1948 (2000).
53. Xin, H.P., Welty, R.J., Hong, Y.G., and Tu, C.W., "Gas-Source MBE Growth of Ga(In)NP/GaP Structures and Their Applications for Red Light-Emitting Diodes," *J. Crystal Growth* 227: 558–561 (2001).
54. Geisz, J.F., and Friedman, D.J., "GaNPAs Solar Cells Lattice Matched to GaP," in *29th IEEE Photovoltaic Specialists Conference* (New York: IEEE, 2002).
55. Geisz, J.F., Reedy, R.C., Keyes, B.M., and Metzger, W.K., "Unintentional Carbon and Hydrogen Incorporation in GaNP Grown by Metal-Organic Chemical Vapor Deposition," *J. Crystal Growth* 259, p. 223 (2003).

CHAPTER 14

GaInNAs and GaInNAsSb Long-Wavelength Lasers

J.S. HARRIS

Solid State and Photonics Lab, Stanford University, Stanford, California

14.1 INTRODUCTION

14.1.1 Long-Wavelength Applications

The incredible growth of the Internet and data transmission are now pushing the bandwidth requirements for metro (MAN), local (LAN), and storage (SAN) area networks to unprecedented performance levels. Low-cost, long-wavelength, single-mode vertical-cavity lasers (VCSELs) that can be directly modulated at 10 Gb/s, that operate uncooled in ambient environments, and that are easily packaged and coupled to fiber are an essential

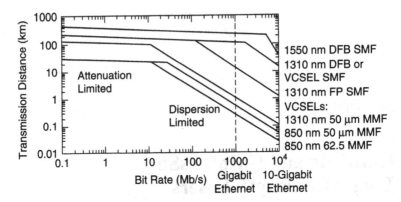

Figure 14.1 Transmission distance vs. laser-modulation frequency for a variety of optical-fiber/laser-diode sources utilized in optical networks.

element to enable this optical revolution. While initial LANs could utilize GaAs-based, 850-nm VCSELs that had most of the above requisite properties, the bit-rate push from 100 Mb/s to 1 Gb/s to 10 Gb/s in the newest systems makes longer-wavelength VCSELs essential. The requirement for such longer-wavelength lasers is illustrated in Figure 14.1. For 850-nm VCSELs at 100 Mb/s, the transmission distance is 7–8 km, certainly suitable for LANs; however, at 10 Gb/s, the transmission distance is ≈50 m, not even useful for a small, intrabuilding LAN [1].

While InGaAsP/InP-based Bragg grating and distributed-feedback (DFB) lasers have been the sources for long-haul, 1.55-μm optical-fiber backbone networks for the past three decades and clearly meet the distance criteria in Figure 14.1, their cost is still far too high to meet the demands of hundreds of million lasers that might be utilized in MANs and LANs in a modern data communication network architecture. In addition to the low-cost VCSELs, there is an additional network challenge in order to realize low-cost, high-bandwidth networks, and that is that a much greater portion of the low-loss fiber bandwidth must be made accessible to enable coarse-wavelength division multiplexing (CWDM) to be utilized. This requirement comes from the necessity to operate the lasers uncooled (i.e., no thermo-electric cooler) and directly modulated. These two requirements translate into much larger bandwidths required per channel (≈25 nm) compared with current dense-wavelength division multiplexing (DWDM) systems (≈0.1 nm) employed in the fiber backbone. The primary driving force for DWDM systems was development of the erbium-doped fiber amplifier (EDFA), which enabled all channels within its gain region to be amplified in parallel. Unfortunately, the EDFA gain region is only about 10% of the available low-loss fiber region, which is illustrated in Figure 14.2 [2]. While there is work ongoing to extend the EDFA gain into the L-band, there are still large

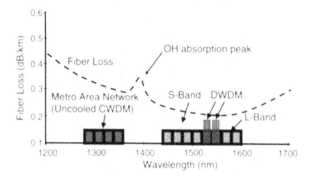

Figure 14.2 Fiber loss vs. wavelength, showing a small portion of low-loss regions used in today's DWDM systems and potential areas for broadband networks.

areas of the fiber not utilized because of a lack of suitable amplifiers. This provides a second, equally compelling driving force to develop better high-power semiconductor lasers in the 1.3–1.6-μm-wavelength region to serve as pump lasers for Raman amplifiers, which can work throughout the entire fiber low-loss region. The sole reason that EDFAs rather than Raman amplifiers dominate the fiber backbone is availability of excellent, reliable, high-power 980-nm InGaAs pump lasers for EDFAs, while there are no comparable pump lasers for Raman amplifiers.

14.2 CANDIDATE LONG-WAVELENGTH MATERIALS SYSTEMS

The above challenges and opportunities have certainly not gone unnoticed to device and materials scientists, and there has been an intense effort to realize both low-cost, long-wavelength VCSELs and high-power pump lasers between 1.3 and 1.6 μm over the past decade [1]. Semiconductor lasers operating in the 1.3–1.6-μm region require materials with band gaps between 0.95 and 0.78 eV. The potential candidate alloys are shown in Figure 14.3. One of the requirements for alloy semiconductors is that they must be reasonably closely lattice-matched to readily available binary substrates (GaAs or InP). The potential choices are thus defined by the intersection of the dashed horizontal lines defining wavelength with the vertical lines below GaAs and InP defining lattice match. For many years, it was believed that there was no suitable alloy adequately lattice-matched to GaAs that would emit at >1.1 μm, so InGaAsP on InP was the only materials system that met the perceived criteria, and 100% of the long-wavelength communications lasers today are fabricated from this system.

This search for new options has led to at least six different choices of materials combinations and approaches [1] by trying to either avoid or accommodate the physical properties limitations of the InGaAsP/InP mate-

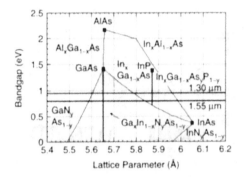

Figure 14.3 Band-gap vs. lattice constant for III-Arsenide alloys showing lines of lattice match to GaAs for Nitride-Arsenide alloys and to InP for Arsenide-Phosphide alloys in the region applicable to long-wavelength fiber systems.

rials system. The major limitations of the InP-based system are described by Kondow et al. [3] and in Chapter 11:

1. There is no good combination of alloys that is lattice matched to produce a large difference in index of refraction with reasonable thermal conductivity for the distributed Bragg reflectors (DBRs) required for VCSELs.
2. T_0, the laser threshold current temperature coefficient, is quite low compared with InGaAs/GaAs, the dominant combination for high-power EDFA pumps.
3. The thermal conductivity of the DBR mirror or cladding layers is inferior to GaAs-based structures, resulting in greater junction heating under operation.

The InGaAsP limitations pushed exploration of several approaches, which can be divided into two "camps" [1]:

1. Those using InGaAsP/InP quantum-well (QW) active regions, but relying on alternative nonepitaxial approaches for the DBR mirrors
2. Those based upon GaAs/AlAs DBR mirror technology, but using a new active gain region of materials closely lattice-matched to GaAs

InGaAsP QW-based VCSELs have been fabricated using metal mirrors [4], wafer-bonded AlAs/GaAs mirrors [5], combined InGaAsP/InP and AlAs/GaAs metamorphic mirrors [6], AlGaAsSb/AlAsSb mirrors [7], and dielectric mirrors [8]. GaAs-based VCSEL approaches included InAs quantum-dot active regions [9] and GaAsSb/InGaAs Type II quantum wells [10], until Kondow et al. [3, 11] discovered, much to everyone's surprise, that GaInNAs was an attractive candidate because its band-gap could be reduced sufficiently to reach the desired wavelength regions, and most importantly, it could be grown on GaAs. This discovery was clearly far from obvious from the known properties of all other III-V ternary and quaternary alloys, where the general rule was that alloys with a smaller lattice constant had an increased band gap. The large electronegativity of N and its small covalent radius cause a very strong negative bowing parameter, and the addition of N to GaAs or GaInAs dramatically decreases the band gap, as discussed in a work by Kondow et al. [3], in Chapters 1–3 of this book, and as illustrated in Figure 14.3. The success of their discovery has catapulted this new material into the lead for the development of a broad range of VCSEL and high-power edge-emitting lasers that I believe will be the foundation of lower-cost optical networks because they not only produce the low-cost VCSELs that have been a key focus, but they enable both Raman and semiconductor optical amplifiers (SOAs) that will provide gain throughout the 1.3–1.6-μm wavelength region, enabling use of the full low-loss fiber bandwidth, which will be crucial to employ lasers that are uncooled and directly modulated. The remainder of this chapter focuses on GaInNAs and more recent work on GaInNAsSb for realizing these lasers.

Research on GaInNAs has revealed several additional factors vis-à-vis InGaAsP/InP that could prove decisive in the race to produce low-cost, long-wavelength VCSELs and high-power Raman pump lasers:

1. For the same band-gap material, the conduction-band well is deeper [3, 11, 12], as described in Chapters 1–3, and the electron effective mass is larger in GaInNAs [12, 13], as described in Chapter 4, thus providing better confinement for electrons and a better match of the valence- and conduction-band densities of states. This leads to a higher T_0, higher operating temperature, higher efficiency, and higher output power [11].

2. Most of the energy-band engineering that is used to minimize heterojunction voltage drops involves the use of intermediate graded layers of $Al_xGa_{(1-x)}As$ or AlAs/GaAs superlattices, all of which are lattice-matched to GaAs and do not require difficult compositional control over both column III and column V constituents in a quaternary layer, such as InGaAsP, to maintain lattice match [1].

3. Compositional control and uniformity of GaInNAs grown by molecular-beam epitaxy (MBE) [14–19] is relatively easy compared with metalorganic vapor-phase epitaxy (MOVPE) [20–29] or with AsP control in InGaAsP [30, 31]. This will translate into better yield and far easier scale-up to larger wafers for lower cost.

4. VCSELs can be straightforwardly fabricated using the well-developed GaAs/AlAs mirror and AlAs oxidation for current and optical aperture confinement technologies.

5. GaInNAs on GaAs provides easy monolithic integration with GaAs
    high-speed electronics, which will be essential to provide low-cost,
    high-speed integrated electrical drivers for direct laser modulation in
    high-speed networks.

14.3 EPITAXIAL GROWTH TECHNIQUES

While the above discussion highlights the advantages of the GaInNAs/GaAs
system over the InGaAsP/InP or GaAsSb/GaAs systems, there are still very
significant challenges before GaInNAs can be used to produce useful,
reliable lasers at 1.3–1.6 μm [11–27]. One of the major challenges is the
completely different set of growth conditions that affect the choice of
epitaxial growth technique. Both MBE [1, 3, 11–20] and MOVPE [21–28]
have been used to grow this material; however, the issues governing choice
are appreciably more complex and challenging than for earlier GaAs-based
systems for 850-nm VCSELs, 980-nm InGaAs EDFA pump lasers, or 1.55-
μm InP-based DFB lasers. MOVPE was the primary choice for all three of
these devices. The major advantages of MOVPE for GaAs-based 850-nm
VCSELs were higher growth rate and particularly much easier growth of
the complex grading/doping interfaces that one utilizes to minimize the
series resistance from the AlAs/GaAs heterojunctions that occur at each
interface of the 25–30 quarter-wave layer pairs of the distributed Bragg
mirrors. While chirp grading by multiple Al/Ga shutter opening/closing is
possible in MBE, this approach was not regarded as "production worthy"
and has limited MBE VCSEL fabrication to research and development in
the past [1].

Edge-emitting InP-based systems were dominated by MOVPE
because, at the outset, valved phosphorus (P)-crackers did not exist, and
early MBE attempts at growing InP-based materials were hazardous
because of the fires that could result when the systems were opened and
exposed to air. The growth temperature, gas flow control with mass-flow
controllers, doping, etc. were all within a range that was compatible with
readily available chemical precursors and growth parameters for MOVPE,
hence it became the dominant optoelectronic epitaxial technology.

The situation for growth of GaInNAs is entirely different compared
with the earlier materials systems. First and foremost is that in order to
incorporate sufficient N, the growth has to be done at much lower growth
temperatures and under metastable growth conditions within the miscibility gap region of the GaInNAs alloy. This is due to the different basic
crystal structures of the constituent alloys and their regions of growth
compatibility. InGaN is a hexagonal (wurtzite) crystal grown at relatively
high temperatures, while InGaAs is cubic (zinc blende) grown at relatively
low temperatures, creating a miscibility gap in the alloys [16–21]. Hence,
as either or both N and growth temperature increase, phase segregation

occurs, and the material breaks up into microscopic regions of InGaAs and InGaN [16–21].

Because growth must be at much lower temperatures than earlier GaAs and InP-based systems, growth by MOVPE is far more challenging. Compared with MOVPE growth of N-based, wide-band-gap systems that use ammonia as the N source, the growth temperature for GaInNAs is too low to achieve reasonable cracking of either ammonia or arsine [25–29]. New sources with difficult precursor reactions and highly nonlinear incorporation ratios greatly complicate the growth compared with work on earlier III-V materials systems. There are also strong precursor reactions between N and Al, thus MOVPE growth is either done in two separate reactors [27, 28], by a two-step growth process [29], or by avoiding the use of Al-containing materials entirely by using InGaP for the wide-band-gap materials [32, 33]. This latter approach has worked well for edge-emitting lasers, but it is not well suited to produce the DBR mirrors or oxide-confining regions that are critical for VCSEL fabrication. In addition, the higher growth temperatures limit the N incorporation where microphase segregation [20, 26] begins and makes it extremely challenging to reach the N compositions needed to produce lasers beyond 1.3 μm. An illustration of this challenge is to look at the literature on GaInNAs lasers and note how many results are reported between 1.25 and 1.28 μm compared with 1.31 μm, where the zero in fiber dispersion occurs, which is the natural target wavelength for high-data-rate networks. Only Fischer's early result [34] stands in contrast to this, with GaInNAs lasers at 1.52 μm.

MBE is certainly not without its challenges in growing these metastable alloys, as discussed in Chapters 10, 11, and 13. Kondow's first work [3, 11] utilized a N plasma source added to a gas-source MBE system. This system provided the insight into the potential for GaInNAs. However, the issues of H incorporation and its effect on the material [35–37] were problematic, as described in Chapter 6. There are also significant challenges created by the very strong sensitivity of composition to growth temperature [25–29] due to the low arsine cracking efficiency. Solid-source MBE with an atomic N plasma source has proven to be the simplest and most effective system to enable growth at the lowest temperatures and over the largest range of N and In compositions [3–20, 34, 38]. Growth temperature is the single most critical parameter controlling growth [16, 17]. When growth temperature exceeds a critical value, MBE growth begins to change from two-dimensional, layer-by-layer growth to three-dimensional island growth with microphase segregation [20, 26]. Figure 14.4 is a TEM (transmission electron microscopy) micrograph showing successful growth of two GaIn-NAs QWs followed by a third, which suddenly phase segregates, resulting in terrible surface morphology and terrible, nonrecoverable luminescence. All of these regions of terrible morphology are In-rich and indicate that some type of In surface segregation is occurring, because all of these regions

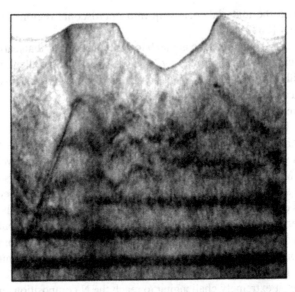

Figure 14.4 TEM cross-section image of a triple-QW, 35% In GaInNAs-GaAs barrier sample where three-dimensional growth and phase segregation have occurred in the third QW.

appear at the top of the QW, never at the starting interface. When this region becomes sufficiently In-rich, phase segregation occurs. Such observations are very common, and the window for good epitaxial growth is quite narrow compared with other III-V alloys. There is a small N-composition dependence on optimum growth temperature, with the temperature range from 420°C to 460°C maintaining two-dimensional epitaxial growth over the greatest range of small N compositions. V/III ratio also has an impact on growth, but much less so than temperature. The issue of strain and surface structure will be revisited later in section 14.3.

We have observed many differences in defects, impurity incorporation, annealing, etc. in GaInNAs alloys compared with other III-V alloy systems [14-17, 37-40], and these are described in Chapter 10. The particular issue of annealing is unique to the GaInNAs and GaInNAsSb systems and is discussed in considerable detail in Section 14.4. Improvements to both materials and laser designs will be required to capitalize on the intrinsic advantages of the GaInNAs materials system to realize the exciting device potential to meet many of the demands for semiconductor lasers in next-generation photonic networks.

The addition of nitrogen to InGaAs quantum wells results in the longest wavelengths achievable on GaAs substrates, and lasers out to 1.52 μm have been demonstrated with InGaNAs-active regions grown on GaAs substrates [34]. Because of the large difference in growth temperatures for GaInN and GaInAs, a very reactive N source is required. We utilize a combined ion-plasma-source and solid-source MBE system where all of the species, except As_2, are atomic and thus impinge on the substrate in a form that maximizes chances for a chemical reaction with N to form the desired alloys.

One of our early and unexpected findings was that the atomic-nitrogen sticking coefficient is constant and near unity, i.e., that the group III growth rate (*not* the N/As flux ratio) controls the nitrogen concentration and N/As mole fraction in the film [16, 17]. At substrate temperatures of 500°C, one does not need to control the As_2 flux, other than to keep the surface As-stabilized. We grow with an As_2 flux of 8 to 20 times the gallium flux, much like the growth of GaAs or InGaAs. This is completely different than any other mixed-column V, III-V alloy semiconductor system [17, 30, 31]. Control of the N/As mole fraction is critical because it changes both the band-gap and the lattice constant. It is truly surprising that there is little sensitivity to the gas-phase mole fraction, and this is completely different than MOVPE-grown InGaAsP [30, 31] or GaInNAs [21–28]. This indicates that MBE growth of GaInNAs will have major advantages over MOVPE in terms of yield and reproducibility.

Another issue is that of background impurities. We have used secondary-ion mass spectroscopy (SIMS) to verify that the impurity concentration (H, O, C, and B) is below 1×10^{17} cm^{-3} in our MBE-grown material [16, 17]. O is a particularly difficult impurity for dilute nitride alloys because it is very difficult to completely eliminate from the N source gas, and it appears to become extremely reactive after passing through the plasma source, increasing its incorporation in all alloys grown while the plasma is running compared with alloys grown while the plasma source is off. We find this particularly critical for growth of AlGaAs, and we never have the plasma source ignited during AlGaAs growth [38]. O contamination can be greatly reduced by utilizing an O filter in the N line, and we often use two in series: one just before the mass-flow controller and one just at the connection to the plasma source. O is a known deep-level nonradiative trap in GaInNAs [42–45], and we have found that it can very substantially reduce photoluminescence (PL) and increase laser threshold currents [38].

Despite the recent laser successes, there is still very significant work remaining to tame this materials system to realize the full wavelength range of high-performance lasers and VCSELs. Most critically, the luminescence properties of GaInNAs deteriorate incredibly rapidly with increasing nitrogen concentration [1, 14–29]. We have undertaken a number of investigations of GaInNAs quantum wells to try to understand the luminescence problems. Thermal annealing increases the PL of GaInNAs QWs by 30–75× over as-grown QWs as well as blueshifting the luminescence peak by 50–80 nm. Typical PL spectra before and after annealing are illustrated in Figure 14.5 [7, 16]. The increase in intensity is presumably due to both the out-diffusion of point defects and an increase in the crystalline quality of the quantum-well material. The wavelength shift could be due to either or both

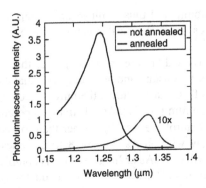

Figure 14.5 Photoluminescence of GaInNAs/GaAs triple-QW sample before and after anneal-ing, illustrating strong improvement in PL intensity and blueshift resulting from the 1 min, 800°C anneal.

nitrogen out-diffusion and group III interdiffusion (GaIn), which could be more significant for InGaNAs compared with InGaAs because of the higher point-defect densities created by the introduction of N. SIMS and nuclear reaction analysis (NRA) measurements suggested that nitrogen out-diffu-sion from the quantum wells was predominantly responsible for this shift [15–20, 39]. Use of GaNAs rather than GaAs barriers reduced the blueshift, but only slightly [15–17, 46]. It thus seemed that there was something beyond compositional changes responsible for the blueshift, and this is discussed in more detail in Section 14.4.

The GaAsN barriers did provide a significant advantage in terms of the growth of multiple QWs (MQW) by providing strain compensation for the QW/barrier pair. The strain compensation occurs because the barrier is under tensile strain while the QW is under compressive strain [15–17, 46–48]. This strain compensation enabled the growth of a far greater thick-ness of high-In composition GaInNAs QWs before strain relaxation occurred. Typically, we could only grow three 7-nm GaInNAs QWs with >30% In with GaAs barriers before the interfaces broke up and microscopic phase segregation and lattice relaxation occurred, as shown in Figure 14.4 [1, 15–17]. By incorporating GaN_zAs_{1-z} barriers with $Ga_{1-x}In_xN_yAs_{1-y}$ QWs, where $z = 1.5\ y$, we have been able to grow up to nine QWs without any deterioration of the QWs. A TEM micrograph illustrating this is shown in Figure 14.6. The inset of the high-resolution TEM shows the ninth GaInNAs QW, clearly illustrating that it is uniform with interfaces as smooth as the first QW. Additionally, the PL from the nine-QW sample is greater than 3× the PL from a similarly grown triple-QW (3QW) sample [1, 16, 17, 47–52].

One of the recent advances in MBE growth of GaInNAs has been the addition of Sb [53–66]. Initial results suggested that Sb was acting only as

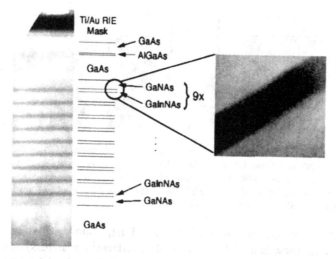

Figure 14.6 TEM cross-section image of nine-quantum-well GaInNAs-GaNAs barrier sample. The inset photo shows a high-resolution image of the ninth quantum well.

a surface surfactant to help maintain the two-dimensional layer-by-layer growth at higher In and N compositions than could be achieved without Sb [53–58]. However, more recent work [59–66] demonstrates that not only does the Sb produce better QWs with smoother interfaces and increased long-wavelength emission (>1.5 μm), but that under the optimum growth conditions, Sb is being incorporated into dilute nitride alloys at levels of 8–10%, well above the incorporation level of N. Hence, one is really forming a quinary alloy of GaInNAsSb. It is not entirely clear what role Sb is playing in the longer wavelength shift of the alloy, but it could be simply decreasing the band-gap or, by changing the degree of strain, the band-gap or relative band-gap offsets. While all of these factors could be important at these concentrations, it appears that these are less important than its effect on the surface kinetics, maintaining two-dimensional growth and much higher PL efficiency [53–66] with higher N and In compositions.

One of the dramatic illustrations of the surfactant effect of Sb is illustrated by comparison of cathodoluminescence (CL) from GaInAs, GaInNAs, and GaInNAsSb QWs. This comparison is shown in Figure 14.7, where the GaInAs QW shows virtually no spatial difference in CL, while the middle image of GaInNAs shows very strong dark regions and suggests that significant microphase segregation is occurring at high In and N composition. By adding Sb to the growth — and with even higher In and N compositions compared with the middle image — the GaInNAsSb on the right is far more homogeneous and much brighter as well.

We have now carried out a fairly extensive investigation of Sb incorporation and its impact on the growth of (GaIn)(NAsSb). Several things

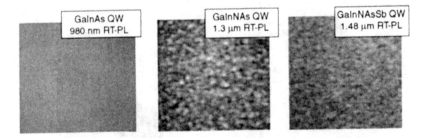

Figure 14.7 Low-temperature (4 K) cathodoluminescence of a standard 980-nm GaInAs QW, 1.3-μm GaInNAs QW, and 1.48-μm GaInNAsSb QW samples showing absolutely no intensity variation in the GaInAs QW, severe variations consistent with the three-dimensional growth and phase segregation described in the TEM image in Figure 14.4, as well as greatly reduced segregation by Sb addition during growth.

have become clear from this investigation. First, a number of quite sophisticated (and new to most III-V materials and device researchers) characterization techniques are required to understand the epitaxial growth and annealing properties of this materials system [58, 64]:

1. Most of the required device alloys are very highly strained and are very thin QWs.
2. The materials must be grown under metastable conditions within the miscibility gap of the alloy.
3. In and Sb are very-high-mass elements, while N is very light and in relatively low concentrations.
4. The N may not be entirely substitutional.

While X-ray diffraction (XRD), PL, SIMS, and TEM have been the primary tools used to characterize III-V alloys in the past, the above issues provide new characterization challenges that require more-sophisticated techniques. The results of several of these measurement and insights into the growth of GaInNAsSb QWs are described.

Studies of Sb incorporation in GaInNAsSb and GaNAsSb show a strong effect of In on Sb incorporation [58]. Compositional analysis was done primarily by SIMS. The structures consisted of 10-nm-thick 3QW (GaIn)(NAsSb) separated by 50-nm-thick Ga(NAsSb) barriers. The only parameter changed was the Sb flux in three steps from 0 to 1.2×10^{-7} Torr to compare the influence of Sb-containing with Sb-free structures and the effect of In. The sample compositions were $30 \pm 2\%$ In and $1.6 \pm 0.5\%$ N in the quaternary QW and $2.3 \pm 0.2\%$ N in the ternary barrier. SIMS analysis showed that adding Sb to the quaternary (GaIn)(NAs) does not change the In content of the QWs. However, Sb increases the N mole fraction in both the quinary (GaIn)(NAsSb) and quaternary Ga(NAsSb), as shown in Figure 14.8. The N mole fraction increases in both the QWs

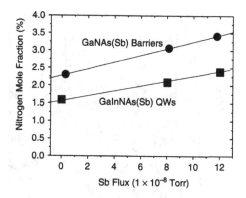

Figure 14.8 Nitrogen mole fraction in GaNAs(Sb) and GaInNAs(Sb) as a function of Sb flux during growth.

and barriers with increasing Sb flux by $\approx$0.06–0.07% per 1×10^{-8} Torr of Sb. The increase is slightly higher in the In-free barriers than in the In-containing QWs, but this may be due to the lower growth rate of the barriers (0.35 μm/h) compared with the QWs (0.5 μm/h). This clearly shows that adding Sb to (GaIn)(NAs) increases the incorporation of N in these material systems.

In order to reach longer wavelengths (>1.3 μm), the In content in the structures must be increased. Figure 14.9 shows the X-ray diffraction (XRD) patterns of structures with In composition of 35–38% for different Sb fluxes [58]. Comparing the Sb-free (absence of Pendellosung fringes) and Sb-containing layers (presence of Pendellosung fringes), it is clear that Sb acts as a surface surfactant to decrease the diffusivity of In on the surface and prevent phase segregation. This allows more highly strained material to be grown with high crystallinity. A second observation from Figure 14.9 is that Sb initially increases the compressive strain in the samples as the Sb flux is increased to 7×10^{-8} Torr, but further increase in Sb flux decreases the strain in the QWs, indicating an increased incorporation of N. This is in agreement with the PL, which redshifts with increasing Sb, indicating a decrease in the band-gap due to both increased Sb and N incorporation. This redshift occurs despite an increase in the compressive strain in the QWs, which should blueshift the emission. The strain effect is overcompensated by the bowing of the band-gap due to increased N incorporation and results in peak PL out to 1.6 μm, as shown in Figure 14.10 [61]. These PL results clearly show the dramatic effect that Sb has in preventing three-dimensional growth and phase segregation when one compares the excellent PL of the 31% In 1.3-μm sample (no Sb, but near the maximum before phase segregation) with the $10^4\times$ lower PL of the 35% In sample and then with the nearly equal intensity PL of the 38% and 39% In samples grown with Sb and increased N incorporation.

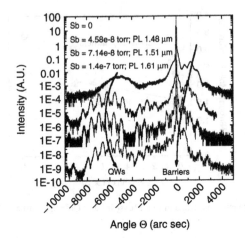

Figure 14.9 High-resolution XRD scans around the GaAs (0,0,4) reflection for (GaIn)(NAsSb)/Ga(NAsSb) MQW structures with 38% In as a function of increasing Sb flux. The top curve without Sb illustrates the same type of three-dimensional growth and phase segregation shown in the TEM of Figure 14.4.

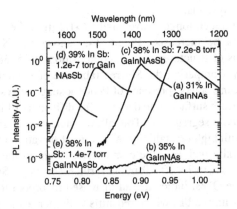

Figure 14.10 Photoluminescence of GaInNAs(Sb) comparing the best 1.3-μm material grown without Sb and the dramatic improvement in PL by adding Sb.

While the effect of Sb incorporation is negligible on In mole fraction and very small on the N mole fraction, the effect of In incorporation on Sb incorporation is very strong. In the low-Sb-composition range (<10%), Sb incorporation in GaAsSb, GaNAsSb, and GaInNAsSb is linear with Sb flux, with little difference between GaAsSb and GaNAsSb. However, when 30–35% In is added in GaInNAsSb, the Sb mole fraction decreases by a factor of about 3× (i.e., from 9% in GaNAsSb to 3% in GaInNAsSb). Since we had not anticipated appreciable Sb incorporation at such low fluxes, a number of lasers were grown and reported using GaNAsSb barriers [51, 52, 59–66]. XRD on these GaNAsSb-barrier lasers shows that the QWs are highly compressively strained, but that the barriers are nearly lattice-matched to the GaAs, hence eliminating the compressive strain we found valuable for growing MQW structures. This dramatic change in Sb incorporation between QW and barrier results in inferior PL and lasers [65], and we have now altered our growth process to close the Sb shutter during growth of the barriers to produce GaNAs barriers [65, 66].

Further understanding of the challenges for GaInNAs growth and the role of Sb have been obtained from a recent examination of samples with the 800-keV atomic-resolution TEM at the Lawrence Berkeley Lab. The samples are polished to expose the (1,0,0) surface rather than the traditional cleaving to expose a (1,1,0) surface. The (0,0,2) diffraction condition is then used in dark field to provide an image that is very sensitive to compositional changes in the group III sublattice [67, 68]. To date, this has only been done on samples grown before we discovered the above large change in Sb incorporation with In, hence the samples consist of GaInNAsSb QWs and GaNAsSb barriers. However, even these first results provide insight into the problems at high In mole fractions, into the effect of Sb, and into the fact that much better control of the growth conditions (particularly substrate temperature) must be maintained, as growth under "nominally constant" conditions results in changing In composition in the QWs. Figure 14.11 shows a strain map from high-resolution TEM of a single QW (SQW) GaInNAs sample. The composition is ≈30% In and ≈1.6% N with peak PL at 1.3 μm, which is still in the two-dimensional growth regime. The strain profile shows very nonuniform In distribution with In strongly segregating near the top surface of the QW. The difference in In composition between lower and upper surfaces of the well is ≈6% (i.e. 24% In at the bottom and 30% In at the top). Such a large gradient in In composition would significantly change the band gap, strain, QW energy, and resulting PL line width and energy. This is for an unannealed sample, which may be homogenized somewhat by the anneal, but this clearly shows the problem of uniformity and In segregation in the QW, which would certainly be best avoided if at all possible.

Figure 14.12 is a TEM (0,0,2) dark-field reflection image of a 3QW GaInNAsSb sample, which is particularly sensitive to composition [67, 68] and shows that the composition across both barriers and QWs is quite uniform in comparison with the strain map in Figure 14.11 for the SQW grown without Sb. Figure 14.13 is an energy-filtered TEM image of the same sample and region as that in Figure 14.12, with the energy-filter window set for electrons scattering from In, and it similarly shows excellent In uniformity in the QWs and that the QW-barrier interfaces are extremely sharp with no diffusion of In into the barrier regions.

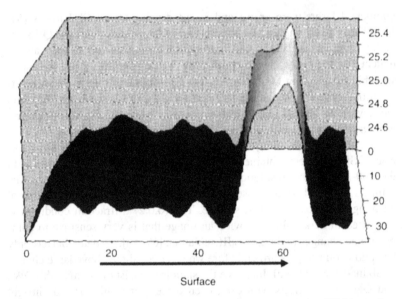

Figure 14.11 Strain map created from mesh overlay of atomic-resolution TEM image for a single, 70-Å GaInNAs/GaAs QW.

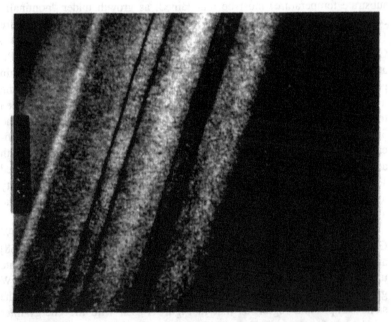

Figure 14.12 Dark-field (0,0,2) diffraction image of a 3QW GaInNAsSb/GaNAsSb sample that is sensitive to III-III and V-V compositional changes, showing QW homogeneity with addition of Sb.

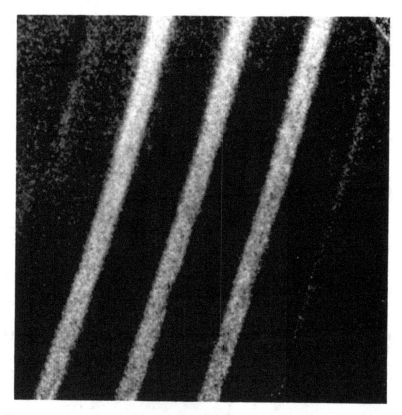

Figure 14.13 Energy-filtered TEM image of same 3QW GaInNAsSb/GaNAsSb sample (Figure 14.2) with the energy filter set for In, illustrating very sharp interfaces and no In diffusion into the QWs.

Figure 14.14 shows an atomic-resolution TEM strain map for a 3QW GaInNAsSb sample with GaNAsSb barriers. The composition is ≈39% In, ≈1.6% N, and ≈2% Sb in the QWs, and ≈2% N and ≈6% Sb in the barriers, with peak PL at 1.5 μm. The strain profile of the sample shows that the first QW has the highest In composition and QW 3 (top) the lowest. The reduction in In is very substantial and increases more from QW 2 to 3 than from QW 1 to 2. The estimated difference in the In composition between QW 1 and 2 is ≈4% (i.e. 39% In in QW 1 and 35% In in QW 2), and the difference between QW 2 and 3 is ≈6%. Figure 14.15 is a blowup of just the top QW and shows that the strain across the QW is very uniform in comparison with the GaInNAs (i.e., no Sb) sample in Figure 14.11. Thus, the In distribution in the wells is quite uniform, but decreasing with each added QW. Figure 14.14 also shows that the Sb composition is not uniform in the barriers and may be changing to balance the increased N incorporation to achieve a lattice

match to the GaAs substrate, as we suggested from the XRD on similar
3QW GaInNAsSb samples discussed previously.

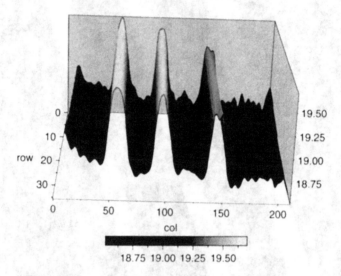

Figure 14.14 Strain map created from mesh overlay of atomic-resolution TEM image for a
70-Å GaInNAsSb/GaNAsSb 3QW.

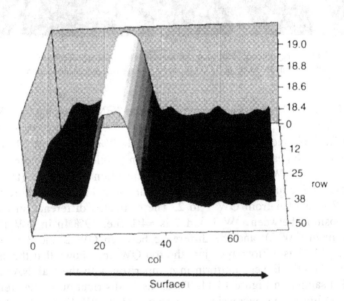

Figure 14.15 Blowup of the third (closest to the surface) QW from the strain map of Figure
14.14.

The above TEM results are quite useful and revealing. However, there are only a few such high-voltage atomic-resolution TEMs in the world, and sample preparation is also difficult and time consuming, which make this technique relatively unsuitable for routine characterization of every sample. Certainly the compositional changes across a SQW and also with progression from QW to QW in a MQW sample are consistent with the significantly broader PL line width that one observes in GaInNAs and GaInNAsSb compared with other III-V alloys. These changes may underlie the successful optimization of GaInNAs QWs based upon PL line width [34, 65, 66] rather than PL intensity, which has been the primary criterion in the past. We are continuing this investigation on samples to find the optimum Sb flux that maintains two-dimensional growth while minimizing other effects on the material and the effect of substrate temperature on In segregation. I believe that these investigations will lead to material with PL line widths and laser thresholds that are comparable with other III-V alloys.

14.4 ANNEALING

Annealing in both GaInNAs and GaInNAsSb shows a completely unique behavior [1, 16, 17, 47, 67–82] compared with all similar III-V semiconductor alloys. There is a dramatic increase (30–75×) in photoluminescence (PL) efficiency and a significant blueshift (50–80 nm) in wavelength which was illustrated in Figure 14.5. Our annealing conditions are 1 min in a rapid thermal annealer with a N ambient and gallium arsenide proximity cap for temperatures ranging between 600 and 900°C, with the optimum generally near 820°C.

There have been a very large number of studies of this annealing behavior, and the initial hypothesis was that the blueshift was due to either indium [47, 48, 77] or nitrogen [1, 16, 17] diffusion from the quantum wells. We have carried out a substantial investigation of this phenomenon, and the annealing behavior appears to be a unique property of alloys with *both* indium and nitrogen. Annealing of GaInAs QWs in GaAs produces absolutely no change in either PL intensity or wavelength. Annealing GaNAs QWs in GaAs produces a small (2–4×) increase in PL intensity and very little wavelength shift, and neither exhibits any discernible difference in PL line width or lattice constant from XRD [58, 64, 65]. However, the situation is entirely different for both GaInNAs and GaInNAsSb QWs on GaAs. Both QWs show similar increases in PL intensity and blueshift, although no change is observed in lattice constant from XRD [51, 63–65]. SIMS measurements before and after annealing of a 3QW GaInNAsSb QW–GaNAsSb barrier sample are shown in Figure 14.16. These profiles are similar to those in references [16, 17, 51] and show a definite decrease in nitrogen composition in the QWs with annealing, but no appreciable decrease or spreading of the In or Sb compositions. This sample is similar

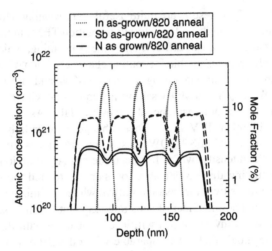

Figure 14.16 SIMS profiles for In, Sb, and N of 3QW GaInNAsSb/GaNAsSb sample before and after 1-min, 800°C anneal.

to those described earlier comparing Sb incorporation with and without In, and it clearly shows the strong decrease in Sb in the QWs. The change in Sb incorporation is far larger than the change in N composition, which is exactly in the ratio of the column III growth rates between QWs and barriers [16].

The similarity of blueshift among many investigators and materials grown by different techniques and annealed under a variety of conditions suggested that there was possibly some very localized change occurring in the material under annealing, resulting in the formation of a more homogeneous alloy with greater equilibrium. Theoretical predictions discussed in Chapter 5 and in the literature [83–85] suggested that the localized unit cell would decrease if N was surrounded by 2, 3, or 4 In atoms in its tetrahedral site. The bond strength of Ga-N is much greater than that of In-N, hence a very real possibility is that because of the low growth temperature, the alloy grows with N being predominantly surrounded by Ga nearest neighbors. Then, during anneal, In and Ga locally exchange lattice sites, resulting in an increase in band-gap of the GaInNAs or GaInNAsSb QWs, but without any change in the QW composition, QW width, or average lattice constant. Measurement of the local N environment is possible, but quite challenging. Several groups have reported results using Raman spectroscopy [67] (described in Chapter 7), IR local-vibrational-mode optical absorption [69, 73] (also described in Chapter 7), X-ray near-edge absorption spectroscopy (XANES) [70, 71], and extended X-ray absorption fine-structure spectroscopy (EXAFS) [70–72]. All of these procedures require great care in sample preparation, measurement, and particularly in interpretation, because some can be done on thin, strained QWs, while others require thicker films that will be lattice relaxed. Clearly, if strain is playing a role in the annealing behavior, one must use thin, strained QWs identical to those used in lasers and PL studies of annealing, rather than thicker, lattice-relaxed layers.

We carried out a series of N K-edge XANES and In K-edge EXAFS measurements to examine the possible changes in the tetrahedral nearest neighbors (NN) in GaInNAs and GaNAs. The samples were grown by solid-source MBE [16, 17] with compositions of $\approx Ga_{0.7}N_{0.03}As_{0.97}$ and $Ga_{0.7}In_{0.3}N_{0.03}As_{0.97}$.

Nitrogen K-edge XANES spectra were taken in fluorescence from both as-grown and annealed samples. The measurements were performed at beam line 4.0.2 of the advanced light source using a high-resolution super-conducting tunnel junction X-ray detector operated at 0.1 K. Indium K-edge EXAFS spectra were measured at beam line BL01B1 of SPring-8 using a liquid N_2-cooled Ge detector.

Since GaInNAs is nominally a zinc-blende random alloy, each N atom sits in the center of a tetrahedron, surrounded by a total of four Ga and/or In atoms, as shown in Figure 14.17. If the arrangement of atoms in the alloy were random, the probability distribution for finding a given number of In atoms surrounding any group V atom is shown in Figure 14.18, for $x = 0.3$. The notation "GaInNAs$^{(n)}$" denotes a model crystal with n In NNs around each N atom (i.e., GaInNAs$^{(0)}$ indicates a crystal where every N atom is bonded to four Ga NN atoms) [71].

The XAS experiments were motivated by atomic-relaxation simulations that showed a decrease in total energy by >10 kcal/mol when two or more In atoms are bonded to N, despite an increase in overall lattice constant leading to a slight increase in compressive strain energy for GaInNAs grown epitaxially on GaAs and as illustrated in Figure 14.19 [71]. The decrease in chemical energy with increasing number of N-In bonds is due to overall decreases in individual bond strains, since the longer N-In bond is stretched less from equilibrium than the N-Ga bond. Furthermore, band-structure calculations show an increase in band-gap of ≈ 150 meV for the thermodynamically favored GaInNAs$^{(3)}$ over GaInNAs$^{(0)}$, consistent with the observed PL blueshift upon anneal [1, 17].

The N K-edge XANES spectra taken from 3000-Å-thick as-grown and annealed GaInNAs samples, as well as from a GaNAs reference sample, exhibit a shift of 0.2–0.3 eV when In is added and an additional 0.1 eV after annealing. Increasing the number of N-In bonds shifts the N K-edge to lower energies. By comparing quantitatively the measured and calculated shifts in the N K-edge XANES spectra, the as-grown material is a nearly random alloy, and the observed shift of 0.2–0.3 eV between GaNAs and as-grown GaInNAs, and the additional 0.1-eV shift after rapid thermal annealing (RTA), is consistent with a distribution of bonds dominated by

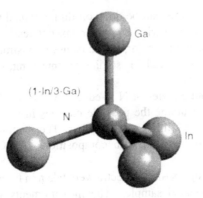

Figure 14.17 Tetrahedral lattice site for N in GaInNAs alloy, illustrating the most likely site of a random, 30% In alloy with 1-In NN and 3-Ga NN (GaInNAs[1]).

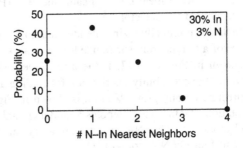

Figure 14.18 The probability of finding a given number of In nearest neighbors to a group V lattice site for a 30% In random alloy.

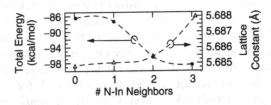

Figure 14.19 Variation of lattice constant and total energy as a function of the N chemical environment.

GaInNAs[1] before annealing and GaInNAs[3] after annealing. Figure 14.20 shows that this corresponds to the random configuration before annealing and the thermodynamic configuration after annealing. This result is not surprising, as the low-temperature MBE growth promotes kinetically dominated epitaxy, while the higher temperature exposure during RTA moves the material toward equilibrium. A previous theoretical study using Monte Carlo simulations predicted a thermodynamically preferred configuration

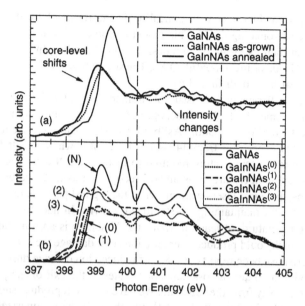

Figure 14.20 (a) Measured N K-edge XANES spectra, and (b) simulated spectra for differing In NN configurations.

favoring N-In$^{(3)}$Ga$^{(1)}$ clusters, but it did not consider the effects of annealing or high In content [84].

In addition to N K-edge XANES, we also measured the NN radial distribution function (RDF) around In from a Fourier transform of In K-edge EXAFS spectra for as-grown and annealed 80-Å GaInNAs quantum-well samples. The RDFs for both as-grown and annealed material show a dominant In-As peak at 2.25 Å and a smaller shoulder at shorter bond lengths. The shoulders correspond to the In-N bond, whose precise length is shown by simulations to depend on the number of Ga atoms around N. Simulated RDFs were then obtained and fit to the experiments, assuming that annealing drives the material from a distribution with mostly GaInNAs$^{(1)}$ to mostly GaInNAs$^{(3)}$. The In-N shoulder in the RDF becomes more prominent after annealing because the number of In-N bonds increases and also because the peak radius shifts further from the center of the overwhelming In-As peak. There is an excellent match between the simulation and experiment, and again we conclude that annealing increases the degree of N-In bonding.

While one might find limitations in either the measurements or interpretation of these results [70, 71], or any of the others [26, 69, 72, 73, 85], and view them as only suggestive rather than conclusive, the similarity in outcome for all of these results on different samples — some prepared by MBE, some by MOVPE, and measured by quite different techniques — provides a rather compelling conclusion that at least half and maybe more of the spectral blueshift with annealing is the result of local In/Ga site exchange surrounding the N. This model is also consistent with the observation that there is virtually no blueshift from annealing InGaAs, GaNAs, GaAs/AlGaAs, or InGaAsP/InP QWs. The strong blueshift is only observed in QWs that have both In and N.

There still remains the issue of dramatic PL intensity increase with annealing, some of which is observed in GaNAs, although not as strongly as in the GaInNAs and GaInNAsSb alloys. The major suspect for this is nonradiative deep-level traps [40–44, 74–82, 86], which result from the very low growth temperature required to incorporate N and In. Growth of GaAs at low temperatures was always recognized as a source of nonradiative traps, and most lasers were grown at the highest possible temperatures consistent with maintaining As-stabilized growth. In addition to the low growth temperature, at least in MBE, the N source is an ion plasma discharge, which could produce energetic ions that damage the growing material. There have been reports [87–89] of reduced ion damage during growth by applying a voltage across deflection plates at the exit of the plasma to deflect ions away from the substrate. We have tried to reproduce such effects and observed mixed results [61–64], which might be summarized as follows: there appears to be some improvement in as-grown PL intensity with voltage applied to the deflection plates, but that after annealing, the difference is relatively small.

14.5 GaInNAs AND GaInNAsSb EDGE-EMITTING LASERS

The first long-wavelength, GaInNAs-on-GaAs, room-temperature, cw laser was reported by Kondow [11] in 1996. The device was a 400-μm-long, broad-area, single-QW laser emitting at 1.18 μm with a threshold current density of 1.83 kA/cm^2 and a T_0 of 126 K. Because the material was grown on GaAs and had a high T_0, there was immediate interest in this system for long-wavelength lasers. With significant development since that time, the threshold current densities have dropped quite dramatically, output powers have increased substantially, and the wavelength has been extended beyond 1.5 μm, making this alloy of significant interest for a variety of lasers across the entire communications-wavelength band [2].

The vast majority of GaInNAs edge-emitting lasers were broad-area lasers fabricated to measure gain, loss, and materials properties as a function of growth and annealing conditions for VCSEL fabrication. Because of the high T_0, Livshits et al. [90] demonstrated the first truly high power potential for the GaInNAs system, with 8-W cw output at 1.3 μm and 10°C chip temperature. More recently, the main focus for higher power lasers has been to push the long-wavelength range of the GaInNAs alloys by first incorporating GaNAs barriers [14, 16–18, 46, 49] and, secondly, by adding Sb to the growth of QW layers [51–57, 59–66]. Sb was initially intended as a surfactant [53–56] to enable the growth of smooth QW interfaces at higher In and N compositions. However, more recent results have shown significant Sb incorporation and formation of a quinary alloy, GaInNAsSb [1, 51, 52, 57, 58], the growth of which was described in Section 14.3.

The design window for edge-emitting lasers is considerably broader and easier to hit compared with VCSELs, hence the approaches and number of results are significantly larger. In particular, the growth issues and precursor reactions for MOVPE can be minimized by the use of InGaP cladding layers next to the GaInNAs/GaAs active quantum-well region, thus avoiding the immediate growth of AlGaAs [27, 32, 33]. There are thus a large number of results reported for both MOVPE and MBE. The review in Chapter 11 by Kondow focuses on 1.3-μm lasers, and that by Potter et al. [91] summarizes progress through 1999, so I will focus on results since that time and particularly on the approaches to achieve longer wavelengths, lower thresholds, and higher power in lasers by addition of Sb to the QW material.

Our initial edge-emitting lasers [38, 92] were fabricated to determine the effects of growth temperature, *in situ* annealing, and composition on threshold current and emission wavelength. Because growth temperature of both active and cladding layers is critical, the growth of these devices is described in detail. The values reported are quite typical for all MBE-grown GaInNAs lasers, as the growth-temperature window is quite narrow ($\approx 25°C$). The structures consisted of a 1.7-μm-thick cladding layer of $Al_{0.33}Ga_{0.67}As$, *n*-doped with Si at a level of 1×10^{18} cm^{-3}, a gain region consisting of 100-nm GaAs, InGaNAs/GaAs quantum-well-active material, and a second 100-nm GaAs layer, followed by an upper cladding layer of 1.5-μm $Al_{0.33}Ga_{0.67}As$, *p*-doped with Be at 5×10^{17} cm^{-3}. A final 10-nm GaAs layer was grown with heavy *p*-doping to facilitate electrical contact formation. The active regions consisted of either one or three InGaNAs/GaAs quantum wells [92]. The quantum-well thickness was varied within the range of 65–90 Å in different samples. The lower waveguide cladding layer was grown at 600°C, the active regions at 420–460°C, and the upper waveguide cladding layer between 580 and 680°C to examine the effect of *in situ* annealing.

The properties of the lasers depend very strongly on the growth temperature of both the active QW region and the upper waveguide cladding layer. Lasers grown under optimized growth conditions yielded threshold current densities (for 20-μm-wide × 800-μm-long devices) as low as 450 A/cm^2 for single quantum-well lasers [59, 63, 92]. The lowest threshold for such lasers is ≈ 350 A/cm^2 [18, 93]. The laser emission wavelengths were in the range of 1230–1250 nm. The laser thresholds decrease as the growth temperature of the active region increases up to 440°C. Beyond 440°C, the threshold currents increase, as well as being quite dependent on quantum-well thickness. The best results were obtained with the top waveguiding layer grown at 680°C, the maximum temperature without Ga desorption and AlGaAs composition changes.

Pushing the emission wavelength beyond 1.3 μm has been a major focus of investigations the past 2–3 years. This was fueled by the first reported success at 1.5 μm by Fischer et al. [34]. While the threshold current was quite high (40 kA/cm^2), this result served as proof that long wavelengths could be achieved. They reported ≈5% N incorporated into the QWs, but this has proven to be a very hard level to achieve without microphase segregation and breakup of the QW interfaces occurring, although they have reported lower thresholds at 1.45 μm [93]. Yang et al. [53] were the first to utilize Sb as a surface surfactant to enable increased N and In incorporation to extend the emission wavelength without breakup of the QW interfaces. The role of Sb is quite complex and almost certainly depends upon the Sb species utilized. Yang [53, 55] and Shimizu [54, 56] both report very small amounts of Sb incorporated (≈1% and <0.5%, respectively) into the QWs, while we have observed 8–9% Sb in GaNAsSb barriers and 2–3% Sb in GaInNAsSb QWs [58–64, 82] with Sb fluxes during growth that are a factor of 4–10× smaller than they report. The major difference in these results is that they evaporate Sb from a solid in a Knudsen cell, producing Sb$_4$, while we utilize an Sb cracker producing predominately Sb and Sb$_2$. This appears to dramatically increase the Sb incorporation [58, 66–71, 82]. We similarly observe a postponement of the two- to three-dimensional growth mode. However, the complete effect of Sb with respect to strain, band-gap offsets, and point defects is completely unknown and may be a very important factor in achieving low-threshold, high-power, long-wavelength lasers. We have recently combined both of these approaches, the incorporation of Sb with GaNAs barriers, to push QW PL emission wavelength out to 1.6 μm (as was shown in Figure 14.10) and lasers out to 1.5 μm [59–64, 82] with dramatically lower threshold current [66].

Separate-confinement heterojunction (SCH) multiple-quantum-well lasers were grown on (1,0,0) n-GaAs substrates. The active region consists of three 7-nm GaInNAsSb quantum wells separated by 20-nm GaNAsSb barriers. The active region is symmetrically embedded in a 120-nm-thick undoped GaAs waveguide. A 1.5-μm Si-doped (5 × 10^{17} cm^{-3}) n-type Al$_{0.33}$Ga$_{0.67}$As cladding layer was grown between the n-substrate and the active layer, and a 1.5-μm Be-doped (7 × 10^{17} cm^{-3}) p-type Al$_{0.33}$Ga$_{0.67}$As cladding layer followed the active layer. The wafers were *ex situ* annealed by rapid thermal annealing (RTA) for 1 min at 780°C to improve material quality [58–64, 82].

Ridge-waveguide lasers with 5-μm-wide × 800-μm-long ridges with cleaved uncoated end facets and three GaInNAsSb QWs and GaNAsSb barriers were fabricated. Figure 14.21a shows the light-output power at

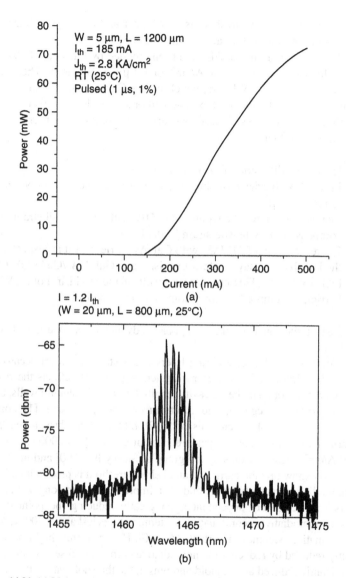

Figure 14.21 (a) Light-output power vs. injection-current (L-I) characteristic for 5-μm-wide ridge-waveguide edge-emitting laser under room-temperature pulsed operation. (b) Emission spectra at 200 mA (≈1.2 I_{th}) under pulsed operation at room temperature.

1.465 μm vs. injection current, with a maximum power exceeding 70 mW and a slope efficiency of 0.21 W/A. Figure 14.21b shows the laser spectra at 1.2× threshold. The minimum threshold current density was 2.8 kA/cm² or 930 A/cm² per quantum well, which was the lowest reported threshold current density at the time for lasers with GaInNAs QWs operating beyond

1.4 μm. However, this threshold is ≈2× higher and the slope efficiency 2× poorer than our best lasers at 1.39 μm [62, 63].

The lowest laser thresholds have been reported by Infineon [18, 94–96], with a threshold current of 350 A/cm² for 1.3-μm SQW lasers. These lasers are similar to our SQW lasers, which utilize GaNAs barriers surrounding GaInNAs QWs. To the best of the author's knowledge, other reported GaInNAs lasers have used GaAs barriers. The N-containing barriers could have several influences [16–18]:

1. Decreased diffusion of N from the QWs
2. Lower QW barrier providing less confinement, hence less N required to reach 1.3 μm
3. Increased compressive strain in the QW with less overall strain due to compensation by tensile strain in the barrier
4. Decreased ratio of $\Delta E_c/\Delta E_v$ with GaNAs barriers, as it is suspected that the hole barrier may be almost vanishing in GaInNAs/GaAs QWs with high (>3%) N (This could be a result of the strain in both QW and barrier, in composition, or a combination of both)

At least at the current time, it appears advantageous to utilize GaNAs barriers.

Comparison of edge-emitting lasers is not straightforward because the range of design and fabrication parameters significantly affects the results, the four most prominent being cavity length, number of quantum wells, broad area vs. ridge waveguide, and facet-mirror coatings. Harris [1] normalized/compared threshold currents of SQW, 800-μm-long, broad-area lasers without facet coatings as a "standard structure," with $J_{th} = 2000 \exp [8 (\lambda - 1.2)]$ A/cm². Measurements by the group at Surrey [97–100] and in Chapter 12 clearly show that the current at and below threshold is predominantly due to nonradiative recombination and that improvements in long-wavelength lasers will be largely dependent upon greatly reducing the point defects causing nonradiative recombination. At temperatures just above 300 K, Auger recombination becomes important [66, 100]. However, this might be significantly reduced by reducing the nonradiative current because $\propto n^3$, but if n is significantly reduced at threshold and pins at the threshold value, then Auger recombination will be much lower. All of these competing recombination mechanisms vary significantly with growth and annealing and, because of their effect on the carrier density at threshold, they can significantly effect the temperature dependence of threshold current and high-temperature operation. This remains one of the most important issues for GaInNAs and GaInNAsSb lasers, because noncooled operation will be critical to achieve both low-cost VCSELs and high-power pump lasers.

Once it was understood that Sb was not just a simple surface surfactant, but was substantially incorporated into the lasers, we focused on improving

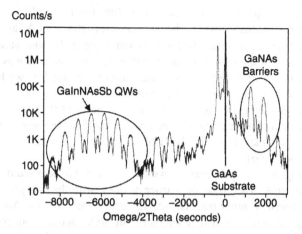

Figure 14.22 High-resolution XRD scans around the GaAs (0,0,4) reflection for (GaIn)(NAsSb)/Ga(NAs) 3QW structures with ≈38% In, ≈1.5%N, and 3% Sb in the QW and ≈2.4% N and ≈9% Sb in the barriers.

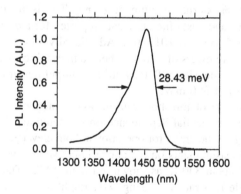

Figure 14.23 Photoluminescence of SQW sample of same composition as the 3QW sample in Figure 14.22.

the QWs and lasers with GaNAs barriers [65]. Many authors have described a "N penalty," which was discussed by Harris [1], showing that the increase in threshold current with wavelength was much higher for lasers operating at wavelengths >1.3 μm compared with those operating between 1.1 and 1.3 μm. As discussed in Section 14.3, the growth under what was thought to be "nominally constant" growth conditions produced QWs with varying levels of In incorporation and relatively broad PL line widths. We have worked to ensure better flux and plasma stability, better O filtering in the N gas line, better substrate temperature control, and growth of AlGaAs layers in a separate, but high-vacuum connected MBE chamber, which allows additional control over igniting and running the plasma source. XRD from such an optimized SQW sample is shown in Figure 14.22. The clear multiple reflections from both the compressively strained QW and tensilely strained barriers is much better than our prior XRD and supports our suggestion that uniformity in these earlier device structures was seriously lacking. Using PL line width as a measure of QW quality, we have been able to both extend the PL wavelength out to 1.6 μm and dramatically improve the intensity, as was shown in Figure 14.9. The most recent improvements have produced a record narrow PL line width with a full width at half maximum (FWHM) of 28.5 meV at 1.46 μm, as shown in Figure 14.23 [66].

Using these improvements, we fabricated SQW edge-emitting lasers with the same structure as the earlier devices, with the exception of using GaNAs barriers rather than GaNAsSb barriers and that the AlGaAs cladding layers were grown in a separate, but high-vacuum connected MBE chamber. The cw, RT light output vs. input current (L-I) for a 20-μm-wide × 760-μm-long ridge waveguide laser is shown in Figure 14.24a. The device lased at 1.490 μm, as shown in Figure 14.24b with a cw threshold of 1.1 kA/cm^2. The cw slope efficiency was 0.34 W/A, corresponding to 40% external quantum efficiency. This is the lowest threshold current density and best efficiency data reported for a dilute nitride laser operating at 1.5 μm [51, 101, 102]. Additionally, devices showed maximum CW output powers of ≈30 mW before thermal rollover and a pulsed threshold current density of 910 A/cm^2 and output power (5-μs pulse and 1% duty cycle) of 300 mW.

Measurements of threshold current as a function of temperature to determine the exponential dependence on temperature (T_0) are important in determining the potential for operation at both elevated temperature and high power. Measurements on the above lasers show a very clear change in behavior at about 330 K, as shown in Figure 14.25. The data cannot be fit with a single T_0 parameter. Using two straight-line fits for 280° T 330° and 330° T 350°, we fine a value of $T_0 = 140$ K for the lower-temperature region and $T_0 = 50$ K for the region above 330 K. These results are from cw operation and particularly at the higher temperature, where thresholds are higher, appreciable junction heating may be occurring, and the 50 K value represents a lowest value. This behavior is similar to that reported by Jin et al. [100] and strongly suggests that two different mechanisms are governing the recombination in the lasers. Earlier results with higher thresholds produced a wide range of T_0 values, at least some of which were taken over a more limited temperature range because the devices no longer lased at higher temperatures. The recombination mechanisms that govern laser operation and their impact on T_0 and high-temperature operation clearly need further work, but progress over the past year suggests that with better understanding of the mechanisms and reduction of threshold current, one might achieve the higher values of T_0 for operation up to 60–80°C, which

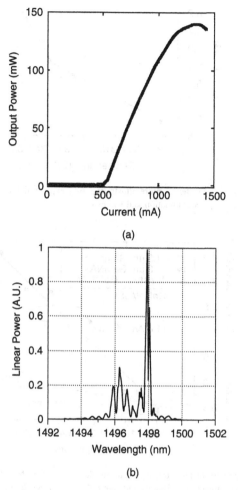

Figure 14.24 (a) Light-output power vs. injection-current (L-I) characteristic for 20-μm-wide × 2450-μm-long ridge-waveguide laser with cw J_{th} = 1.1 kA/cm². (b) Emission spectra at 600 mA, cw and RT.

would represent a huge advance in terms of maintaining sufficient gain and output power for uncooled laser operation in MAN, SAN, and LAN networks.

Several of the above devices were HR coated (R ≈ 98.5) on one mirror facet. A 20 × 1000-μm device was pushed to the maximum current we could deliver in pulsed mode and produced 672 mW power out of a single facet. This device had a similar 1.15-kA/cm² threshold current and 0.35-W/A slope efficiency. The most remarkable aspect of these lasers is that their threshold current density follows closely an extrapolation of the 1.1–1.3-μm laser thresholds with wavelength, which is 5× lower than all

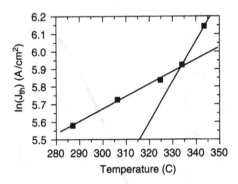

Figure 14.25 Log (J_{th}) vs. temperature illustrating two exponential regions and different mechanisms dominating recombination and values of T_0.

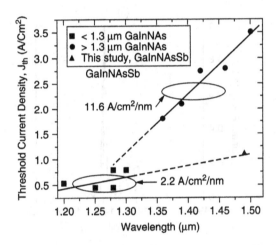

Figure 14.26 Summary of laser threshold current density for GaInNAs ($\lambda \leq 1.3$ µm) GaIn-NAsSb ($1.3 \leq \lambda \leq 1.5$ µm) lasers. Squares indicate $\lambda \leq 1.3$-µm-range lasers with linear fit (solid line), circles indicate previous $1.3 \leq \lambda \leq 1.5$-µm-range lasers with linear fit (solid line), and the triangles represent the low J_{th} result in Figure 14.24. Extrapolation of the ≤ 1.3-µm-range data (dashed line) fit this low-threshold 1.5-µm point. (Adapted from Ref. [1].)

previously reported lasers operating at wavelengths >1.3 µm. This is illustrated in Figure 14.26 and clearly suggests that by using Sb and optimized growth and annealing conditions, there is no "N penalty."

14.6 SUMMARY

The discovery of 1.3–1.6-µm active quantum-well material that can be grown on GaAs to capitalize on the superior AlAs/GaAs materials and processing technology has been a real breakthrough and has fueled a complete reevaluation of long-wavelength lasers. I believe that GaInNAsSb on GaAs will be the fundamental technology for wide-bandwidth MAN/LAN/SAN applications and optical switching and routers.

The major challenge of this materials system has been to understand the differences compared with other III-V alloy systems and to produce low-threshold lasers at any desired wavelength between 1.3 and 1.6 µm. The most recent results incorporating Sb to form a quinary alloy, GaIn-NAsSb, appear to overcome many of the prior problems with phase segregation, and I believe that GaInNAsSb will be the active gain material of choice because it has significantly higher gain for VCSELs, it is closer to the existing QW technologies than InAs QDs, and it has fundamental energy-band advantages over its other competitors. GaInNAsSb also has an inherent lateral uniformity advantage over other active QW materials choices. However, this advantage is obtained only by solid-source MBE. As illustrated from the TEM strain maps, there are still challenges for vertical uniformity. However, recent improvements based upon these observations suggest that with proper feedback and control during QW growth, these problems can be overcome by MBE.

While MBE has been utilized for production of very-low-cost, edge-emitting CD-lasers, it has not been utilized in the production of VCSELs, although it has been the tool of choice for most of the research and development of VCSELs. The newest versions of production MBE systems, with their greater versatility in the number of liquid-metal sources, could easily change the role of MBE. As described in this chapter, not only the large wafer capability, but most importantly for VCSELs, the vertical configuration, which allows up to eight or even ten column III metal sources, enables very simple step grading of the mirrors and higher growth rates without oval defects. This eliminates the greatest challenges that have faced VCSEL production by MBE. This advance in equipment combined with the significantly easier growth of GaInNAsSb by MBE will likely make MBE the choice for production of both VCSELs and high-power edge-emitting lasers. Progress has been fast and furious, and the future for this materials system and the potential for its inclusion as a major part of the optical networks is indeed bright.

Acknowledgments

The work at Stanford has been the result of efforts by a number of graduate students, including Seth Bank, Homen Yuen, Wonill Ha, Mark Wistey, Vincent Gambin, Tihomir Gugov, Vince Lordi, Sylvia Spruytte, Chris Coldren, Mike Larson, Lynford Goddard, as well as postdoctoral workers Kerstin Volz, Dr. Seongsin Kim of Stanford, and Drs. Danielle Chamberlin and Jeff Rosner of Agilent Technologies.

References

1. Harris Jr., J.S., "GaInNAs Long-Wavelength Lasers: Progress and Challenges," *Semicond. Sci. Technol.* 17: 880–891 (2002).
2. Kaiser, P., "Photonic Network Trends and Impact on Optical Components," in *2001 Digest of LEOS Summer Topical Meetings: WDM Components*, IEEE, New York, 3; private communication (2001).
3. Kondow, M., Uomi, K., Niwa, A., Kitatani, T., Watahiki, S., and Yazawa, Y., "GaInNAs: A Novel Material for Long-Wavelength-Range Laser Diodes with Excellent High-Temperature Performance," *Jpn. J. Appl. Phys.* 35: 1273–1275 (1996).
4. Soda, H., Iga, K., Kitahara, C., and Suematsu, Y., "GaInAsP/InP Surface Emitting Injection Lasers," *Jpn. J. Appl. Phys.* 18: 2329–2330 (1979).
5. Jayaraman, V., Geske, J.C., MacDougal, M.H., Peters, F.H., Lowers, T.D., and Char, T.T., "Uniform Threshold Current, Continuous-Wave, Singlemode 1300-nm Vertical Cavity Lasers from 0 to 70 Degrees C," *Electron. Lett.* 34: 1405–1407 (1998).
6. Yuen, W., Li, G.S., Nabiev, R.F., Boucart, J., Kner, P., Stone, R., Zhang, D., Beaudoin, M., Zheng, T., He, C., Yu, K., Jansen, M., Worland, D.P., and Chang-Hasnain, C.J., "High-Performance 1.6-μm Single-Epitaxy Top-Emitting VCSEL," *Electron. Lett.* 36: 1121–1123 (2000).
7. Hall, E., Almuneau, G., Kim, J.K., Sjolund, O., Kroemer, H., and Coldren, L.A., "Electrically Pumped, Single-Epitaxial VCSELs at 1.55 μm with Sb-Based Mirrors," *Electron. Lett.* 35: 1337–1338 (1999).
8. Uchiyama, S., Yolouchi, N., and Ninomiya, T., "Continuous-Wave Operation up to 36 Degrees C of 1.3-μm GaInAsP-InP Vertical-Cavity Surface-Emitting Laser," *IEEE Photonics Tech. Lett.* 9: 141–142 (1997).
9. Lott, J.A., Ledentsov, N.N., Ustinov, V.M., Alferov, Z.I., and Bimberg, D., "InAs-InGaAs Quantum Dot VCSELs on GaAs Substrates Emitting at 1.3 μm," *Memoirs Inst. Scientific Indust. Res. Osaka* 57: 80–87 (2000).
10. Blum, O., and Klem, J.F., "Characteristics of GaAsSb Single-Quantum-Well Lasers Emitting near 1.3 μm," *IEEE Photonics Tech. Lett.* 12: 771–773 (2000).
11. Kondow, M., Nakatsuka, S., Kitatani, T., Yazawa, Y., and Okai, M., "Room-Temperature Continuous-Wave Operation of GaInNAs/GaAs Laser Diode," *Electron. Lett.* 32: 2244–2245 (1996).
12. Hetterich, M., Dawson, M.D., Egorov, A.Y., Bernklau, D., and Riechert, H., "Electronic States and Band Alignment in GaInNAs/GaAs Quantum-Well Structures with Low Nitrogen Content," *Appl. Phys. Lett.* 76: 1030–1032 (2000).
13. Hai, P.N., Chen, W.M., Buyanova, I.A., Xin, H.P., and Tu, C.W., "Direct Determination of Electron Effective Mass in GaNAs/GaAs Quantum Wells," *Appl. Phys. Lett.* 77: 1843–1845 (2000).
14. Harris Jr., J.S., "Tunable Long-Wavelength Vertical-Cavity Lasers: the Engine of Next Generation Optical Networks?" *IEEE J. Selected Top. Quantum Electron.* 6: 1145–1160 (2000).
15. Spruytte, S.G., Coldren, C.W., Marshall, A.F., Larson, M.C., and Harris, J.S., "MBE Growth of Nitride-Arsenide Materials for Long-Wavelength Optoelectronics," in *1999 GaN Proceedings of Fall MRS Meeting* (MRS Research Society, Pittsburgh, PA, 1999), W8.4.
16. Spruytte, S.G., Larson, M.C., Wampler, W., Coldren, C.W., Krispin, P., Petersen, H.E., Picraux, S., Ploog, K., and Harris, J.S., "Nitrogen Incorporation in Group III-Nitride-Arsenide Materials Grown by Elemental Source Molecular Beam Epitaxy," *J. Crystal Growth* 227–228: 506–515 (2001).
17. Spruytte, S.G., "MBE Growth of Nitride-Arsenides for Long-Wavelength Opto-Electronics," Ph.D. thesis (Stanford, CA: Stanford University, 2001).

18. Riechert, H., Ramakrishnan, A., and Steinle, G., "Development of InGaAsN-Based 1.3-μm VCSELs," *Semicond. Sci. Technol.* 17: 892–897 (2002).
19. Harmand, J.C., Ungaro, G., Largeau, L., and LeRoux, G., "Comparison of Nitrogen Incorporation in Molecular-Beam Epitaxy of GaAsN, GaInAsN, and GaAsSbN," *Appl. Phys. Lett.* 77: 2482–2484 (2000).
20. Spruytte, S.G., Coldren, C.W., Marshall, A.F., Larson, M.C., Harris, J.S., "MBE growth of nitride-arsenide materials for long wavelength opto-electronics", *MRS Internet J. Nitride Semicon. Res.* 5, suppl.1: U407–U412 (2000).
21. Jin, C., Qiu, Y., Nikishin, S.A., and Temkin, H., "Nitrogen Incorporation Kinetics in Metalorganic Molecular Beam Epitaxy of GaAsN," *Appl. Phys. Lett.* 74: 3516–3518 (1999).
22. Kawaguchi, M., Gouardes, E., Schlenker, D., Kondo, T., Miyamoto, T., Koyama, F., and Iga, K., "Low Threshold Current Density Operation of GaInNAs Quantum Well Lasers Grown by Metalorganic Chemical Vapour Deposition," *Electron. Lett.* 36: 1776–1777 (2000).
23. Sato, S., and Satoh, S., "Metalorganic Chemical Vapor Deposition of GaInNAs Lattice Matched to GaAs for Long-Wavelength Laser Diodes," *J. Crystal Growth* 192: 381–385 (1998).
24. Mereuta, A., Saint-Girons, G., Bouchoule, S., Sagnes, I., Alexandre, F., Le Roux, G., Decobert, J., and Ougazzaden, A., "(InGa)(NAs)/GaAs Structures Emitting in 1–1.6-μm Wavelength Range," *Optical Mater.* 17: 185–188 (2001).
25. Stolz, W., "Alternative N-,P- and As-Precursors for III/V-Epitaxy," *J. Crystal Growth* 209: 272–278 (2000).
26. Hasse, A., Volz, K., Schaper, A.K., Koch, J., Hohnsdorf, F., and Stolz, W., "TEM Investigations of (GaIn)(NAs)/GaAs Multi-Quantum Wells Grown by MOVPE," *Crystal Res. Technol.* 35: 787–792 (2000).
27. Johnson, R., Honeywell, private communication (2003).
28. Takeuchi, T., Chang, Y.L., Tandon, A., Bour, D., Corzine, S., Twist, R., Tan, M., and Luan, H.C., "Low Threshold 1.2-μm InGaAs Quantum Well Lasers Grown under Low As/III Ratio," *Appl. Phys. Lett.* 80: 2445–2447 (2002).
29. Pan, Z., Miyamoto, T., Schlenker, A.D., Sato, S., Koyama, B.F., and Iga, K., "Low Temperature Growth of GaInNAs/GaAs Quantum Wells by Metalorganic Chemical Vapor Deposition Using Tertiarybutylarsine," *J. Appl. Phys.* 84: 6409–6411 (1998).
30. Stringfellow, G.B., *Organometallic Vapor-Phase Epitaxy: Theory and Practice* (Boston: Academic Press, 1989), 123.
31. LaPierre, R.R., Robinson, B.J., and Thompson, D.A., "Group V Incorporation in InGaAsP Grown on InP by Gas Source Molecular Beam Epitaxy," *J. Appl. Phys.* 79: 3021–3027 (1996).
32. Tansu, N., and Mawst, L.J., "Low-Threshold Strain-Compensated InGaAs(N) (λ = 1.19–1.31 μm) Quantum-Well Lasers," *IEEE Photonics Tech. Lett.* 14: 444–446 (2002).
33. Jikutani, N., Sato, S., Takahashi, T., Itoh, A., Kaminishi, M., and Satoh, S., "Threshold Current Density Analysis of Highly Strained GaInNAs Multiple Quantum Well Lasers Grown by Metalorganic Chemical Vapor Deposition," *Jpn. J. Appl. Phys.* 41: 1164–1167 (2002).
34. Fischer, M., Reinhardt, M., and Forchel, A., "GaInAsN/GaAs Laser Diodes Operating at 1.52 μm," *Electron. Lett.* 36: 1208–1209 (2000).
35. Polimeni, A., Baldassarri, G.H., Bissiri, H.M., Capizzi, M., Fischer, M., Reinhardt, M., and Forchel, A., "Effect of Hydrogen on the Electronic Properties of $In_xGa_{1-x}As_{1-y}N_y$/GaAs Quantum Wells," *Phys. Rev. B* 63: 201304/1-4 (2001).
36. Buyanova, I.A., Izadifard, M., Chen, W.M., Polimeni, A., Capizzi, M., Xin, H.P., and Tu, C.W., "Hydrogen-Induced Improvements in Optical Quality of GaNAs Alloys," *Appl. Phys. Lett.* 82: 3662–3664 (2003).

37. Polimeni, A., Baldassarri, G.H., Bissiri, M., Capizzi, M., Frova, A., Fischer, M., Reinhardt, M., and Forchel, A., "Role of Hydrogen in III-N-V Compound Semiconductors," *Semicond. Sci. Technol.* 17: 797–802 (2002).

38. Coldren, C.W., Spruytte, S.G., Harris, J.S., and Larson, M.C., "Group III Nitride-Arsenide Long Wavelength Lasers Grown by Elemental Source Molecular Beam Epitaxy," *J. Vac. Sci. Technol. B* 18: 1480–1483 (2000).

39. Spruytte, S.G., Coldren, C.W., Marshall, A.F., and Harris, J.S., "Compositional Evolution and Structural Changes during Anneal of Group III-Nitride-Arsenide Alloys," in *MRS Spring 2000 Meeting Proceedings,* Materials Research Society, Pittsburgh, PA (2003).

40. Krispin, P., Spruytte, S.G., Harris, J.S., and Ploog, K.H., "Origin and Annealing of Deep-Level Defects in p-Type GaAs/Ga(As,N)/GaAs Heterostructures Grown by Molecular Beam Epitaxy," *J. Appl. Phys.* 89: 6294–6298 (2001).

41. Krispin, P., Spruytte, S.G., Harris, J.S., and Ploog, K.H., "Electrical Depth Profile of p-Type GaAs/Ga(As, N)/GaAs Heterostructures Determined by Capacitance-Voltage Measurements," *J. Appl. Phys.* 88: 4153–4158 (2000).

42. Thinh, N.Q., Buyanova, I.A., Chen, W.M., Xin, H.P., and Tu, C.W., "Formation of Nonradiative Defects in Molecular Beam Epitaxial GaN_xAs_{1-x} Studied by Optically Detected Magnetic Resonance," *Appl. Phys. Lett.* 79: 3089–3091 (2001).

43. Polyakov, A.Y., Smirnov, N.B., Govorkov, A.V., Botchkarev, A.E., Nelson, N.N., Fahmi, M.M.E., Griffin, J.A., Khan, A., Mohammad, S.N., Johnstone, D.K., Bublik, V.T., Chsherbatchev, K.D., Voronova, M.I., and Kasatochkin, V.S., "Studies of Deep Centers in Dilute GaAsN and InGaAsN Films Grown by Molecular Beam Epitaxy," *Solid State Electron.* 46: 2155–2160 (2000).

44. Polyakov, A.Y., Smirnov, N.B., Govorkov, A.V., Botchkarev, A.E., Nelson, N.N., Fahmi, M.M.E., Griffin, J.A., Khan, A., Mohammad, S.N., Johnstone, D.K., Bublik, V.T., Chsherbatchev, K.D., and Voronova, M.I., "Interface Properties and Deep Levels in InGaAsN/GaAs and GaAsN/GaAs Heterojunctions," *Solid State Electron.* 46: 2141–2146 (2000).

45. Skowronski, M., "Complexes of Oxygen and Native Defects in GaAs," *Phys. Rev. B* 46: 9476–9481 (1992).

46. Egorov, A.Y., Bernklau, D., Borchert, B., Illek, S., Livshits, D., Rucki, A., Schuster, M., Kaschner, A., Hoffmann, A., Dumitras, G., Aman, M.C., and Riechert, H., "Growth of High Quality InGaAsN Heterostructures and Their Laser Application," *J. Crystal Growth* 227–228: 545–552 (2001).

47. Pavelescu, E.M., Jouhti, T., Peng, C.S., Li, W., Konttinen, J., Dumitrescu, M., Laukkanen, P., and Pessa, M., "Enhanced Optical Performances of Strain-Compensated 1.3-μm GaInNAs/GaNAs/GaAs Quantum-Well Structures," *J. Crystal Growth* 241: 31–38 (2002).

48. Pavelescu, E.M., Peng, C.S., Jouhti, T., Konttinen, J., Li, W., Pessa, M., Dumitrescu, M., and Spanulescu, S., "Effects of Insertion of Strain-Mediating Layers on Luminescence Properties of 1.3-μm GaInNAs/GaNAs/GaAs Quantum-Well Structures," *Appl. Phys. Lett.* 80: 3054–3056 (2002).

49. Ha, W., Gambin, V., Wistey, M., Bank, S., Kim, S., and Harris, J.S., "Long Wavelength GaInNAs Ridge Waveguide Lasers with GaNAs Barriers," in *LEOS 2001 Conference Proceedings* IEEE, New York (2001).

50. Spruytte, S.G., Larson, M.C., Wampler, W., Coldren, C.W., and Harris, J.S., "Molecular Beam Epitaxial Growth of Group III-Nitride-Arsenides for Long Wavelength Optoelectronics," in *Proceedings of International Symposium on Compound Semiconductors* Institute of Physics, Bristol, U.K. (2000), 61–66.

51. Ha, W., Gambin, V., Wistey, M., Bank, S., Kim, S., and Harris Jr., J.S., "Long-Wavelength GaInNAs(Sb) Lasers on GaAs," *IEEE J. Quantum Electron.* 38: 795–800 (2002).

52. Harris Jr., J.S., and Gambin, V., "GaInNAs: a New Material in the Quest for Communications Lasers," in *MRS Spring 2000 Meeting Proceedings* Materials Research Society, Pittsburgh, PA, April.

53. Yang, X., Jurkovic, M.J., Heroux, J.B., and Wang, W.I., "Molecular Beam Epitaxial Growth of InGaAsN:Sb/GaAs Quantum Wells for Long-Wavelength Semiconductor Lasers," *Appl. Phys. Lett.* 75: 178–180 (1999).
54. Shimizu, H., Kumada, K., Uchiyama, S., and Kasukawa, A., "1.2-µm Range GaInAs SQW Lasers Using Sb as Surfactant," *Electron. Lett.* 36: 1379–1381 (2000).
55. Yang, X., Heroux, J.B., Mei, L.F., and Wang, W.I., "InGaAsNSb/GaAs Quantum Wells for 1.55-µm Lasers Grown by Molecular Beam Epitaxy," *Appl. Phys. Lett.* 78: 4068–4070 (2001).
56. Shimizu, H., Kumada, K., Uchiyama, S., and Kasukawa, A., "High-Performance CW 1.26-µm GaInNAsSb-SQW Ridge Lasers," *IEEE J. Selected Top. Quantum Electron.* 7: 355–364 (2001).
57. Harmand, J.C., Ungaro, G., Ramos, J., Rao, E.V.K., Saint-Girons, G., Teissier, R., Le Roux, G., Largeau, L., and Patriarche, G., "Investigations on GaAsSbN/GaAs Quantum Wells for 1.3–1.55-µm emission," *J. Crystal Growth* 227: 553–557 (2001).
58. Volz, K., Gambin, V., Ha, W., Wistey, M.A., Yuen, H., Bank, S., and Harris, J.S., "The Role of Sb in the MBE Growth of (GaIn)(NAsSb)," *J. Crystal Growth* 251: 360–366 (2003).
59. Ha, W., Gambin, V., Wistey, M., Bank, S., Yuen, H., Kim, S., and Harris Jr., J.S., "Multiple-Quantum-Well GaInNAs-GaNAs Ridge-Waveguide Laser Diodes Operating Out to 1.4 µm," *IEEE Photonics Tech. Lett.* 14: 591–593 (2002).
60. Ha, W., Gambin, V., Wistey, M., Bank, S., Yuen, H., Kim, S., and Harris Jr., J.S., "Long Wavelength GaInNAsSb/GaNAsSb Multiple Quantum Well Lasers," *Electron. Lett.* 38: 277–278 (2002).
61. Gambin, V., Ha, W., Wistey, M., Yuen, H., Bank, S., Kim, S., and Harris Jr., J.S., "GaInNAsSb for 1.3–1.6-µm Long Wavelength Lasers Grown by Molecular Beam Epitaxy," *IEEE J. Selected Top. Quantum Electron.* 8: 1260–1267 (2002).
62. Bank, S., Ha, W., Gambin, V., Wistey, M., Yuen, H., Goddard, L., Kim, S., and Harris Jr., J.S., 1.5-µm GaInNAs(Sb) Lasers Grown on GaAs by MBE," *J. Crystal Growth* 251: 367–371 (2003).
63. Ha, W., "Long Wavelength GaInNAs and GaInNAsSb Lasers on GaAs," Ph.D. thesis (Stanford, CA: Stanford University, 2002).
64. Gambin, V., "Long Wavelength Luminescence from GaInNAsSb on GaAs," Ph.D. thesis (Stanford, CA: Stanford University, 2002).
65. Yuen, H.B., Bank, S.R., Wistey, M.A., Ha, W., Moto, A., and Harris, J.S., "An Investigation of GaNAs(Sb) for Strain Compensated Active Regions at 1.3 and 1.55 µm," in *Abstracts of 45th Electronic Materials Conference*, The Metallurgical Society, Warrendale, PA (2003).
66. Bank, S., Wistey, M., Yuen, H., Goddard, L.L., Ha, W., Harris Jr., J.S., "A Low Threshold CW GaInNAsSb/GaAs Laser at 1.49 µm," *Electron. Lett.* 39, 1445–46 (2003).
67. Albrecht, M., Grillo, V., Remmele, T., Strunk, H.P., Egorov, A.Y., Dumitras, G., Riechert, H., Kaschner, A., Heitz, R., and Hoffmann, A., "Effect of Annealing on the In and N Distribution in InGaAsN Quantum Wells," *Appl. Phys. Lett.* 81: 2719–2721 (2002).
68. Chauveau, J.M., Trampert, A., Pinault, M.A., Tournie, E., Du, K., and Ploog, K.H., "Correlations between Structural and Optical Properties of GaInNAs Quantum Wells grown by MBE," *J. Crystal Growth* 251: 383–387 (2003).
69. Alt, H.C., Egorov, A.Y., Riechert, H., Wiedemann, B., Meyer, J.D., Michelmann, R.W., and Bethge, K., "Local Vibrational Mode Absorption of Nitrogen in GaAsN and InGaAsN Layers Grown by Molecular Beam Epitaxy," *Physica B* 302–303: 282–90 (2001).
70. Gambin, V., Lordi, V., Ha, W., Wistey, M., Takizawa, T., Uno, K., Friedrich, S., and Harris, J.S., "Structural Changes on Annealing of MBE Grown (Ga,In)(N,As) as Measured by X-ray Absorption Fine Structure," *J. Crystal Growth* 251: 408–411 (2003).

71. Lordi, V., Gambin, V., Friedrich, S., Funk, T., Takizawa, T., Uno, K., and Harris, J.S., "Nearest-Neighbor Configuration in (GaIn)(NAs) Probed by X-ray Absorption Spectroscopy," *Phys. Rev. Lett.* 90: 145505/1-4 (2003).

72. Ciatto, G., Boscherini, F., D'Acapito, F., Mobilio, S., Baldassari, G., Polimeni, H.H., Capizzi, M., Gollub, D., and Forchel, A., "Atomic Ordering in (InGa)(AsN) Quantum Wells: an In K-Edge X-ray Absorption Investigation," *Nucl. Instrum. Meth. Phys. Res. B* 200: 34–39 (2003).

73. Kurtz, S., Webb, J., Gedvilas, L., Friedman, D., Geisz, J., and Olson, J., "Structural Changes during Annealing of GaInAsN," *Appl. Phys. Lett.* 78: 748–750 (2001).

74. Spruytte, S., Wampler, W., Krispin, P., Coldren, C., Larson, M., Ploog, K., Harris, J.S., "Incorporation of Nitrogen in Nitride-Arsenides: Origin of Improved Luminescence Efficiency after Anneal," *J. Appl. Phys.* 89: 4401–4406 (2001).

75. Buyanova, I.A., Pozina, G., Hai, P.N., Thinh, N.Q., Bergman, J.P., Chen, W.M., Xin, H.P., and Tu, C.W., "Mechanism for Rapid Thermal Annealing Improvements in Undoped GaN_xAs_{1-x}/GaAs Structures Grown by Molecular Beam Epitaxy," *App. Phys. Lett.* 77: 2325–2327 (2000).

76. Kageyama, T., Miyamoto, T., Makino, S., Koyama, F., and Iga, K., "Thermal Annealing of GaInNAs/GaAs Quantum Wells Grown by Chemical Beam Epitaxy and Its Effect on Photoluminescence," *Jpn. J. Appl. Phys.* 38: L298–300 (1999).

77. Wei, L., Pessa, M., Ahlgren, T., and Decker, J., "Origin of Improved Luminescence Efficiency after Annealing of Ga(In)NAs Materials Grown by Molecular-Beam Epitaxy," *Appl. Phys. Lett.* 79: 1094–1096 (2001).

78. Kitatani, T., Nakahara, K., Kondow, M., Uomi, K., and Tanaka, T., "Mechanism Analysis of Improved GaInNAs Optical Properties through Thermal Annealing," *J. Crystal Growth* 209: 345–349 (2000).

79. Pornarico, A., Lomascolo, M., Cingolani, R., Egorov, A.Y., and Riechert, H., "Effects of Thermal Annealing on the Optical Properties of InGaNAs/GaAs Multiple Quantum Wells," *Semicond. Sci. Technol.* 17: 145–149 (2002).

80. Xin, H.P., Kavanagh, K.L., Kondow, M., and Tu, C.W., "Effects of Rapid Thermal Annealing on GaInNAs/GaAs Multiple Quantum Wells," *J. Crystal Growth* 201–202: 419–422 (1999).

81. Makino, S., Miyamoto, T., Kageyama, T., Ikenaga, Y., Arai, M., Koyama, F., and Iga, K., "Composition Dependence of Thermal Annealing Effect on 1.3-μm GaInNAs/GaAs Quantum Well Lasers Grown by Chemical Beam Epitaxy," *Jpn. J. Appl. Phys.* 40: L1211–1213 (2001).

82. Gambin, V., Ha, W., Wistey, M., Kim, S., and Harris, J.S., "GaInNAs Material Properties for Long Wavelength Opto-Electronic Devices," in *MRS Fall 2001 Meeting Proceedings* Materials Research Society, Pittsburgh, PA, (2001).

83. Magri, R., and Zunger, A., "Real-Space Description of Semiconducting Band Gaps in Substitutional Systems," *Phys. Rev. B* 44: 8672–8684 (1991).

84. Kim, K., and Zunger, A., "Spatial Correlations in GaInAsN Alloys and Their Effects on Band-Gap Enhancement and Electron Localization, *Phys. Rev. Lett.* 86: 2609–2611 (2001).

85. Klar, P.J., Gruning, H., Koch, J., Schafer, S., Volz, K., Stolz, W., Heimbrodt, W., Saadi, A.M.K., Lindsay, A., and O'Reilly, E.P., "(Ga, In)(N, As)-Fine Structure of the Bandgap due to Nearest-Neighbor Configurations of the Isovalent Nitrogen," *Phys. Rev. B* 64: 121203/1-4 (2001).

86. Buyanova, I.A., Chen, W.M., Monemar, B., Xin, H.P., and Tu, C.W., "Effect of Growth Temperature on Photoluminescence of GaNAs/GaAs Quantum Well Structures," *Appl. Phys. Lett.* 75: 3781–3783 (1999).

87. Li, L.H., Pan, Z., Zhang, W., Lin, Y.W., Wang, X.Y., Wu, R.H., and Ge, W.K., "Effect of Ion-Induced Damage on GaNAs/GaAs Quantum Wells Grown by Plasma-Assisted Molecular Beam Epitaxy," *J. Crystal Growth* 223: 140–144 (2001).

88. Pan, Z., Li, L., Zhang, W., Wang, X., Lin, Y., and Wu, R., "Growth and Characterization of GaInNAs/GaAs by Plasma-Assisted Molecular Beam Epitaxy," *J. Crystal Growth* 227–228: 516–520 (2001).
89. Li, L.H., Pan, Z., Zhang, W., Wang, X.Y., and Wu, R.H., "Quality Improvement of GaInNAs/GaAs Quantum Wells Grown by Plasma-Assisted Molecular Beam Epitaxy," *J. Crystal Growth* 227–228: 527–531 (2001).
90. Livshits, D.A., Egorov, A.Y., Riechert, H., "8-W Continuous Wave Operation of InGaAsN Lasers at 1.3 μm," *Electron. Lett.* 36: 1381–1382 (2000).
91. Potter, R., Mazzucato, S., Balkan, N., Adams, M.J., Chalker, P.R., Joyce, T.B., and Bullough, T.J., "The Effect of In/N Ratio on the Optical Quality and Lasing Threshold in $Ga_xIn_{1-x}As_{1-y}N_y$/GaAs Laser Structures," *Superlattices Microstructures* 29: 169–186 (2001).
92. Coldren, C.W., Larson, M.C., Spruytte, S.G., and Harris, J.S., "1200-nm GaAs-Based Vertical Cavity Lasers Employing GaInNAs Multiquantum Well Active Regions," *Elect. Lett.* 36: 951–952 (2000).
93. Fischer, M.O., Reinhardt, M., and Forchel, A., "Room-Temperature Operation of GaIn-AsN-GaAs Laser Diodes in the 1.5-μm Range," *IEEE J. Selected Top. Quantum Electron.* 7: 149–151 (2001).
94. Egorov, A.Y., Bernklau, D., Borchardt, B., Illek, S., Livshits, D.A., Rucki, A., Schuster, M., Kaschner, A., Hoffmann, A., Dumitras, G., Amann, M.C., and Riechert, H., "Growth of High Quality InGaAsN Heterostructures and Their Laser Application," *J. Crystal Growth* 227–228: 545–552 (2001).
95. Steinle, G., Mederer, F., Kircherere, M., Michalzik, R., Kristen, G., Egorov, A.Y., Riechert, H., Wolf, H.D., and Ebeling, K., "Data Transmission up to 10 Gbit/s with 1.3-μm Wavelength InGaAsN VCSELs," *Electron. Lett.* 37: 632–634 (2001).
96. Höhnsdorf, F., Koch, J., Leu, S., Stolz, W., Borchert, B., and Druminski, M., "Reduced Threshold Current Densities of (GaIn)(NAs)/GaAs Single Quantum Well Lasers for Emission Wavelengths in the Range 1.28–1.38 μm," *Electron. Lett.* 35: 571–572 (1999).
97. Jin, S.R., Sweeney, S.J., Knowles, G., Adams, A.R., Higashi, T., Riechert, H., and Thijs, P.J.A., "Optical Investigation of Recombination Processes in GaInNAs, InGaAsP and AlGaInAs Quantum-Well Lasers Using Hydrostatic Pressure," in *IEEE 18th International Semiconductor Laser Conference Proceedings* (New York: IEEE, 2002).
98. Sweeney, S.J., Jin, S.R., Fehse, R., Adams, A.R., Higashi, T., Riechert, H., and Tbijs, P.J.A., "A Comparison of the Thermal Stability of InGaAsP, AlGaInAs and GaInNAs Quantum-Well Lasers for 1.3-μm Operation," in *IEEE 18th International Semiconductor Laser Conference. Conference Proceedings* (New York: IEEE, 2002).
99. Fehse, R., Sweeney, S.J., Adams, A.R., O'Reilly, E.P., Egorov, A.Y., Riechert, H., and Illek, S., "Insights into Carrier Recombination Processes in 1.3-μm GaInNAs-Based Semiconductor Lasers Attained Using High Pressure," *Electron. Lett.* 37: 92–93 (2001).
100. Jin, S.R., Sweeney, S.J., Tomic, S., Adams, A.R., and Riechert, H., "Unusual Increase of the Auger Recombination Current in 1.3-μm GaInNAs Quantum-Well Lasers under High Pressure," *Appl. Phys. Lett.* 82: 2335–2337 (2003).
101. Gollub, D., Moses, S., Fischer, M., and Forchel, A., "1.42-μm Continuous-Wave Operation of GaInNAs Laser Diodes," *Electron. Lett.* 39: 777–778 (2003).
102. Li, L.H., Sallet, V., Patriarche, G., Largeau, L., Bouchoule, S., Merghem, K., Travers, L., and Harmand, J.C., "1.5-μm Laser on GaAs with GaInNAsSb Quinary Quantum Well," *Electron. Lett.* 39: 519–520 (2003).

Index

R

Raman
  amplifier, 397
  scattering
    disorder-activated, 206
    method, 199
    mode
      Ga-N, 202
      local vibrational (LVM), 202
      resonant, 200, 208, 214, 218
      phonon line width, 200, 214, 216
      profile, 208–211
      second order, 202
      selection rules, 206
Recombination
  Auger, 345, 348–350
  coefficient, 353
    Auger, 345, 355, 356
    monomolecular, 345, 355, 356
    radiative, 345, 346
  depletion region, 385–387
  effect on solar cell performance,
    384–387
  mechanism, 256, 262
  monomolecular, 345, 348
  non-radiative, 345, 380
    Ga(In)NAs, 268–274
    Ga(Al)NP, 274, 275
  radiative 345, see also
    Photoluminescence
  under pressure, 361
Relaxation
  atomic, 5,6
Ridge mesa, 319
Rutherford backscattering channeling
    (RBS), 270
  Sb incorporation, 406, 407

S

Scattering
  alloy, 67, 83–85
  cross-section, 83–84, 86
  Raman, see Raman scattering
  time, 96
Solar cell, 85
  field-aided collection, 373, 381, 385
  lattice matched to Si, 388
  multijuction, 374–378, 387
  open-circuit voltage, 387, 389
Solar spectrum, 374

Solubility
  N equilibrium, 226
  N epitaxial, 228
  N surface-enhanced, 227
  N thermodynamic, 225
Spin-orbit splitting, 33
Sticking coefficient, 403
Strain
  compensation, 404
  compressive, 407, 409
  distribution, 411
  map, 409–412
  tensile, 131, 134
Supercell
  calculations, 6
  face-centered cubic, 71,72
  disordered,73
  simple cubic, 71,71
Superlattice, 294
Surfactant, 405
Symmetry point group, 213

T

Tight-binding calculations, 66, 69, 86, 141
Transition
  E_0, 32
  E_-, 33
  E_+, 33
  optical, 32
  compositional dependence, 41
  pressure dependence, 35, 38
  oscillator strength, 267
  type I – type II, 146–151
Transmission electron microscopy, 401, 409
Transport
  magneto-, 102
  properties
    GaInNAs, 285
Trap, 269, 270 373, 385–387, 403
Two-band model, 137–138
Type conversion, 379, 383

V

Vacancy, see defect
Valence band, 34
  discontinuity, 313
  dispersion, 78, 80
  light holes, 78
  offset, 81, 132, 136, 299

Printed in the United States
by Baker & Taylor Publisher Services